THE ANALYSIS OF LINEAR MODELS

THE ANALYSIS OF LINEAR MODELS

R. R. HOCKING
Texas A & M University

Brooks/Cole Publishing Company
Monterey, California

Brooks/Cole Publishing Company
A Division of Wadsworth, Inc.

Printed in the United States of America
10 9 8 7 6 5 4 3 2 1

Library of Congress Cataloging in Publication Data

Hocking, R. R. (Ronald R.), [date]
The Analysis of linear models.

Bibliography: p.
Includes index.
1. Linear models (Statistics) I. Title.
QA276.H56 1984 519.5 84-14606

ISBN 0-534-03618-X

Sponsoring Editor: *Craig Barth*
Editorial Assistant: *Eileen Galligan*
Marketing Representative: *Ragu Raghavan*
Production Editor: *Candyce Cameron*
Manuscript Editor: *Marilu Uland*
Permissions Editor: *Carline Haga*
Interior and Cover Design: *Katherine Minerva*
Art Coordinators: *Rebecca Tait, Michele Judge*
Interior Illustration: *Tim Keenan*
Typesetting: *Jonathan Peck Typographers, Ltd., Santa Cruz, California*
Printing and Binding: *The Maple-Vail Book Manufacturing Group, York, Pennsylvania*

To My Wife, Olga,
and My Children, Judy, Laura, and Raymond.
And, To a Special Person, Hoh.

PREFACE

The motivation for writing a textbook must be based either on the feeling that you can do a better job of presenting known material or on the knowledge that you have something new to contribute to the literature. In The Analysis of Linear Models, my presentation of the material is different from what we have seen in recent years and perhaps goes back to the basic ideas of Fisher and Snedecor. It is my feeling, however, that in the process of "mathematizing" the analysis of linear models, we have obscured the simple concepts and objectives of the analysis. This fact became apparent to me several years ago when I was attempting to explain the output of various ANOVA computer programs to a consulting client. In the process, I realized that many of the difficulties I encountered in communicating results to nonstatisticians stemmed from their lack of understanding of the classical over-parameterized linear model.

In this book, I return to first principles both for model formulation and for analysis with the hope that this will provide a simple and thorough understanding of the topic for new students and a different insight for those with previous exposure to linear models.

With regard to new material, I suspect that much of what I say is known to many statisticians, but not all of it appears in the literature. The material on the analysis of unbalanced data and especially on designs with missing data should resolve some of the confusion in that area. In addition, the notation developed to describe the balanced, mixed model should unify the development of that topic. The general area of variance-component analysis has been greatly reworked, and though the presentation here does not resolve all of the problems, it does give a summary of current methods and may provide some useful insight into alternative methods for modeling and analysis.

This book divides naturally into three main areas. The first three chapters provide an introduction to linear models and develop the necessary theory for the analysis of the model with scalar covariance matrix. In Chapters 4 through 7, that theory is applied to a special class of models that are traditionally known as *Analysis of Variance* models. The approach I take both for model formulation and for analysis is based simply on means, and the term *cell means model* is suggested as an appropriate description of the model. The features of these middle chapters are

1. The cell means model is made conceptually simple and avoids such unnecessary concepts as nonfull rank models, nonestimable parameters, and nontestable hypotheses.
2. The development is easily extended to the case of unbalanced data.

It is emphasized that the simplicity of the cell means approach is in describing the model and the analysis. It is recognized that the actual computation of estimates, test statistics, and confidence bounds may, in some cases, be more conveniently performed in terms of a transformed model.

Chapters 8 through 10 contain a discussion of the general linear model characterized by having a nonscalar covariance matrix. Here the model formulation also differs slightly from the usual presentation in that the variance components are viewed as covariances or components of a covariance term. This interpretation sheds some light on certain classical controversies, such as the problem of negative estimates of variance components.

I have assumed that students have had prior exposure to the topics of linear regression and the analysis of variance at the level of a good introductory course in statistical methods. In addition, a course in matrix algebra and a course in mathematical statistics are assumed. The essential ideas on these topics are reviewed in Appendixes A and B. Extensive use of matrix algebra is made in the development of the theory and methodology. The statistical theory required consists of an understanding of the concepts of (a) the distribution of random variables, (b) point and interval estimates of parameters, and (c) tests of parametric hypotheses. We rely primarily on maximum likelihood estimation and likelihood ratio tests.

Chapters 1 through 7 contain the analysis for linear models with scalar covariance matrix. Depending on the background of the student, this material may be suitable for a one-semester, three-credit course. Chapter 1 contains several simple examples to illustrate a variety of linear models and serves to introduce some notation. The distribution theory for linear and quadratic functions of normal variables is given in Chapter 2. In Chapter 3, we give a detailed development of the analysis of the linear model, and in Chapter 4 the results are specialized to the cell means model.

Chapters 5 through 7 contain illustrations of the concepts in Chapter 4. The one-way classification model is discussed in Chapter 5. In Chapter 6, the two-factor model is discussed in detail. This chapter also contains important ideas that generalize to more complex models. The concept of an Effective Model is introduced to deal with the missing cell problem. Computational problems are addressed, and a canonical model is introduced, to simplify both the notation and the computations. The canonical model also provides the link to the nonfull rank model often used for this analysis. In Chapter 7, the ideas are illustrated for more complex models.

In Chapters 8 through 10, we discuss variance component models. These three chapters form the basis for a second course in linear models. Chapter 8 reviews the current approaches for parameter estimation and hypothesis testing. A detailed presentation of the balanced, mixed model is given in Chapter 9. Relations are established among the various methods of analysis. Unbalanced models also are introduced. In Chapter 10, we discuss classical approaches to the problem of estimating variance components. A feature of this chapter is the plausible explanation for the problem of negative estimates of variance components, which suggests a scheme for data and model inspection that should be introduced into the analysis.

One of the advantages of the cell means model is that the analysis can be described with relatively unsophisticated mathematics. In some cases, the developments involve matrix manipulations that are described compactly so as to keep the main issues in focus. The student is urged to fill in the details as needed to ensure understanding. The exercises at the end of each chapter consist partly of such verifications, but they also include some more-demanding problems.

The ideas in the text are illustrated by examples, some of which merely illustrate simple points while others provide relatively detailed data analyses. The primary purpose of the book is to provide a solid foundation for the applied statistician who is concerned with the formulation and analysis of linear models.

Acknowledgements

I would like to acknowledge the many contributions of former students and associates, especially Michael Speed, Michael Kutner, David Smith, Olga Hackney, Leigh Murray, and Anne Coleman. My wife, Olga Pendleton, deserves special recognition for encouraging me to write this book, for valuable technical contributions, and for countless hours of proofreading. Our son, Raymond, born during the writing of this book, showed infinite patience and understanding.

I would also like to thank those who improved the quality of this book with their helpful feedback and criticism. The reviewers include Lyle Broemeling, Oklahoma State University; G. Rex Bryce, Brigham Young University; Franklin Graybill, Colorado State University; Ramon C. Littell, University of Florida; Roland Sink, Pasadena City College; James Stapleton, Michigan State University; and Jessica Utts, University of California at Davis. And, finally, at Brooks/Cole Publishing Company, thanks to Craig Barth, Eileen Galligan, Candy Cameron, and Katherine Minerva.

R. R. Hocking

CONTENTS

LIST OF EXAMPLES

CHAPTER 1

CHAPTER 2

CHAPTER 3

none

CHAPTER 4

CHAPTER 5

none

CHAPTER 6

CHAPTER 7

CHAPTER 8

CHAPTER 9

CHAPTER 10

THE ANALYSIS OF LINEAR MODELS

CHAPTER ONE
INTRODUCTION TO LINEAR MODELS

To motivate the analyses to be developed in this book, we begin with a brief review of what are known as linear statistical models. The purpose of this introduction is to emphasize the flexibility and generality of these linear models in the description of data from a wide variety of experimental and observational settings. The analysis of these models will be developed in subsequent chapters, in which we shall focus primarily on a special class of the general model.

1.1. SIMPLE LINEAR MODELS

The term *linear model* arises initially in situations in which the mean of a random variable, y, may be expressed as a linear function of unknown parameters, $\theta_j, j = 0, 1, \ldots, m$. Specifically, it is assumed that we may write the mean of y as

$$E(y) = \theta_0 + \sum_{j=1}^{m} \theta_j x_j. \tag{1.1}$$

The variables, x_j, in (1.1) may take many forms which will be illustrated in the following by examples. For the purpose of this book we shall assume that these variables are nonrandom. [See Graybill (1976) for a discussion of the case in which the x_j are random and (1.1) denotes the mean of y conditional on the x_j.]

In the simplest situation, y may be the output or response from an experiment in which the inputs x_j are controlled. In other cases, the input variables are not specified in advance but observed at the same time as the response. Additional input variables also may be computed as functions of observed variables. In other applications, the actual variables may be qualitative, and the x_j are simply indicator variables. We shall see by the examples given below that the linear structure on the mean indicated in (1.1) has very broad application.

Continuing with the specification of the model, we may make distributional assumptions about the error or residual, e, defined as

$$e = y - E(y). \tag{1.2}$$

Clearly, e has mean zero. We further assume that the variance of e is σ^2, a parameter which does not depend on the x_j. For making inferences, it is usually assumed that e follows a univariate normal distribution. Thus, we may summarize the assumptions on e by writing

$$e \sim N(0, \sigma^2) \tag{1.3}$$

where ($\sim$) is read "is distributed as" and the notation $N(a, b)$ denotes the univariate normal distribution with mean a and variance b.

Typically, we will have N observations, y_i, $i = 1, \ldots, N$, corresponding to inputs $(x_{i1} \ldots x_{im})$, $i = 1, \ldots, N$. Let

$$Y = \begin{bmatrix} y_1 \\ \vdots \\ y_N \end{bmatrix} \tag{1.4}$$

denote the N vector of responses, and let

$$\theta = \begin{bmatrix} \theta_0 \\ \theta_1 \\ \vdots \\ \theta_m \end{bmatrix} \tag{1.5}$$

denote the $m + 1$ vector of parameters. Further, let X denote the $N \times (m + 1)$ matrix of inputs, including a column of ones. That is,

$$X = \begin{bmatrix} 1 & x_{11} & \cdots & x_{1m} \\ \vdots & \vdots & \vdots & \vdots \\ 1 & x_{N1} & \cdots & x_{Nm} \end{bmatrix}. \tag{1.6}$$

If we let $E(Y)$ denote the vector of expected values, $E(y_i)$—the assumption of the linear mean structure—(1.1) is written in matrix form as

$$E(Y) = X\theta. \tag{1.7}$$

To complete the description of the simple linear model, we add the assumption that the errors associated with these observations are uncorrelated. Thus, if we define

$$e_i = y_i - E(y_i), \tag{1.8}$$

we have

$$\begin{aligned} \text{Var}(e_i) &= \sigma^2 & i &= 1, \ldots, N \\ \text{Cov}(e_i, e_j) &= 0 & i \neq j &= 1, \ldots, N, \end{aligned} \tag{1.9}$$

where Var and Cov denote variance and covariance, respectively.

If we let e denote the N-vector of e_i, then

$$e = Y - E(Y) \tag{1.10}$$

and we see that the mean vector and the covariance matrix of e are given by

$$\begin{aligned} E(e) &= 0 \\ \text{Var}(e) &= \sigma^2 I. \end{aligned} \tag{1.11}$$

The term *covariance matrix* refers to the $N \times N$ matrix of variances and covariances in (1.9). If we add the assumption of normality, it follows that the e_i are independently normally distributed. Equivalently, the vector e follows an N-variate normal distribution with mean and covariance matrix given by (1.11). Thus, we may write

$$e \sim N(0, \sigma^2 I). \tag{1.12}$$

Here, the notation, $N(\mu, V)$, denotes the multivariate normal distribution with mean vector, μ, and covariance matrix, V. (See Section 2.2 for a discussion of this distribution.)

The model is also described by writing

$$Y \sim N(X\theta, \sigma^2 I) \tag{1.13}$$

to compactly summarize the mean structure and the distributional assumptions.

Alternatively, we may write

$$Y = X\theta + e \tag{1.14}$$

where e is assumed to satisfy (1.11) or, more generally, (1.12). The use of (1.14) is favored in the literature and will be used in this book.

The literature on linear models is roughly dichotomized into *regression models* and what we shall call *means models*. The distinction is not a sharp one, however, as many models combine both features. In the remainder of this section, we provide simple examples illustrating the distinction between these two general types of linear models as well as situations in which both features are combined.

1.1.1. Regression Models

To illustrate the types of models included under this heading, we give three examples. There is quite a substantial literature in this area, so we shall not go into great detail. The reader is referred to recent books by Daniel and Wood (1980), Draper and Smith (1981), Gunst and Mason (1980), Neter, Wasserman, and Kutner (1982), Seber (1977), and Weisberg (1980). The review paper by Hocking (1983b) contains several other references.

EXAMPLE 1.1

A study was conducted to investigate the apparent decline in the mathematics skills of incoming college freshmen. In Table 1.1, we show the average scores for male and female students taking a national exam for each of the years 1967 through 1978. Focusing for now on just the male students, we may view the average score as the response variable y and the year as the input variable x. A reasonable model for this data is to assume that the expected value of y is linear in x. Thus, the assumed model is

$$E(y) = \theta_0 + \theta_1 x. \tag{1.15}$$

Here, the parameter θ_1 denotes the rate of change of $E(y)$ with respect to time and would be of primary interest to the analyst.

Note that, for this example, the matrix X defined in (1.6) would consist of the column of ones and the first column in Table 1.1.

We could also postulate the same model structure for the female students but allow for the possibility of different parameters in (1.15).

■

TABLE 1.1 Average scores in mathematics.

Year	*Score (male)*	*Score (female)*
67	514	467
68	512	470
69	513	470
70	509	465
71	507	466
72	505	461
73	502	460
74	501	459
75	495	449
76	497	446
77	497	445
78	494	444

EXAMPLE 1.2

A manufacturer of a stainless-steel product uses sheets of processed stainless steel and has noticed that the production level (number of units per sheet) is variable. It is believed that the production level (PROD), y, depends on three factors: x_1 = width of the sheets (WID), x_2 = density (DENS), and x_3 = tensile strength (STR). Records for the past twenty days are shown in Table 1.2. In this case, the assumed model is

$$E(y) = \theta_0 + \theta_1 x_1 + \theta_2 x_2 + \theta_3 x_3. \tag{1.16}$$

For this model, we may view θ_i, $i = 1, 2, 3$ as the rate of change of $E(y)$ with respect to x_i, $i = 1, 2, 3$, keeping all other variables fixed. For this example, the X matrix consists of the column of ones and the columns labeled x_1, x_2, and x_3 in Table 1.2.

■

TABLE 1.2 Steel-production data.

Observation	*WID* x_1	*DENS* x_2	*STR* x_3	*PROD* y
1	19.8	128	86	763
2	20.9	110	72	650
3	15.1	95	62	554
4	19.8	123	82	742
5	21.4	77	52	470
6	19.5	107	72	651
7	25.2	123	84	756
8	26.2	95	83	563
9	26.8	116	76	681
10	28.8	100	64	579
11	22.0	110	80	716
12	24.2	107	71	650
13	24.9	125	81	761
14	25.6	89	61	549
15	24.7	103	71	641
16	26.2	103	67	606
17	21.0	110	77	696
18	29.4	133	83	795
19	21.6	96	65	582
20	20.0	91	62	559

EXAMPLE 1.3

The quality of a particular product is known to be a function of two variables, x_1 = temperature and x_2 = pressure, which may be varied in the production process. Runs are made with nine different combinations of temperature and pressure, and the quality of the product is measured. The data in coded form are shown in Table 1.3. Here we shall assume that average quality is a quadratic function of x_1 and x_2. Thus,

$$E(y) = \theta_0 + \theta_1 x_1 + \theta_2 x_2 + \theta_{11} x_1^2 + \theta_{22} x_2^2 + \theta_{12} x_1 x_2. \tag{1.17}$$

Note that the subscripting has been modified to conform with standard notation. For this model, the input variables in the general sense consist of x_1 and x_2 as defined and the computed variables, x_1^2, x_2^2, and x_1x_2. Thus, the X matrix consists of the column of ones, the x_1 and x_2 columns from Table 1.3, and three additional columns consisting of the computed variables for all nine observations.

■

TABLE 1.3 Production-quality data.

Observation	x_1	x_2	y
1	−1	−1	149
2	−1	1	141
3	1	−1	139
4	1	1	146
5	0	0	167
6	−1.5	0	128
7	1.5	0	137
8	0	−1.5	130
9	0	1.5	134

1.1.2. Means Models

The models we include under this heading are often referred to in the literature as *experimental design models* or *analysis of variance models,* but neither term is completely appropriate. Such models often arise in situations that do not correspond to designed experiments. Further, the analysis of variance constitutes only a portion of the analysis of such models, and even that name may be inappropriate for the analysis. The term *means model* is an attempt to define a general label for such models, but, more than that, it corresponds to a formulation of the model that is somewhat different from what is usually presented in the literature.

The simplest model of this type is that in which $E(y)$ is simply a scalar, say μ. Actually, this model is a special case of (1.1), with $\mu = \theta_0$ and $\theta_j = 0, j = 1, \ldots, m$. Assuming N observations on a population with this simple structure, we may write the model in the matrix form in (1.14) as

$$Y = J\mu + e. \tag{1.18}$$

Here, and in the remainder of this book, the letter J corresponds to a vector of ones of appropriate length. In this case, J clearly has length N. In cases where the length of J is not clear, we shall indicate its length by a subscript, for example J_N.

This model may seem trivial, but it illustrates a simple fact about the models to be considered under this heading and to be emphasized in this book. That is, we shall be concerned with the analysis of means of populations. Generally, these means are just constants that do not depend on other variables. In regression models, we have seen that the mean of y is a function of the input variables x_j.

To illustrate more interesting means models and to see how they fit into the linear models framework, we give the following two examples.

EXAMPLE 1.4
An experiment was conducted to study the effect of long-term freezing on the rising of bread dough. One factor of interest in this study was the type of flour used. The experiment consisted of preparing four batches of dough with each of three types of flour and freezing for a specified time. The data are shown in Table 1.4, where the response denotes the volume increase four hours after removal from the freezer.

TABLE 1.4 Dough-rising data (freeze time 1).

		Flour Type		
		1	2	3
Batch	1	1.1	2.7	3.1
	2	1.8	2.9	3.2
	3	1.0	3.3	3.3
	4	1.2	2.8	3.2

To develop a model for this experiment, we let μ_i, $i = 1, 2, 3$ denote the mean volume increase for flour type i, and let y_{ij}, $i = 1, 2, 3$, $j = 1, 2, 3, 4$ denote the response for the jth batch using flour type i. The use of the multiple subscripts for such models is not essential but greatly simplifies the description. Assuming that the twelve batches were prepared independently, we may view these data as resulting from a random sample of four observations from each of three populations. Making the standard assumptions on the error distribution, we may write the model algebraically as

$$y_{ij} = \mu_i + e_{ij} \qquad \begin{array}{l} i = 1, 2, 3 \\ j = 1, 2, 3, 4 \end{array} \tag{1.19}$$

where it is assumed that the e_{ij} are independently distributed, $N(0, \sigma^2)$.

To see how this model fits into the linear model format, we introduce the indicator variables, w_i, $i = 1, 2, 3$, defined as

$$w_i = \begin{cases} 1 \text{ if the observation is on flour type } i \\ 0 \text{ otherwise.} \end{cases} \tag{1.20}$$

We may then write (1.19) as

$$y_{ij} = \mu_1 w_1 + \mu_2 w_2 + \mu_3 w_3 + e_{ij}. \tag{1.21}$$

While this expression for the model may seem to be unnecessarily more complicated than (1.19), we now recognize the sense in which this model fits into the linear model format. Comparing (1.21) with (1.1) with $\theta_0 = 0$, $\theta_i = \mu_i$, $i = 1, 2, 3$, and $x_i = w_i$, we see that the models are mathematically equivalent. The merits of expressing the mean structure in this way will become apparent as we develop the analysis.

To express the model in matrix form analogous to (1.14), let the vector of responses Y be written in the natural way as

$$Y' = (y_{11}\, y_{12}\, y_{13}\, y_{14}\, y_{21}\, y_{22}\, y_{23}\, y_{24}\, y_{31}\, y_{32}\, y_{33}\, y_{34}). \tag{1.22}$$

Further, define $\mu' = (\mu_1 \mu_2 \mu_3)$ as the parameter vector and W as the matrix of indicator variables. With the ordering in (1.22), W is given by

$$W = \begin{bmatrix} 1 & 0 & 0 \\ 1 & 0 & 0 \\ 1 & 0 & 0 \\ 1 & 0 & 0 \\ 0 & 1 & 0 \\ 0 & 1 & 0 \\ 0 & 1 & 0 \\ 0 & 1 & 0 \\ 0 & 0 & 1 \\ 0 & 0 & 1 \\ 0 & 0 & 1 \\ 0 & 0 & 1 \end{bmatrix}. \tag{1.23}$$

With e denoting the vector of errors subscripted as in (1.22), we write the matrix form of the model as

$$Y = W\mu + e. \tag{1.24}$$

This form of the means model is quite general and will be used in the following chapters. The coefficient matrix W and the parameter vector μ will be used to distinguish the means model from the regression model.

It will be convenient, however, to express W in more compact notation. Assuming that we are examining p populations (treatments), W will have p columns. With n observations on each population, these columns will each have n ones, as in (1.23). Using the notation described in Appendix A.I.1, we write

$$W = D_p(J_n). \tag{1.25}$$

This notation denotes a diagonal display of p column vectors of ones, each of length n. Alternatively, we may use the Kronecker product notation in Appendix A.I.8 and write

$$W = I_p \otimes J_n. \tag{1.26}$$

This latter notation will be particularly useful in the developments in later chapters. ■

EXAMPLE 1.5

A study was conducted to examine college students' awareness of current events. A random sample of students was taken from each of three colleges—namely the colleges of engineering, business, and liberal arts. The response measured is the score on an examination. To model this data, we let y_{ij} denote the score for the jth student from the ith college. If we let μ_i denote the mean response for a student from the ith college, we may postulate the same model used in Example 1.4. That is, the model is given algebraically by (1.19) or in matrix form by (1.24). Thus, we see that the means model may apply equally well to data from a designed experiment, as in Example 1.4, or to data from an observational or survey study, as in this example.

It often happens in such studies that we do not have the same number of observations on each population. This could happen by chance or by design. For example, the number of students examined from each college might be proportional to the size of the college. Even in carefully designed experiments, the number of observations per population might differ since some results may be lost for various reasons. This situation, sometimes referred to as the case of unbalanced data, does not appreciably complicate the model or the analysis. The only distinction in the model statement is that the indicator matrix, W in (1.23), will have strings of ones of unequal length. The Kronecker product notation is not suited for this case, but we can use the notation for diagonal matrices. Thus, we write

$$W = D_p(J_{n_i}) \tag{1.27}$$

to denote the diagonal display of the vectors, J_{n_i}, $i = 1, \ldots, p$. ■

EXAMPLE 1.6

To extend the bread-dough experiment from Example 1.4, consider a situation in which two different freezing times were examined. In particular, each type of flour was studied for each of the two freezing times. We thus identify six distinct populations defined by the combinations of flour type and freeze time. Assume that the data in Table 1.4 correspond to one of the freeze times and that the data in Table 1.5 correspond to the second freeze time. To model this experiment, we could proceed as in Example 1.4, letting μ_i, $i = 1, \ldots, 6$ denote the means for the six populations (treatment combinations) and use the model given in (1.19) or, equivalently, (1.24). For notational purposes, we shall see that it is convenient to use two subscripts to define the means. We write μ_{ij}, $i = 1, 2, 3$ and $j = 1, 2$ to denote the mean response for the ith flour and the jth freeze time. Similarly, we let y_{ijk}, $i = 1, 2, 3$, $j = 1, 2$ and $k = 1, 2, 3, 4$ denote the kth response to the (i, j)th treatment combination. The model is now written in algebraic form as

$$y_{ijk} = \mu_{ij} + e_{ijk} \qquad \begin{array}{l} i = 1, 2, 3 \\ j = 1, 2 \\ k = 1, 2, 3, 4. \end{array} \tag{1.28}$$

In matrix form, we let

$$\mu = (\mu_{11}\ \mu_{12}\ \mu_{21}\ \mu_{22}\ \mu_{31}\ \mu_{32})'$$

and (1.29)

$$W = I_6 \otimes J_4$$

and write the model as in (1.24) with Y defined in the obvious way.

TABLE 1.5 Dough-rising data (freeze time 2).

		Flour Type		
		1	2	3
Batch	1	1.6	1.7	1.9
	2	2.8	1.5	2.2
	3	2.1	1.3	2.0
	4	1.2	2.8	1.2

Apart from this notational difference, the model is identical to that in Example 1.4. We shall see, however, that the analysis of this model is more complex because of the potential questions suggested by the inclusion in the study of two factors: type of flour and freeze time.

We also note that it is possible that the number of observations per treatment combination is different. If we let n_{ij} denote the number of observations for the (i, j)th treatment combination, the only change is in the W matrix, which may be written, using the notation in Appendix A.I.1, as

$$W = D_a[D_b(J_{n_{ij}})]. \tag{1.30}$$

Here, we have generalized to let a and b denote the ranges on i and j.

■

1.1.3. Constrained Models

We shall have occasion to consider situations in which the parameters of the linear model are required to satisfy certain constraints. In this book, we shall consider only the case of linear equality constraints. In general, we shall allow for a set of linear constraints written as

$$G\theta = g \tag{1.31}$$

where G is a known matrix of size $q \times p$ of rank q. (Note $p = m + 1$ for regression models.)

Constraints on the parameters arise in the following two ways:

1. It is known, *a priori,* that the parameters satisfy the relations.
2. We may hypothesize that such relations are true and wish to test this hypothesis. We shall see that this requires fitting the model with and without the constraints.

It is natural to ask why this extension of the model should be treated separately. It seems that the obvious approach would be to impose the constraints on the model and analyze the resulting, unconstrained model. Thus, in Example 1.1, which is concerned with the production of stainless steel products, it may be known that $\theta_2 = 5\theta_3$. If so, we may simply write the model (1.16) as

$$E(y) = \theta_0 + \theta_1 x_1 + \theta_3(5x_2 + x_3). \tag{1.32}$$

Note that the dimensionality of the model (that is, the number of parameters) is reduced by one. In general, with q linearly independent constraints as in (1.31), we could solve for q of the parameters in terms of the remaining $p - q$ parameters, substitute these relations into the model, and consider the resulting, "reduced model." In regression models, this is almost invariably the way constraints are handled.

When dealing with means models, though, it is sometimes more convenient for notational purposes to carry the constraints along as a part of the model. In addition, we shall see that certain theoretical developments are simpler if we work with the constrained model, rather than with the unconstrained model reduced by the constraints. We shall also see that problems with "missing cells" are made clearer by keeping the constraints as a separate part of the model.

EXAMPLE 1.7

To illustrate a typical constraint with the means model, recall the experiment involving the study of flour types and freeze times. Suppose it is *known* that the difference in the effects of the two freeze times is the same for all three types of flour. That is, the parameters satisfy the constraints

$$\begin{aligned} \mu_{11} - \mu_{12} &= \mu_{21} - \mu_{22} \\ &= \mu_{31} - \mu_{32}. \end{aligned} \tag{1.33}$$

In terms of the vector, μ, defined in (1.29), we may write these relations in the form of (1.31) as

$$\begin{bmatrix} 1 & -1 & 0 & 0 & -1 & 1 \\ 0 & 0 & 1 & -1 & -1 & 1 \end{bmatrix} \mu = \begin{bmatrix} 0 \\ 0 \end{bmatrix}. \tag{1.34}$$

Thus, a complete statement of this constrained model is

$$\begin{aligned} Y &= W\mu + e \\ \text{subject to} \quad G\mu &= g \end{aligned} \tag{1.35}$$

where W and μ are defined in (1.29) and G and g are defined in (1.34). We may choose to use the constraints in (1.34) to eliminate two of the parameters. Solving for μ_{11} and μ_{21} and substituting into the original model statement, we obtain the reduced model. In algebraic form, this model is written as

$$\begin{aligned} y_{ijk} &= \mu_{ij} + e_{ijk} \qquad (i, j) \neq (1, 1) \text{ or } (2, 1) \\ y_{11k} &= \mu_{12} + \mu_{31} - \mu_{32} + e_{11k} \\ y_{21k} &= \mu_{22} + \mu_{31} - \mu_{32} + e_{21k}. \end{aligned} \tag{1.36}$$

In matrix form, we write the reduced model as

$$Y = W_R \mu_R + e. \tag{1.37}$$

Here, $\mu_R = (\mu_{12}\, \mu_{22}\, \mu_{31}\, \mu_{32})'$ and

$$W_R = \begin{bmatrix} J & 0 & J & -J \\ J & 0 & 0 & 0 \\ 0 & J & J & -J \\ 0 & J & 0 & 0 \\ 0 & 0 & J & 0 \\ 0 & 0 & 0 & J \end{bmatrix} \tag{1.38}$$

where J and 0 denote vectors of ones and zeros of length four, respectively.

It is evident that the notational simplicity lies with the model statement (1.35) but that the reduced model reflects the true dimension of the problem. This will be seen to have computational advantages. Both forms of the model will be utilized in the developments in the following chapters.

■

1.1.4. Combined Models

It was implied earlier that the distinction between regression and means models is not sharp in that some models have both features. A model that combines both features is developed in the following example, using the data from Example 1.1 involving mathematics scores, by year, for both male and female students.

EXAMPLE 1.8
To motivate the combined model, consider first two simple models that use all of the data but perhaps ignore important information.

In the first case, we might fit the simple regression model (1.15) to the data in Table 1.1, ignoring the sex difference and simply assuming that we have two observations for each year.

In the second instance, the interest might be in a comparison of the performance of the two sexes. Thus, we might ignore the yearly effect and fit the means model (1.19) with $i = 1, 2$ and $j = 1, \ldots, 12$.

It is clear that the effect of time might be masked by differences in the sexes if we use the regression model. Alternatively, the difference in the performance of males and females might be hidden if we use the means model.

To consider both effects, we suggest two possible models. First, consider the model written as

$$E(y_{ij}) = \gamma_i + \theta_1 x_{ij} \qquad \begin{aligned} i &= 1, 2 \\ j &= 1, \ldots, 12. \end{aligned} \tag{1.39}$$

Here, y_{ij} denotes the response in year j for males ($i = 1$) or females ($i = 2$), and $x_{1j} = x_{2j}$, $j = 1, 2, \ldots, 12$, denotes the jth coded year value. An examination of this model reveals that we are assuming that the effect of time, θ_1, is the same for either sex but that we are allowing for a difference, $\gamma_1 - \gamma_2$, in the level of performance. This is most easily seen by plotting $E(y_{ij})$ versus x_{ij} for specific values of γ_1, γ_2, and θ_1 and noting that the plots are parallel lines.

As a second model, we might write

$$E(y_{ij}) = \gamma_i + \theta_1 x_{ij} + \theta_2 z_{ij} \tag{1.40}$$

where we define the new variable z_{ij} as

$$z_{ij} = \begin{bmatrix} x_{ij} & i = 1 \\ 0 & i = 2. \end{bmatrix} \tag{1.41}$$

In this case, we are allowing for differences in both the level of performance and the effect of time. This is easily seen by plotting $E(y_{ij})$ versus x_{ij} for a specific set of values for γ_1, γ_2, θ_1, and θ_2. Again, we obtain a pair of straight lines, but in this case the slope is $\theta_1 + \theta_2$ for the males and θ_1 for the females.

Both models are encountered in practice. As noted, (1.40) has the advantage of generality in that it allows for different slopes. In this case, the difference in performance of the two sexes can be stated only as a function of the year. On the other hand, if the assumption $\theta_2 = 0$ is valid, then (1.39) is appropriate and $\gamma_1 - \gamma_2$ measures the difference in performance. It should be emphasized that this simpler interpretation does not justify the preference of (1.39) over (1.40).

■

Such models, especially (1.39), are often referred to as *analysis of covariance* models or, alternatively, *regression with indicator variables*. The former terminology is generally used when the primary interest is in the qualitative factor (in this case sex) and the quantitative factor (time) is introduced to improve the model. The second terminology is used when the emphasis is reversed. A discussion of the analysis of such models is provided in Chapter 7. The paper by Cox and McCullagh (1982) provides a review of the use of such models from the view of the analysis of covariance. Here again, the description of the model seems misleading since we are not concerned with covariance in the usual sense. The terminology arose since the quantitative variables added to the means model were referred to as covariates, or concomitant variables.

1.2. GENERAL LINEAR MODELS

To this point, we have focused on the linear structure of the mean vector in our model development. The covariance structure on the observations has been particularly simple in that we have assumed $\text{Var}(Y) = \sigma^2 I$. In many cases, the assumption of constant variance and zero covariance is inappropriate. In the last three chapters of this book, we shall consider models with a more general covariance structure. In particular, we shall assume that

$$\text{Var}(Y) = V = \sum_{t=0}^{T} \phi_t V_t \tag{1.42}$$

where the matrices, V_t, are assumed to be known symmetric matrices. The unknown parameters, ϕ_t, sometimes called variance components, are constrained so that V is positive definite.

Our first observation is that this structure is not restrictive, since any positive definite matrix can be written in this form. The models considered will be such that the number of parameters, $T + 1$, is small so that we will have sufficient data for analysis.

To illustrate the structure, consider the following simple example.

EXAMPLE 1.9

Suppose we are interested in evaluating the effect of three different fertilizers used in growing cotton. In this case, the basic experimental unit is a one-acre plot planted in cotton. The plan is to apply the three fertilizers, each to n plots, and observe the yield. Our first reaction is that the model will be the one used in Example 1.4. That is, we let μ_i denote the mean yield for fertilizer i and let y_{ij} denote the response from the jth plot using the ith fertilizer. Thus, the mean structure is

$$E(y_{ij}) = \mu_i \qquad i = 1, 2, 3 \qquad j = 1, \ldots, n. \tag{1.43}$$

We have available for the study three plots on each of four different fields. The experimental procedure is to apply each fertilizer to one plot in each field. The model (1.43) still applies to the mean structure, but now the subscript j identifies the particular field for $j = 1, \ldots, 4$.

In this situation, we no longer feel comfortable in assuming that the responses are uncorrelated. Instead, we allow the responses within the same field to be correlated. In particular, suppose we assume that all responses are subject to the same variance, all responses in the same field are subject to the same covariance, and observations in different fields are uncorrelated. Thus, we may write

$$\begin{aligned} \mathrm{Var}(y_{ij}) &= \sigma^2 \\ \mathrm{Cov}(y_{ij}, y_{i^*j}) &= \gamma \text{ for } i \neq i^* \\ \mathrm{Cov}(y_{ij}, y_{i^*j^*}) &= 0 \text{ for } j \neq j^* \end{aligned} \tag{1.44}$$

and the matrix V is defined. A parameterization that allows for a convenient notation for describing V is to let

$$\begin{aligned} \gamma &= \phi_b \\ \sigma^2 &= \phi_0 + \phi_b. \end{aligned} \tag{1.45}$$

In this case, we may write

$$V = \phi_0 I_N + \phi_b (J_3 J_3' \otimes I_4). \tag{1.46}$$

Here, V has dimension $N = 12$, and we are assuming that the observations are ordered in the usual way as described in (1.22). It is easily verified that the covariance structures in (1.44) and (1.46) are identical with the parameters related by (1.45).

■

The complete description of our general linear model, assuming normality, is given by writing

$$Y \sim N(X\theta, V). \tag{1.47}$$

Here V is given by (1.42), and any linear constraints on θ are either adjoined to the model statement or have already been imposed on the model.

Alternatively, we may write

$$Y = X\theta + e \tag{1.48}$$

where

$$e \sim N(0, V). \tag{1.49}$$

In the literature on means models, the case $V = \sigma^2 I$ is referred to as the fixed effect model. The general means model with nonscalar covariance matrix is referred to as the mixed model, the implication being that we are interested in making inferences on both the means and the variance components. The special case in which all the means are the same is called the random model. In this terminology, the model of Example 1.9 would be a mixed model. We shall use this terminology for ease of reference with the literature on such models.

1.3. DISCUSSION

The examples in the preceding sections should serve to illustrate linear models in general and to focus on the distinction between regression and means models. The mathematical equivalence of these two classes of models and combinations of them allows for the development of a single body of statistical theory for performing basic inference—that is, the development of point and interval estimates of the parameters and the construction of statistics for testing linear hypotheses on these parameters.

The development of the analysis for the linear model with scalar covariance matrix is presented in Chapter 3. No distinction is made between applications to regression and to means models. In Chapter 4 the results that are unique to means models are spelled out, and some additional concepts are discussed. The theory is presented for the general case of unbalanced data. Simplifications that are a consequence of balanced data also are noted. Much of the statistical theory depends upon the distribution of linear and quadratic functions of normal variables, and this distribution theory is presented in Chapter 2.

Chapters 5 through 7 contain examples to illustrate the analysis of means models. Chapter 6, while restricted to the two-way classification model, contains essential results on hypothesis formulation, computing methods, and reparameterized models. The discussion of this latter concept will help to clarify certain confusions which have arisen in the recent literature on means models.

A unique feature of the presentation in this book is that the coefficient matrix, X, in all linear models will be assumed to have full column rank; that is, the rank of X is equal to the number of columns. This includes the case in which X is associated with a regression model, the case in which $X = W$ in the unconstrained means models, and the case in which $X = W_R$ in the reduced version of the constrained means model. In the regression case, the assumption is satisfied if $N \geq m + 1$ in (1.6), unless extraneous input variables that are linear functions of other input variables are included in the model statement. For the unconstrained means model, W—as defined in general by (1.27)—has full column rank if all populations (for example, treatments or treatment combinations) have been observed (that is, all $n_i > 0$). If any $n_i = 0$, that population is simply not included in the analysis. In the constrained case, we shall see that W_R may have full column rank even though certain populations have not been sampled. This is a consequence of the known linear relations on the means. This situation, sometimes called the missing-cell problem, will be discussed in detail in Chapters 4 and 6. The lone exception to our full column rank assumption occurs in cases where the number and location of the missing cells lead us to what we call an unconnected model. Although we shall reduce this problem to one of full rank, it does require special treatment.

Apart from the general results presented in Chapter 3, this book is primarily devoted to the means model. There are several reasons for the decision to so restrict the discussion. The primary reason is that the initial motivation for this book was to clarify some misconceptions associated with analysis of the means model, especially in cases with unbalanced and/or missing data. The book was extended to include variance component or mixed models for essentially the same reason.

The second reason for the restriction is that, although regression and means models may be mathematically equivalent, there are important practical differences with regard to analysis and interpretation. Typically, the experiments leading to means models are well defined, and the basic structure of the model is clear. The problem is essentially a classical one of making inferences about the parameters—namely the means and variance components.

In most regression analyses, this is rarely the case. The analyst does not know the exact form of the model, and the problem is often one of model building. Even if the analyst has recorded all relevant input variables, there is often a question of the functions of these variables, which should be included in the model. Such decisions are usually made after several passes at the analysis. While the results in Chapter 3 are essential to the decisions that go into the modeling, they are of questionable use for inference on the final model.

In fact, the fitting of regression models may be more art than science. Many interesting problems arise in the process of developing the model. For example, even though X has full column rank, there may be near linear dependencies in the columns of X. The resulting problems were emphasized in the important papers by Hoerl and Kennard (1970a, b) and have been the subject of much discussion in the literature. Other problems also arise, since the values of the input variables are often not controlled by the researcher. It often happens that a small number of observations has an unreasonable influence on the results. The study of influential observations is discussed at length in Belsley, Kuh, and Welsch (1980) and Cook and Weisberg (1982).

While regression modeling is an important problem, the decision to exclude these topics from this book was based on the fact that there are already several good texts in this area, including those mentioned earlier in this chapter.

Finally, it is emphasized that the distinction is not as sharp as I have implied. Outlier and residual analysis recommended for regression analysis is equally important for the analysis of means. In the discussion of variance component estimation, for example, we will use an important graphical diagnostic to investigate influential observations. The student of linear models is urged to study both regression analysis and the analysis of means. It is hoped that the material in this book, however, will resolve many of the questions in the analysis of means and variance components.

CHAPTER 1 EXERCISES

1. Plot the data for the male students in Example 1.1. Sketch a straight line fitting the data, and determine the slope and intercept of this line. Do the same for the female students. Does it appear that the slopes of these two lines are the same? If you force the slopes to be the same, interpret the difference in the inputs. Does it appear that a straight line is adequate for either data set? Does a common slope seem reasonable?

2. For the data in Example 1.2, plot the following:
   (a) PROD vs WID
   (b) PROD vs DENS
   (c) PROD vs STR
   (d) DENS vs STR.
   Which variables appear to be most important with regard to describing production? Based on these plots, estimate—graphically—the rate of change of production with respect to either DENS or STR. Are there any observations that seem unusual? What do you infer from the DENS vs STR plot?
3. Determine the sample mean and variance for each of the flour types in Example 1.4. Determine the average variance. Repeat this for the data from Table 1.5. Compute the average of the sample variance for the six populations.
4. Compute the differences in the sample means corresponding to the constraint (1.33) in Example 1.7. Do you think that the no-interaction constraint might be valid for this experiment?
5. Plot the combined data from Example 1.1. Sketch the locus of means, $E(y)$, for the following four models:
   (a) $E(y) = \mu_i$
   (b) $E(y) = \theta_0 + \theta_1 x$
   (c) $E(y) = \gamma_i + \theta_1 x$
   (d) $E(y) = \gamma_i + \theta_1 x + \theta_2 z$
   where z is defined in (1.41). (Use approximate values for the parameters.)
6. Write the X matrix for the models (1.39) and (1.40).
7. Write out the covariance matrix for Example 1.9 using the notation in (1.44) and (1.46).

CHAPTER TWO
THE DISTRIBUTION OF LINEAR AND QUADRATIC FUNCTIONS OF NORMAL VECTORS

2.1. INTRODUCTION

The analysis of linear models leads to a study of the distribution of linear and quadratic functions of the response vector Y and of the ratios of such functions. In this chapter, we shall determine these distributions under the assumption that Y follows a multivariate normal distribution. The results developed here will be utilized in later chapters when we examine the distributional properties of estimators, develop the statistics for testing hypotheses, and determine confidence intervals for parameter functions.

2.2. THE MULTIVARIATE NORMAL DISTRIBUTION

We referred to the multivariate normal distribution in Chapter 1. In this section, we will recall the definition and some essential properties of this distribution. [For a detailed discussion, see, for example, Kshirsagar (1972).] We will introduce the concept of moments of vectors of random variables and also some related algebra.

2.2.1. Moments of Random Vectors

Since we will be dealing with vectors and matrices whose elements are random variables, it is convenient to introduce a notation for the expected value of such quantities. If M is a matrix whose elements, m_{ij}, are random variables, we define the expected value of M as the matrix whose elements are the expected values of the m_{ij}. Thus,

$$E(M) = [E(m_{ij})]. \tag{2.1}$$

(Note that we already used this notation in Chapter 1 to describe the expected value of the vector Y.)

With this definition, we may now establish the following results, which are the natural extensions of moments for scalars.

(a) Let Y be a vector with components y_i, such that $E(y_i) = \mu_i$ and hence $E(Y) = \mu$. The covariance matrix, $V = \text{Var}(Y)$—that is, the matrix whose elements, v_{ij}, represent the variances and covariances of the elements of Y—is given by

$$\begin{aligned} V &= \text{Var}(Y) \\ &= \{E[(y_i - \mu_i)(y_j - \mu_j)]\} \\ &= E[(Y - \mu)(Y - \mu)']. \end{aligned} \tag{2.2}$$

(b) If A is a matrix of constants and a is a vector of constants, then the linear function $Z = AY + a$ has the following moments:

$$E(Z) = A\mu + a = \delta, \tag{2.3}$$

and

$$\begin{aligned} \text{Var}(Z) &= \{E[(z_i - \delta_i)(z_j - \delta_j)]\} \\ &= E[(Z - \delta)(Z - \delta)'] \\ &= E[A(Y - \mu)(Y - \mu)'A'] \\ &= AVA'. \end{aligned} \tag{2.4}$$

(c) Let X be another linear function of Y—say, $X = BY + b$—and let $E(X) = B\mu + b = \eta$. Then the matrix whose elements are the covariances of the elements of X and those of Z is given by

$$\begin{aligned} \text{Cov}(X, Z) &= \{E[(x_i - \eta_i)(z_j - \delta_j)]\} \\ &= E[(X - \eta)(Z - \delta)'] \\ &= E[B(Y - \mu)(Y - \mu)'A'] \\ &= BVA'. \end{aligned} \tag{2.5}$$

(Note that these results do not depend on normality.)

2.2.2. The Density and Moment Generating Functions

We shall use the notation

$$Y \sim N_p(\mu, V) \tag{2.6}$$

to denote the fact that the p-vector Y follows a multivariate normal distribution with mean vector μ and covariance matrix V. This is a convenient notation for the random vector whose components have joint density

$$f(Y) = (2\pi)^{-p/2}|V|^{-1/2}\exp[-(Y - \mu)'V^{-1}(Y - \mu)/2] \tag{2.7}$$

where the components of Y satisfy $-\infty < y_i < \infty$. The moment generating function of Y is

$$m_Y(t) = E[\exp(t'Y)] = \exp[t'\mu + (t'Vt)/2]. \tag{2.8}$$

A fundamental property of normal vectors is the preservation of normality under linear transformations. That is, if A is an arbitrary $k \times p$ matrix and a an arbitrary k-vector, then

$$Z = AY + a \tag{2.9}$$

is distributed as $N_k(A\mu + a, AVA')$ in the sense that the moment generating function of Z is

$$m_Z(t) = \exp[t'(A\mu + a) + (t'AVA't)/2]. \tag{2.10}$$

If $k \leq p$ and the rank of A is k, then the density of Z is given by (2.7) with appropriate mean vector and covariance matrix. Otherwise, the covariance matrix, AVA', is singular, and Z is said to have a singular normal distribution. The simple interpretation of this is that Z consists of k_1 components which have a density of the form (2.7), where k_1 is the rank of A. The remaining $k - k_1$ components are linear functions of these k_1 components.

2.3. NONCENTRAL DISTRIBUTIONS

In our discussion of hypothesis testing, we shall be interested in the distribution of the test statistic under both the null and the alternative hypotheses. The distributions in the latter case are frequently called *noncentral distributions,* for reasons that will soon be apparent. In particular, we shall discuss the noncentral chi-square, the noncentral t, and the noncentral F distributions.

2.3.1. The Noncentral Chi-Square Distribution

From elementary distribution theory it is known that, if y is univariate, $N(0, 1)$, then y^2 follows a chi-square distribution with one degree of freedom. If y_i, $i = 1, \ldots, p$ are independent, $N(0, 1)$—that is, $Y \sim N_p(0, I)$—then

$$z = Y'Y \tag{2.11}$$

is distributed as (central) chi-square with p degrees of freedom. We now consider the distribution of z when $Y \sim N_p(\mu, I)$.

DEFINITION 2.1
If $Y \sim N_p(\mu, I)$, then $z = Y'Y$ is called a noncentral chi-square variable with p degrees of freedom and noncentrality parameter $\lambda = \mu'\mu/2$. Notationally, we shall write

$$z \sim \chi^2(p, \lambda). \tag{2.12}$$

■

THEOREM 2.1
If $z \sim \chi^2(p, \lambda)$, the moment generating function of z is given by

$$m_z(t) = (1 - 2t)^{-p/2}\exp[2t\lambda/(1 - 2t)] \tag{2.13}$$

for $t < 1/2$. The density function of z is given by

$$f(z) = \sum_{j=0}^{\infty} (e^{-\lambda}\lambda^j/j!)\chi^2(p + 2j, 0; z) \tag{2.14}$$

where $\chi^2(k, 0; z)$ denotes the density function for the central chi-square distribution with k degrees of freedom.

Proof: We shall just sketch the essential ideas. The moment generating function is obtained directly as

$$m_z(t) = E[\exp(tz)] = E[\exp(tY'Y)]. \tag{2.15}$$

The last expectation in (2.15), taken with respect to the density of Y, yields (2.13) after straightforward manipulations.

To establish the density function, it is sufficient to show that the moment-generating function of the density, (2.14), agrees with (2.13). From (2.14), using the moment-generating function for the central chi-square distribution, we have

$$\begin{aligned} m_z(t) &= \sum_{j=0}^{\infty} (e^{-\lambda}\lambda^j/j!)(1 - 2t)^{-[(p/2)+j]} \\ &= (1 - 2t)^{-p/2} \sum_{j=0}^{\infty} e^{-\lambda}[\lambda/(1 - 2t)]^j/j! \\ &= (1 - 2t)^{-p/2}e^{-\lambda}e^{\lambda/(1-2t)}. \end{aligned} \tag{2.16}$$

This expression is easily reduced to (2.13), and the proof is complete.

■

The noncentral chi-square distribution also has an interesting geometric interpretation. Using (2.11), the cumulative distribution of z may be written as

$$\begin{aligned} F_z(z^*) &= \text{Prob}(z \leq z^*) \\ &= \text{Prob}(Y'Y \leq z^*) \\ &= \text{Prob}\left(\sum y_i^2 \leq z^*\right). \end{aligned} \tag{2.17}$$

Thus, $F_z(z^*)$ describes the probability content of a sphere centered at the origin with squared radius z^*. If $E(Y) = \mu$, then the spherical contours of the normal density are centered at μ. Figure 2.1 illustrates the situation for the case $p = 2$.

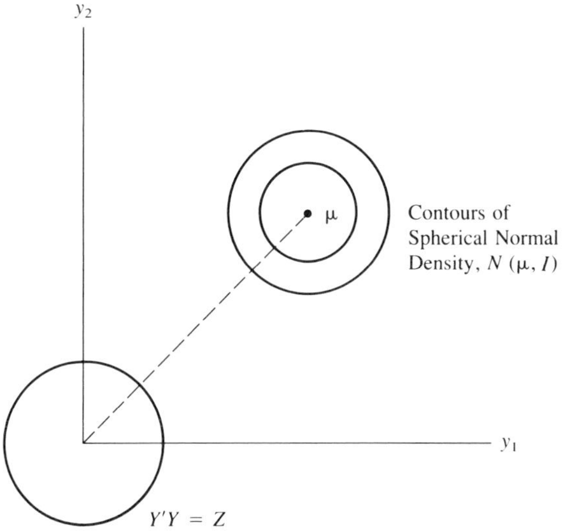

FIGURE 2.1
Noncentral chi-square probabilities

By the symmetry of the spherical normal density, the probability content of this sphere depends only on the distance from the origin to the center of the distribution. Thus, this distance or, equivalently, its square, $\mu'\mu$, is an essential parameter of the distribution. Historically, the parameter $\lambda = \mu'\mu/2$ has been used and is appropriately called the *noncentrality parameter*. In view of this interpretation, the cumulative distribution function of z could be obtained by integrating the normal density over the sphere centered at the origin.

We shall need a notation for the percentage points of this distribution. The upper 100α percentage point will be denoted by $\chi^2(\alpha; p, \lambda)$. That is

$$F_z[\chi^2(\alpha; p, \lambda)] = 1 - \alpha. \tag{2.18}$$

There should be no confusion over the use of the notation $\chi^2(p, \lambda)$ to denote the random variable, of $\chi^2(p, \lambda; z)$ to denote the density, and of $\chi^2(\alpha; p, \lambda)$ to denote the percentage points. In the central case, $\lambda = 0$, we shall write either $\chi^2(p)$ or $\chi^2(p, 0)$.

Additional properties of the noncentral chi-square density are contained in the following corollary.

Corollary 2.1.1

(a) If z_i, $i = 1, \ldots, k$ are independent $\chi^2(p_i, \lambda_i)$, then

$$z = \sum_{i=1}^{k} z_i \sim \chi^2(p, \lambda) \tag{2.19}$$

with $p = \Sigma p_i$ and $\lambda = \Sigma \lambda_i$.

(b) The first two moments of the distribution are

$$E[\chi^2(p, \lambda)] = p + 2\lambda \tag{2.20}$$

and

$$\text{Var}[\chi^2(p, \lambda)] = 2p + 8\lambda. \tag{2.21}$$

Proof: The first result follows directly from the moment-generating function, and the second is a consequence of the moments of the central chi-square distribution and the expression (2.14) for the density, or it may be obtained directly from the moment generating function.

■

An immediate extension of Theorem 2.1 to the general multivariate normal vector, Y, is given by the following corollary.

Corollary 2.1.2
If $Y \sim N_p(\mu, V)$, then

$$z = Y'V^{-1}Y \sim \chi^2(p, \lambda) \tag{2.22}$$

with $\lambda = \mu'V^{-1}\mu/2$.

Proof: From Appendix A.I.9 there exists a nonsingular matrix C such that $V = C'C$. Let $Y = C'X$, and note that

$$\begin{aligned} X &\sim N[(C')^{-1}\mu, I] \\ z &= Y'V^{-1}Y = X'X. \end{aligned} \tag{2.23}$$

Thus, by Theorem 2.1, z is distributed as $\chi^2(p, \lambda)$, with

$$\lambda = \mu'C^{-1}C'^{-1}\mu/2 = \mu'V^{-1}\mu/2. \tag{2.24}$$

■

2.3.2. The Noncentral t Distribution

A natural extension of the Student's t distribution is given in the following definition.

DEFINITION 2.2
If y and z are independent random variables with $y \sim N(\delta, 1)$ and $z \sim \chi^2(p, 0)$, then the ratio

$$t = y/(z/p)^{1/2} \tag{2.25}$$

is called a noncentral t variable with noncentrality parameter δ.

■

Note that the special case $\delta = 0$ yields the (central) Student's t statistic. The central t statistic arises in simple applications where we are testing the hypothesis H_0: $\mu = 0$ in a normal population. The test statistic has the form (2.25) and follows the central t distribution if $\mu = 0$. If $\mu = \delta \neq 0$, then the test statistic follows the noncentral t distribution. This distribution can be used to compute the power of this test as a function of μ.

As with the noncentral chi-square distribution, we shall use the notation $t(p, \delta)$ to denote the random variable and use $t(\alpha; p, \delta)$ to denote the upper 100α percentage point of the distribution.

We shall not provide further detail on this distribution, but you may refer to Owen (1968) for the evaluation of percentage points and a discussion of the theory and application of this distribution.

2.3.3. The Noncentral F Distribution

The noncentral t distribution arises in a special case of our discussion of hypothesis testing. More generally, we shall encounter test statistics that are ratios of independent chi-square random variables. In particular, the denominator will always be a central chi-square variable, while the numerator may be noncentral.

DEFINITION 2.3

If $z_1 \sim \chi^2(p_1, \lambda)$ and $z_2 \sim \chi^2(p_2, 0)$, where z_1 and z_2 are independent, then the ratio

$$u = (z_1/p_1)/(z_2/p_2) \tag{2.26}$$

is called a noncentral F variable with noncentrality parameter λ. We will use the notation $F(p_1, p_2, \lambda)$ to refer to this random variable and we will let $F(\alpha; p_1, p_2, \lambda)$ denote the upper 100α percentage point.

■

To relate the noncentral t and F distributions, recall that, in the central case, if $t \sim t(p, 0)$, then $t^2 \sim F(1, p, 0)$. In the noncentral case if $t \sim t(p, \delta)$, then $t^2 \sim F(1, p, \lambda)$ with $\lambda = \delta^2/2$.

We shall use the noncentral F distribution to determine the power of tests of hypotheses on linear functions of the linear model parameters. Under our null hypotheses, the test statistics will follow a central F distribution, $F(p_1, p_2, 0) = F(p_1, p_2)$. Thus, we will be interested in the power defined by

$$\prod(\alpha; p_1, p_2, \lambda) = \text{Prob}[u > F(\alpha; p_1, p_2)]. \tag{2.27}$$

It has been shown by Patnaik (1949) that an approximation to the noncentral F distribution is given by the approximate relation

$$F(p_1, p_2, \lambda) \mathrel{\dot{\sim}} [(p_1 + 2\lambda)/p_1]F(p^*, p_2) \tag{2.28}$$

where

$$p^* = (p_1 + 2\lambda)^2/(p_1 + 4\lambda). \tag{2.29}$$

It follows that we may use central F tables to approximate the noncentral F values. Thus, we write (2.27) as

$$\prod (\alpha; p_1, p_2, \lambda) \doteq \text{Prob}\{F(p^*, p_2) > [p_1/(p_1 + 2\lambda)]F(\alpha; p_1, p_2)\}. \tag{2.30}$$

Another useful result is to observe the behavior of (2.27) as a function of λ, p_1, p_2, and α. Ghosh (1973) established the following relations:

$$\prod (\alpha; p_1, p_2, \lambda)$$

is

(1) monotone increasing in λ
(2) monotone increasing in α (2.31)
(3) monotone increasing in p_2
(4) monotone decreasing in p_1.

For each of the statements in (2.31) it is assumed that the remaining parameters are fixed.

2.4. THE DISTRIBUTION OF QUADRATIC FORMS

The test statistics we encounter in the analysis of linear models will depend on quadratic forms of normal random variables; hence we are interested in the distributional properties of such statistics. In particular, we shall examine the distribution of

$$q = Y'AY \tag{2.32}$$

where A is symmetric and $Y \sim N(\mu, V)$. Special cases of (2.32), in which q is a noncentral chi-square variable, namely $A = V = I$ and $A = V^{-1}$, have already been noted. We shall be interested in determining general requirements on A for which q is so distributed.

The first and second moments of q will be used repeatedly in our work, and we will use the moment generating function to prove some of the distributional results. The following theorem provides these facts.

THEOREM 2.2
If $Y \sim N_p(\mu, V)$, then the quadratic form $q = Y'AY$ has moment generating function

$$m_q(t) = |I - 2tAV|^{-1/2}\exp\{-\mu'[I - (I - 2tAV)^{-1}]V^{-1}\mu/2\} \tag{2.33}$$

for $t < t_0$, the smallest root of the determinant in (2.33).

The proof of this theorem follows in the usual way by evaluating $E[\exp(tY'AY)]$. ■

Corollary 2.2.1

Let $q_1 = Y'A_1Y$, $q_2 = Y'A_2Y$, and $T = BY$ where B is an $r \times p$ matrix. Then,

$$\begin{aligned}
&\textbf{(1)}\ E(q_1) = tr(A_1V) + \mu'A_1\mu \\
&\textbf{(2)}\ \mathrm{Var}(q_1) = 2tr(A_1VA_1V) + 4\mu'A_1VA_1\mu \\
&\textbf{(3)}\ \mathrm{Cov}(q_1, q_2) = 2tr(A_1VA_2V) + 4\mu'A_1VA_2\mu \\
&\textbf{(4)}\ \mathrm{Cov}(Y, q_1) = 2VA_1\mu \\
&\textbf{(5)}\ \mathrm{Cov}(T, q_1) = 2BVA_1\mu.
\end{aligned} \tag{2.34}$$

Proof: Parts (1) and (2) may be obtained directly from $m_q(t)$ using the rules for differentiation of determinants and matrices given in Appendix A.II.1. Alternatively, we may obtain (1) as follows:

$$\begin{aligned}
E(Y'AY) &= E[tr(AYY')] \\
&= tr\{E[A(Y - \mu)(Y - \mu)'] + A\mu\mu'\} \\
&= tr(AV) + \mu'A\mu.
\end{aligned} \tag{2.35}$$

From this development we observe that this result is actually independent of the assumption of normality. The expression for the variance, (2), does depend on the normality assumption.

The third result follows by applying (2) to the sum of A_1 and A_2. Thus,

$$\begin{aligned}
\mathrm{Cov}[Y'(A_1 + A_2)Y] = {} & 2tr[(A_1 + A_2)V(A_1 + A_2)V] \\
& + 4\mu'(A_1 + A_2)V(A_1 + A_2)\mu.
\end{aligned} \tag{2.36}$$

Also,

$$\mathrm{Cov}(Y'A_1Y + Y'A_2Y) = \mathrm{Var}(q_1) + \mathrm{Var}(q_2) + 2\mathrm{Cov}(q_1, q_2). \tag{2.37}$$

Equating the right sides of these equations and solving for $\mathrm{Cov}(q_1, q_2)$ establishes the result.

To establish (4), note that this expression denotes the vector of covariances of q_1 with the elements of Y. We have

$$\begin{aligned}
\mathrm{Cov}(y_i, q_1) &= \mathrm{Cov}\left(y_i, \sum_k \sum_j a_{kj} y_k y_j \right) \\
&= 2 \sum_k \sum_j v_{ik} a_{kj} \mu_j.
\end{aligned} \tag{2.38}$$

This follows since, for the normal distribution, first and third moments are zero. Expressing (2.38) for $i = 1, \ldots, p$ as a vector yields (4). The final result, (5), follows immediately from (4).

■

We now give several theorems on the distribution of quadratic forms. For motivation, we examine the results of Theorem 2.2 and Corollary 2.2.1. It is apparent that the behavior of the matrix AV is important. Comparing the mean and variance of q with the first two moments of the noncentral chi-square distribution, we note that $tr(AV)$ plays the role of degrees of freedom. Hence, if q is to be chi-square, this must be integral. Further, the variance expression suggests $AVAV = AV$. These two facts suggest that AV is idempotent. (See Appendix A.I.6.) Pursuing the analogy suggests $\lambda = \mu'A\mu/2$. These observations will be confirmed in Theorem 2.3. Continuing with this search for clues, we see by (2.34) that suggested conditions for independence of pairs of quadratic forms and linear and quadratic forms are $A_1VA_2 = 0$ and $BVA = 0$. These observations will be confirmed in Theorems 2.4 and 2.5.

THEOREM 2.3

If $Y \sim N_p(\mu, V)$, then $q = Y'AY \sim \chi^2(r, \lambda)$, for *all* μ, with r denoting the rank of A and $\lambda = \mu'A\mu/2$ if and only if AV is idempotent.

Proof: Suppose AV is idempotent, and let $V = C'C$ where C is nonsingular. Then, $B = CAC'$ is idempotent of rank $r = r(A)$. From Appendices A.I.6 and A.I.7, there exists an orthogonal matrix P such that

$$P'BP = \begin{bmatrix} I_r & 0 \\ 0 & 0 \end{bmatrix}. \tag{2.39}$$

Letting $Z = P'(C')^{-1}Y$, we see that

$$Z \sim N_p[P'(C')^{-1}\mu, I] \tag{2.40}$$

and

$$q = Y'AY = Z'P'CAC'PZ = Z_1'Z_1 \tag{2.41}$$

where Z_1 represents the first r components of Z. Further, Z_1 has mean vector,

$$E(Z_1) = (I_r 0)P'(C')^{-1}\mu. \tag{2.42}$$

Hence, using Theorem 2.1,

$$q \sim \chi^2(r, \lambda) \tag{2.43}$$

and

$$\begin{aligned}\lambda &= (\mu'C^{-1}PP'BPP'(C')^{-1}\mu/2 \\ &= (\mu'A\mu)/2.\end{aligned} \tag{2.44}$$

Conversely, given that $q \sim \chi^2(r, \lambda)$, compare the moment-generating function for the noncentral chi-square distribution from Theorem 2.1 with that for a general quadratic form from Theorem 2.2 and realize that they must agree for all μ, and hence for $\mu = 0$. This implies that

$$(1 - 2t)^r = |I - 2tAV| = \prod_{i=1}^{p} (1 - 2t\theta_i). \tag{2.45}$$

Here, θ_i are the eigenvalues of AV, and hence $(1 - 2t\theta_i)$ are the eigenvalues of $(I - 2tAV)$. Equating coefficients in the two polynomials in t, we see that the θ_i must be either zero or one and, hence, from Appendix A.I.6, AV is idempotent.

■

THEOREM 2.4
If $Y \sim N_p(\mu, V)$, then $q_1 = Y'A_1Y$ and $q_2 = Y'A_2Y$ are independent, for *all* μ, if and only if $A_1VA_2 = A_2VA_1 = 0$.

Proof: Assuming $A_1VA_2 = A_2VA_1 = 0$, let $V = C'C$ and let

$$Q = CA_1C' \qquad T = CA_2C'. \tag{2.46}$$

Note that QT is symmetric ($QT = 0$) and hence, by Appendix A.I.7, there is an orthogonal matrix, P, such that $P'QP$ and $P'TP$ are diagonal. But, since $QT = 0$, the nonzero diagonal elements of $P'QP$ and $P'TP$ must occur in different positions, say,

$$P'QP = \begin{bmatrix} D_1 & 0 \\ 0 & 0 \end{bmatrix} \qquad P'TP = \begin{bmatrix} 0 & 0 \\ 0 & D_2 \end{bmatrix}. \tag{2.47}$$

Letting $Z = P'(C')^{-1}Y$, we see that

$$q_1 = Y'A_1Y = Z_1'D_1Z_1 \tag{2.48}$$

and

$$q_2 = Y'A_2Y = Z_2'D_2Z_2 \tag{2.49}$$

where Z has been partitioned to conform with D_1 and D_2. Since, by (2.40), Z_1 and Z_2 are independent, the same holds for q_1 and q_2. Note that the same argument holds if we assume $A_2VA_1 = 0$.

Conversely, since q_1 and q_2 are independent, it follows that $\text{Cov}(q_1, q_2) = 0$. Using Corollary 2.2.1, we have

$$2tr(A_1VA_2V) + 4\mu'A_1VA_2\mu = 0 \tag{2.50}$$

for all μ. Letting $\mu = 0$ implies that

$$tr(A_1VA_2V) = 0 \tag{2.51}$$

and hence for all μ,

$$\mu'A_1VA_2\mu = 0 \tag{2.52}$$

which implies $A_1VA_2 = 0$. Reversing the roles of A_1 and A_2 implies that $A_2VA_1 = 0$.

■

THEOREM 2.5
If $Y \sim N_p(\mu, V)$, $q = Y'AY$, and $T = BY$, then q and T are independent, for *all* μ, if and only if $BVA = 0$.

Proof: Suppose $BVA = 0$, let $V = C'C$ and let P be an orthogonal matrix which diagonalizes $Q = CAC'$. Letting $Z = P'(C')^{-1}Y$, we have

$$q = Y'AY = Z'DZ \tag{2.53}$$

and

$$T = BY = GZ \tag{2.54}$$

where $D = P'QP$ and $G = BC'P$. We may then write

$$0 = BVA = GDP'(C')^{-1}, \tag{2.55}$$

or equivalently, $GD = 0$. Partitioning as usual, we have

$$G = (G_1|G_2) \qquad D = \begin{bmatrix} D_1 & 0 \\ 0 & 0 \end{bmatrix} \tag{2.56}$$

where $r(D_1) = r(A)$. But $GD = 0$ implies $G_1D_1 = 0$, hence

$$T = G_2Z_2 \qquad q = Z_1'D_1Z_1 \tag{2.57}$$

and the independence follows from the independence of Z_1 and Z_2.
Conversely, from Corollary 2.2.1 we have

$$0 = \text{Cov}(T, q) = 2BVA\mu \tag{2.58}$$

for all μ and hence $BVA = 0$.

■

Note: Theorems 2.3, 2.4 and 2.5 all require that the stated distributional property hold for *all* μ in order to establish the matrix conditions (AV idempotent, $A_1VA_2 = 0$, $BVA = 0$). Actually it is only required that the properties hold for a specific vector μ. The proofs of these stronger theorems are more difficult and are omitted since our primary interest is in the sufficiency of the conditions. For an interesting historical account and proof of Theorem 2.4, see Gundberg (1984).

THEOREM 2.6 (Cochran's Theorem)
Let $Y \sim N_p(\mu, V)$, A_i, $i = 1, \ldots, m$ be symmetric, $r(A_i) = r_i$, and

$$A = \sum A_i \tag{2.59}$$

with $r(A) = r$. If AV is idempotent, and

$$r = \sum r_i \tag{2.60}$$

then $q_i = Y'A_iY$ are mutually independent with

$$q_i \sim \chi^2(r_i, \mu'A_i\mu/2). \tag{2.61}$$

Proof: Transform, as usual, to the case $V = I$ and appeal to result (5) on idempotent matrices in Appendix A.I.6.

■

EXAMPLE 2.1

To illustrate Theorems 2.2 through 2.6, let $Y \sim N(\mu J, \sigma^2 I)$ where μ is a scalar and J is an n-vector of ones. Let y_i denote the elements of Y and let

$$\bar{y}_. = J'Y/n \tag{2.62}$$

$$\begin{aligned} q_1 &= n\bar{y}_.^2 \\ &= Y'(JJ'/n)Y \end{aligned} \tag{2.63}$$

$$\begin{aligned} q_2 &= \sum(y_i - \bar{y}_.)^2 \\ &= Y'(I - JJ'/n)Y. \end{aligned} \tag{2.64}$$

Then, $\bar{y}_.$ and q_2 are independent, q_1 and q_2 are independent, and

$$q_1 \sim \sigma^2\chi^2[1, n\mu^2/(2\sigma^2)] \tag{2.65}$$

$$q_2 \sim \sigma^2\chi^2(n - 1, 0). \tag{2.66}$$

Further,

$$E(q_1) = \sigma^2 + n\mu^2 \tag{2.67}$$

$$E(q_2) = (n - 1)\sigma^2. \tag{2.68}$$

The distributional results on q_1 and q_2 follow from Theorems 2.3 and 2.4. Alternatively, they are obtained directly from Theorem 2.6 since, with $A = I$, we have

$$I = JJ'/n + (I - JJ'/n) \tag{2.69}$$

and

$$n = 1 + (n - 1). \tag{2.70}$$

This result is particularly useful when decomposing quadratic forms into sums of quadratic forms. In this case, the "total sum of squares," $Y'Y$, is expressed as the sum of the "residual sum of squares," q_2, and the "sum of squares due to the mean," q_1.

■

CHAPTER 2 EXERCISES

1. If $X \sim N_p(\mu, I)$, find the distribution of $Y = AX + a$. Use this result to find the distribution of $\bar{y} = J'Y/p$.
2. If $Y \sim N_p(\mu, V)$, verify that the moment generating function of Y is given by (2.8).
3. If $Y \sim N_p(\mu, V)$ and $Z = AY + a$, verify that the moment generating function of Z is given by (2.10).
4. Use (2.15) to establish the moment generating function of $z = Y'Y$ when $Y \sim N_p(\mu, I)$.
5. Verify the results of Corollary 2.1.1. Determine the moments in part (2) from the density function and from the moment generating function.
6. Let $u \sim F(p_1, p_2, \lambda)$.
   (a) Determine $E(u)$.
   (b) Find the density of u.
7. Let $Y \sim N_n(\mu J, \sigma^2 I)$, and let H be the Helmert matrix defined as follows: The first row of H is $J'/\sqrt{n}$. For $r = 2, \ldots, n$, the rth row of H has its first $r - 1$ components equal to $[r(r-1)]^{-1/2}$, the rth component equals $-(r-1)/[r(r-1)]^{1/2}$, and the remaining components are zero.
   (a) Show that H is an orthogonal matrix.
   (b) Determine the distribution of $Z = HY$.
   (c) Use this transformation to determine the joint distribution of $\bar{y}$ and $S^2 = Y'(I - JJ'/n)Y/\sigma^2$.
8. Let $Y \sim N_p(\mu, V)$ where $V = (1 - r)I_p + rJ_pJ_p'$.
   (a) For $p = 2$ and $p = 3$, determine the distribution of $Z = AY$ where

$$A = \begin{bmatrix} 1 & 1 \\ 1 & -1 \end{bmatrix} \text{ for } p = 2 \qquad A = \begin{bmatrix} 1 & 1 & 1 \\ 1 & -1 & 0 \\ 1/2 & 1/2 & -1 \end{bmatrix} \text{ for } p = 3.$$

   (b) Generalize this idea for $p > 3$.
9. Derive the distribution of the noncentral chi-square distribution as follows:
   (a) Let $z = Y'Y$ where $Y \sim N_p(\mu, I)$, and let $X = AY$ where A is an orthogonal matrix whose first row is $\mu'/\gamma^{1/2}$ with $\gamma = \mu'\mu$.
   (b) Let $v = x_1^2$ and $u = X'X - v$. Note that $u \sim \chi^2(p - 1)$, and show that v is independent of u with density

$$f(v) = (2\pi v)^{-1/2}\exp[-(v + \gamma)/2] \sum_{r=0}^{\infty} (v\gamma)^r/(2r)!$$

   for $v \geq 0$.
   (c) Find the joint distribution of $z = u + v$ and $t = u/(u + v)$, and then find the marginal distribution of z. To relate to (2.14), note that $\lambda = \gamma/2$ and

$$\Gamma[k + (1/2)] = \Gamma(2k + 1)\sqrt{\pi}/[2^{2k}\Gamma(k + 1)].$$

10. Let $Y \sim N_n(\mu J, \sigma^2 I)$ denote a random sample of size n from $N(\mu, \sigma^2)$. For testing the hypothesis H_0: $\mu = 0$ against the alternative H_A: $\mu \neq 0$, we may use the test statistic

$$F = \bar{y}^2/(s^2/n)$$

where $\bar{y} = J'Y/n$ and $s^2 = Y'(I - JJ'/n)Y/(n - 1)$.

(a) Use Exercise 7 to determine the distribution of F.

(b) Use the approximation (2.28) to compute the power of this test for $\alpha = 0.05$ and $\mu/(\sigma/\sqrt{n}) = 1, 2, 5$.

11. Verify the moment generating function in (2.33).

12. Verify (2.38).

13. Illustrate (2.45) for the case $r = 2$, $p = 3$.

14. Suppose $Y \sim N(\mu, \sigma^2 I)$ and let $q_i = Y'A_iY$, $i = 1, 2, 3$, where

$$A_1 = (1/3)JJ' \qquad A_2 = (1/2)\begin{bmatrix} 1 & -1 & 0 \\ -1 & 1 & 0 \\ 0 & 0 & 0 \end{bmatrix}$$

$$A_3 = (1/6)\begin{bmatrix} 1 & 1 & -2 \\ 1 & 1 & -2 \\ -2 & -2 & 4 \end{bmatrix}$$

Determine the distribution of the q_i. Check for independence of these quadratic forms.

15. Suppose $Y \sim N(W\mu, \sigma^2 I)$ where $W = I_a \otimes J_n$. Let $q_i = Y'A_iY$, $i = 1, 2$, where

$$A_1 = I - (1/n)WW'$$
$$A_2 = (1/n)WW' - (1/an)JJ'.$$

(a) Determine the distribution of q_1 and q_2.

(b) Check for independence.

(c) Determine the distribution of $(q_2/r_2)/(q_1/r_1)$ where r_1 and r_2 are the ranks of A_1 and A_2.

16. Suppose $Y \sim N(\mu J, V)$ where

$$V = \phi_0 I + \phi_1(I_a \otimes J_n J_n')$$
$$\phi_0 > 0 \qquad \phi_0 + n\phi_1 > 0.$$

(a) Determine the distribution of q_1 and q_2 as defined in Exercise 15.

(b) Determine the distribution of the ratio in (c) of Exercise 15.

17. In Exercise 15, define $M_i = A_i/\sigma^2$, $i = 1, 2$, and $M = M_1 + M_2$. Apply Theorem 2.6 to determine the joint distribution of q_1 and q_2.

18. In Exercise 16, define $M_1 = A_1/\phi_0, M_2 = A_2/(\phi_0 + n\phi_1)$, and $M = M_1 + M_2$. Apply Theorem 2.6 to determine the joint distribution of q_1 and q_2.

19. Suppose $Y \sim N(\mu, V)$ and $q = Y'AY \sim \chi^2(r, \lambda)$.

(a) Develop an expression for the expected value of q in terms of the noncentrality parameter.

(b) Determine the expected value of q_1 and q_2 in Exercises 15 and 16.

CHAPTER THREE
ESTIMATION AND HYPOTHESIS TESTING FOR SIMPLE LINEAR MODELS

3.1. INTRODUCTION

In Chapter 1, we used the term *linear model* to describe situations in which the expected value of a response is a linear function of a set of unknown parameters. The model also allowed for linear equality constraints on these parameters. The term *simple linear model* refers to the special case in which observations on the model are assumed to be uncorrelated with constant variance. For the purpose of making statistical inferences, we further assume that the observations follow a normal distribution.

In matrix notation, this simple linear model is summarized by writing the N-vector of observations, Y, as

$$Y = X\theta + e \tag{3.1}$$

where

$$e \sim N(0, \sigma^2 I_N). \tag{3.2}$$

The matrix, X, is assumed to be known and of dimension $N \times p$ of rank $r(X) = p$, and θ denotes the p-vector of unknown parameters in the model. The variance, σ^2, is also unknown. If there are known linear relations on θ, we write them as

$$G\theta = g. \tag{3.3}$$

The matrix, G, is assumed to be $q \times p$ of rank q.

In Chapter 1, we distinguished between *regression models* and *means models* but noted that they are mathematically equivalent in the sense of the general definition of linear models. The objective of this chapter is to develop the general results for point and interval estimation of these parameters and to develop the test statistics for testing linear hypotheses on θ. These results will apply to both regression and means models, but our illustrations will focus on the former. In Chapter 4, we will restrict the discussion to means models, noting certain simplifications that arise and developing additional results that are unique to means models. In this chapter and in Chapter 4, the theory is described separately for *unconstrained* and *constrained models*. It has been noted that any constrained model can be reduced to an unconstrained model. Conversely, the constrained analysis includes the unconstrained situation as a special case. We find it instructive to treat the two situations separately. This approach will be especially useful when we consider means models.

3.2. ESTIMATION OF PARAMETERS

For parameter estimation, we shall use the method of maximum likelihood discussed in Appendix B.I.2. In the case of constrained models, the maximization is subject to those same constraints. The natural logarithm of the likelihood function for the model given by (3.1) and (3.2) is

$$\begin{aligned} \text{Ln L}(\theta, \sigma^2) &= -(N/2)\text{Ln}(2\pi) - (N/2)\text{Ln } \sigma^2 \\ &\quad - (Y - X\theta)'(Y - X\theta)/(2\sigma^2). \end{aligned} \tag{3.4}$$

Point estimates of θ and σ^2 are obtained by maximizing (3.4), subject to the constraints $G\theta = g$ if appropriate.

3.2.1. The Unconstrained Model

The likelihood equations are obtained by differentiating (3.4) with respect to θ and σ^2. Applying the rules given in Appendix A.II.2 for differentiation of functions with respect to vectors, and simplifying, we find that the likelihood equations are given by

$$X'X\hat{\theta} = X'Y \tag{3.5}$$

and

$$\hat{\sigma}^2 = Q(\hat{\theta})/N \tag{3.6}$$

where

$$Q(\theta) = (Y - X\theta)'(Y - X\theta). \tag{3.7}$$

Under the assumption that X has full column rank, we obtain the solution

$$\hat{\theta} = (X'X)^{-1}X'Y \tag{3.8}$$

$$\hat{\sigma}^2 = Y'(I - X(X'X)^{-1}X')Y/N. \tag{3.9}$$

The nonsingularity of $X'X$ establishes the uniqueness of the solution, and it is easily verified that $(\hat{\theta}, \hat{\sigma}^2)$ maximizes the likelihood function.

To establish the properties of these estimators, note, from (3.8), that $\hat{\theta}$ is a linear function of the normal vector, Y; hence, applying the results of Section 2.2.2, it follows that $\hat{\theta}$ is normally distributed with mean and covariance matrix

$$E(\hat{\theta}) = (X'X)^{-1}X'E(Y) = \theta \tag{3.10}$$

and

$$\mathrm{Var}(\hat{\theta}) = \sigma^2(X'X)^{-1}. \tag{3.11}$$

The distribution of $Q(\hat{\theta})$ is established by noting that the matrix of the quadratic form in (3.9) is idempotent. Application of Theorem 2.3 shows that

$$Q(\hat{\theta})/\sigma^2 \sim \chi^2(N - p). \tag{3.12}$$

The independence of $\hat{\theta}$ and $Q(\hat{\theta})$ is a direct consequence of Theorem 2.5. From (3.9) and (3.12) we see that

$$E(\hat{\sigma}^2) = [(N - p)/N]\sigma^2. \tag{3.13}$$

This suggests the use of the unbiased estimator

$$s^2 = Q(\hat{\theta})/(N - p). \tag{3.14}$$

The optimality properties of these estimators follow by using Appendices B.I.4 and B.I.5 to show that $\hat{\theta}$ and $Q(\hat{\theta})$ are a complete set of sufficient statistics for θ and σ^2. The Rao-Blackwell Theorem and the Lehmann-Scheffé Theorem, Appendices B.I.5 and B.I.6, may be used to establish that all functions of $\hat{\theta}$ and $Q(\hat{\theta})$ are uniformly minimum variance, unbiased estimates of their expected values.

The properties of the estimators are summarized in the following theorem.

THEOREM 3.1

For the unconstrained model

$$Y = X\theta + e$$

with

$$e \sim N(0, \sigma^2 I),$$

the minimum variance, unbiased estimators of θ and σ^2 are given by

$$\hat{\theta} = (X'X)^{-1}X'Y$$

and

$$s^2 = Y'[I - X(X'X)^{-1}X']Y/(N - p).$$

Further, $\hat{\theta}$ and s^2 are independently distributed as follows:

$$\hat{\theta} \sim N[\theta, \sigma^2(X'X)^{-1}]$$

and

$$s^2 \sim [\sigma^2/(N-p)]\chi^2(N-p).$$

■

3.2.2. The Constrained Model

The estimation problem for the constrained model is solved by maximizing the likelihood function or, equivalently, the log likelihood, (3.4), subject to the constraints. That is, we consider the problem

$$\begin{aligned} &\text{maximize Ln L}(\theta, \sigma^2) \\ &\text{subject to} \quad G\theta = g. \end{aligned} \tag{3.15}$$

There are two general approaches to solving this problem. The first, which was indicated in Example 1.7, will be called the *model reduction method.* For this method, the constraints are solved for q of the parameters in terms of the remaining $r = p - q$ parameters, and the result is substituted into the likelihood function to yield an unconstrained model with r parameters.

To demonstrate this approach, assume that θ and G are partitioned so that the constraints are written as

$$G_1\theta_1 + G_2\theta_2 = g, \tag{3.16}$$

where G_1 is $q \times q$ of rank q. Solving for θ_1 yields

$$\theta_1 = G_1^{-1}g - G_1^{-1}G_2\theta_2. \tag{3.17}$$

Partitioning X in the same way as $X = (X_1 | X_2)$ and substituting into the model, we obtain

$$Y - X_1G_1^{-1}g = (X_2 - X_1G_1^{-1}G_2)\theta_2 + e. \tag{3.18}$$

For simplicity, define

$$\begin{aligned} Y_R &= Y - X_1G_1^{-1}g \\ X_R &= X_2 - X_1G_1^{-1}G_2 \\ \theta_R &= \theta_2 \end{aligned} \tag{3.19}$$

and note that X_R has rank r. Then, the reduced, unconstrained model, (3.18), is given by

$$Y_R = X_R\theta_R + e. \tag{3.20}$$

The maximum likelihood analysis of this model is identical to that given in Section 3.2.1, with likelihood equations given by (3.5) through (3.7). Thus, the minimum variance, unbiased estimators for θ_R and σ^2 are given by

$$\hat{\theta}_R = (X_R'X_R)^{-1}X_R'Y_R \tag{3.21}$$

and

$$s_R^2 = Q_R(\hat{\theta}_R)/(N - r) \tag{3.22}$$

where

$$Q_R(\theta_R) = (Y_R - X_R\theta_R)'(Y_R - X_R\theta_R). \tag{3.23}$$

Proceeding exactly as in Section 3.2.1, we see that these estimators are independently distributed with

$$\hat{\theta}_R \sim N[\theta_R, \sigma^2(X_R'X_R)^{-1}] \tag{3.24}$$

and

$$s_R^2 \sim [\sigma^2/(N - r)]\chi^2(N - r). \tag{3.25}$$

Letting $\tilde{\theta}_2 = \hat{\theta}_R$, the remaining components of θ are estimated by substituting $\tilde{\theta}_2$ in (3.17) to obtain

$$\tilde{\theta}_1 = G_1^{-1}g - G_1^{-1}G_2\tilde{\theta}_2. \tag{3.26}$$

Notationally, we may combine (3.21) and (3.26) to write the estimator for θ as

$$\tilde{\theta} = \begin{bmatrix} \tilde{\theta}_1 \\ \tilde{\theta}_2 \end{bmatrix} = \begin{bmatrix} G_1^{-1}g \\ 0 \end{bmatrix} + \begin{bmatrix} -G_1^{-1}G_2 \\ I_r \end{bmatrix} \hat{\theta}_R. \tag{3.27}$$

Note that

$$E(\tilde{\theta}) = \theta \tag{3.28}$$

and

$$\operatorname{Var}(\tilde{\theta}) = \sigma^2 \begin{bmatrix} -G_1^{-1}G_2 \\ I_r \end{bmatrix} (X_R'X_R)^{-1} \begin{bmatrix} -G_1^{-1}G_2 \\ I_r \end{bmatrix}'. \tag{3.29}$$

This covariance matrix is $p \times p$ of rank r; hence, $\tilde{\theta}$ has a singular, normal distribution, as described in Section 2.1. This singularity need not concern us, since, for analysis, the actual dimension of this problem is r, and we are concerned only with $\theta_R = \theta_2$, or the linear functions of θ_R defined by (3.17).

The substitution approach has the advantage that it is conceptually simple and that the essential computations are done in the actual dimension, r, of the problem. The disadvantage is that we must face up to the partitioning of θ, G, and X, and we also lose any simple structure in X. While this poses no difficulty for a given problem, it does complicate general theoretical results.

As an alternative solution, we consider the method of LaGrange Multipliers (see Appendix A.II.4) for the solution of the constrained optimization problem (3.15). In this case the LaGrangian function is given by

$$F(\theta, \sigma^2, \delta) = \text{Ln L}(\theta, \sigma^2) + \delta'(G\theta - g)/\sigma^2 \tag{3.30}$$

where δ is a q-vector of unknown multipliers. Differentiating with respect to θ, σ^2, and δ yields, after simplification, the stationary equations

$$X'X\tilde{\theta} - G'\delta = X'Y \tag{3.31}$$

$$\tilde{\sigma}^2 = Q(\tilde{\theta})/N \tag{3.32}$$

$$G\tilde{\theta} = g \tag{3.33}$$

where $Q(\theta)$ is defined in (3.7). To solve for $\tilde{\theta}$, use (3.31) to obtain

$$\tilde{\theta} = (X'X)^{-1}(X'Y + G'\delta). \tag{3.34}$$

Substituting this expression for $\tilde{\theta}$ into (3.33) and solving for δ yields

$$\delta = [G(X'X)^{-1}G']^{-1}[g - G(X'X)^{-1}X'Y]. \tag{3.35}$$

Returning to (3.34) with this expression for δ, the constrained estimator is given in the form

$$\tilde{\theta} = A\hat{\theta} + Bg \tag{3.36}$$

where $\hat{\theta}$ is the unconstrained estimator

$$\hat{\theta} = (X'X)^{-1}X'Y \tag{3.37}$$

and

$$\begin{aligned} A &= I - BG \\ B &= (X'X)^{-1}G'[G(X'X)^{-1}G']^{-1}. \end{aligned} \tag{3.38}$$

The estimate of σ^2 is then obtained by substituting (3.36) into (3.32). An unbiased estimate of σ^2 is given by

$$s^2 = Q(\tilde{\theta})/(N - r). \tag{3.39}$$

Since the two approaches to constrained optimization are equivalent, the estimates of θ and σ^2 given by (3.27) and (3.22) are identical to those given by (3.36) and (3.39). It follows that the distributional properties developed for the estimators obtained by the model reduction method apply to the estimators obtained by the method of LaGrange Multipliers. It is instructive, however, to develop these properties directly. To do so, we first note some properties of the matrix A given in (3.38). These properties are summarized in the following lemma, which is easily established.

Lemma 3.1

For the matrix A, given by (3.38), we have

(a) A is idempotent,

(b) $XA(X'X)^{-1}X'$ is idempotent and symmetric,

(c) $(X'X)^{-1}A' = A(X'X)^{-1} = A(X'X)^{-1}A'$, and

(d) the rank of $[I - XA(X'X)^{-1}X']$ is $N - r$.

■

Looking first at the estimator for θ given by (3.36), we may use (3.1) to write

$$\begin{aligned} \tilde{\theta} &= A\hat{\theta} + Bg \\ &= \theta + A(X'X)^{-1}X'e \end{aligned} \tag{3.40}$$

where $e \sim N(0, \sigma^2)$. From Section 2.1, it follows that

$$\tilde{\theta} \sim N[\theta, \sigma^2 A(X'X)^{-1}]. \tag{3.41}$$

To examine $Q(\tilde{\theta})$, we use (3.40) to write

$$Y - X\tilde{\theta} = [I - XA(X'X)^{-1}X']e. \tag{3.42}$$

Hence, using the idempotency of the coefficient matrix for e in (3.42), we may write

$$Q(\tilde{\theta}) = e'[I - XA(X'X)^{-1}X']e. \tag{3.43}$$

From Theorem 2.3 and Lemma 3.1,

$$Q(\tilde{\theta}) \sim \sigma^2\chi^2(N - r). \tag{3.44}$$

The independence of $\tilde{\theta}$ and $Q(\tilde{\theta})$ is easily established using Theorem 2.5. Further, $\tilde{\theta}$ and $Q(\tilde{\theta})$ are a complete set of sufficient statistics for θ and σ^2. This is shown by writing

$$Q(\theta) = Q(\tilde{\theta}) + (\tilde{\theta} - \theta)'X'X(\tilde{\theta} - \theta) \tag{3.45}$$

in the likelihood function and applying Appendices B.I.4 and B.I.5.

The properties of the constrained estimators are summarized in the following theorem.

THEOREM 3.2
For the constrained model,

$$Y = X\theta + e$$
$$\text{subject to} \quad G\theta = g,$$

the minimum variance, unbiased estimators of θ are given by either

$$\tilde{\theta} = \begin{bmatrix} \tilde{\theta}_1 \\ \tilde{\theta}_2 \end{bmatrix} \qquad \text{or} \qquad \tilde{\theta} = A\hat{\theta} + Bg$$

and

$$\tilde{\theta} \sim N[\theta, \operatorname{Var}(\tilde{\theta})]$$

with $\operatorname{Var}(\tilde{\theta})$ being the $p \times p$ matrix of rank r given in either (3.29) or (3.41).

The minimum variance, unbiased estimator for σ^2 is given by either

$$s^2 = Q_R(\hat{\theta}_R)/(N - r) \qquad \text{or} \qquad s^2 = Q(\tilde{\theta})/(N - r).$$

Further, s^2 is independent of $\tilde{\theta}$ with

$$s^2 \sim [\sigma^2/(N - r)]\chi^2(N - r).$$

■

3.2.3. Least Squares Estimation

Under the assumption of normally distributed errors, we were able to develop the maximum-likelihood estimators for θ and σ^2 and the distribution of these estimators. It is of interest to relate these results to those obtained by the method of least squares.

Assume that the model is given by (3.1) and (3.3), apart from the normality assumption on the error vector. We do assume $E(e) = 0$ and $\operatorname{Var}(e) = \sigma^2 I$. The least squares estimators are obtained by solving the problem

$$\begin{aligned} &\text{minimize } Q(\theta) \\ &\text{subject to} \quad G\theta = g. \end{aligned} \tag{3.46}$$

This problem may be solved using either the model reduction or the LaGrange Multiplier procedure, and the estimator of θ is seen to be identical to the maximum likelihood estimator, $\tilde{\theta}$, given by Theorem 3.2. The method of least squares does not provide an estimate of σ^2, but intuition suggests an estimator based on $Q(\tilde{\theta})$, and a natural choice is $s^2 = Q(\tilde{\theta})/N - r$ where $r = p - q$. The mean and variance of $\tilde{\theta}$ and the mean of s^2 follow directly. The optimality of $\tilde{\theta}$ as an estimate of θ is restricted to the class of linear unbiased estimators. These results are summarized in the following theorem, generally known as the Gauss-Markov Theorem.

THEOREM 3.3

Assuming $E(e) = 0$ and $\text{Var}(e) = \sigma^2 I$, the least squares estimators, $\tilde{\theta}$ and s^2, have the following properties:

(a) $E(\tilde{\theta}) = \theta$, $\text{Var}(\tilde{\theta}) = \sigma^2 A(X'X)^{-1}$.
(b) $E(s^2) = \sigma^2$.
(c) For any vector a, $a'\tilde{\theta}$ is the unbiased estimator of $a'\theta$, which has minimum variance among all linear unbiased estimators of $a'\theta$.

Proof: Parts (a) and (b) are readily established. To verify part (c), consider an arbitrary linear function of Y to be used as an estimator of $a'\theta$, say,

$$(a'\theta)^* = c'Y + d. \tag{3.47}$$

Then

$$E[(a'\theta)^*] = c'X\theta + d \tag{3.48}$$

and

$$\text{Var}[(a'\theta)^*] = c'c\sigma^2. \tag{3.49}$$

Now determine c and d so that $(a'\theta)^*$ is unbiased for $a'\theta$ and has minimum variance. Let

$$d = f'g \tag{3.50}$$

for some vector f. Then, since $G\theta = g$, we may write the requirement for unbiasedness as

$$a'\theta = c'X\theta + f'G\theta. \tag{3.51}$$

Since this must hold for any vector θ, the condition to be satisfied is

$$a' = c'X + f'G. \tag{3.52}$$

Our estimator is then determined by minimizing (3.49) with respect to c and f, subject to the constraint (3.52), with d then determined from (3.50). The method of LaGrange Multipliers is applied to the problem

$$\begin{aligned} &\text{minimize } c'c \\ &\text{subject to} \quad a' - c'X = f'G. \end{aligned} \tag{3.53}$$

The stationary equations are

$$X\rho = c \tag{3.54}$$

$$G\rho = 0 \tag{3.55}$$

$$a - X'c = G'f \tag{3.56}$$

where ρ is the vector of multipliers. To solve for c, f, and ρ, substitute c from (3.54) into (3.56) and obtain

$$\rho = (X'X)^{-1}(a - G'f). \tag{3.57}$$

Using (3.57) in (3.55) yields

$$\begin{aligned} f &= [G(X'X)^{-1}G']^{-1}G(X'X)^{-1}a \\ &= B'a \end{aligned} \tag{3.58}$$

where B is defined in (3.38). Finally, using (3.57) and (3.58) in (3.50) and (3.54) yields

$$\begin{aligned} d &= a'Bg \\ c &= X(X'X)^{-1}A'a. \end{aligned} \tag{3.59}$$

Thus, the minimum variance unbiased estimator of $a'\theta$ is given by

$$\begin{aligned} (a'\theta)^* &= c'Y + d \\ &= a'(A\hat{\theta} + Bg) \\ &= a'\tilde{\theta} \end{aligned} \tag{3.60}$$

with $\tilde{\theta}$ given by (3.36).

■

The impact of Theorem 3.3 is that we could have obtained our estimators directly by application of least squares or, alternatively, as indicated in the proof, by seeking a linear, unbiased, minimum variance estimator. For the purpose of making inferences on θ and σ^2, we need the assumption of normality to determine the distribution of the estimators and subsequent test statistics or to construct confidence regions.

The likelihood equations for θ given in (3.5) are often referred to as the *normal equations*. This terminology is based on a geometric interpretation of least squares. We shall not elaborate on this but will feel free to use this term, which has become standard in the literature.

3.3. TESTS OF LINEAR HYPOTHESES ON θ

The distributional properties of the estimators provide us with the necessary information for making inferences about θ. This information is used in two different ways that allow the researcher to interpret the results. Confidence regions on θ or confidence intervals on linear functions of θ provide a convenient mechanism for making inferences. Alternatively, the inferences can be based on tests of hypotheses on linear functions of θ.

It is logically easier to consider the hypothesis testing problem first, so we shall defer regional and interval estimation to Section 3.4. This order is also consistent with the application of these methods. One of the first steps in an analysis is to attempt to obtain a general indication of the nature of the effects being studied. This is frequently done in the form of tests of general linear hypotheses on θ. A more detailed analysis in terms of confidence intervals may be suggested as a result of the preliminary tests.

The discussion in this section follows the same format as that in Section 3.2, where we treated the unconstrained and constrained models separately.

3.3.1. The Unconstrained Model

In the general situation, we wish to test a linear hypothesis of the form

$$H_0\colon H\theta = h \tag{3.61}$$

against the two-sided alternative that at least one component of the vector of equalities fails to hold. Here, H is a $v \times p$ matrix of rank $v \leq p$ reflecting the linear functions of θ that are of interest to the analyst. In most applications, the vector h is equal to zero, but for completeness we shall consider the general case, $h \neq 0$. In some instances, it may be appropriate to consider one-sided alternatives—that is, hypotheses of the form $H\theta \leq h$—against the alternative that at least one of the inequalities does not hold. These cases are rare, except when $v = 1$, so we will not discuss them. One-sided inferences can be made from the confidence intervals if desired.

The test statistic will be derived from the likelihood-ratio principle described in Appendix B.II.1. In this case, the likelihood-ratio statistic is given by

$$\gamma = \frac{\begin{array}{c}\max \mathrm{L}(\theta, \sigma^2) \\ \text{subject to} \quad H\theta = h\end{array}}{\max \mathrm{L}(\theta, \sigma^2)}. \tag{3.62}$$

The computations required for evaluating γ have been done in Section 3.2. In particular, the denominator is maximized by $\hat{\theta}$ and $\hat{\sigma}^2$ as given by (3.8) and (3.9). The constrained maximization for the numerator is the same as that developed in Section 3.2.2 for the constrained model, yielding $\tilde{\theta}$ and $\tilde{\sigma}^2$, as given by (3.27) or (3.36) and (3.22) or (3.39). Recall that the maximization in that case was accomplished by two methods, model reduction and LaGrange Multipliers. The former is generally preferred in practice, as the dimensionality of the resulting matrices is reduced; the latter is more convenient for theoretical developments.

By substituting these estimators into (3.62) and simplifying, we get

$$\gamma^{2/N} = Q(\hat{\theta})/Q(\tilde{\theta}) \tag{3.63}$$

where, from Section 3.2,

$$\begin{aligned} \hat{\theta} &= (X'X)^{-1}X'Y \\ \tilde{\theta} &= A\hat{\theta} + Bh \end{aligned} \tag{3.64}$$

$$\begin{aligned} B &= (X'X)^{-1}H'[H(X'X)^{-1}H']^{-1} \\ A &= I - BH \end{aligned} \tag{3.65}$$

$$Q(\theta) = (Y - X\theta)'(Y - X\theta). \tag{3.66}$$

Recalling from Section 3.2 that $\hat{\theta}$ and $\tilde{\theta}$ minimize $Q(\theta)$ in the unconstrained and constrained models, we see that $Q(\tilde{\theta}) \geq Q(\hat{\theta})$ and, hence, $0 \leq \gamma \leq 1$.

To develop the test statistic, it is convenient to express the difference, $Q(\tilde{\theta}) - Q(\hat{\theta})$, as a quadratic form. To do this, write

$$Y - X\tilde{\theta} = Y - X\hat{\theta} + XB(H\hat{\theta} - h) \tag{3.67}$$

and note that

$$(Y - X\hat{\theta})'XB = 0. \tag{3.68}$$

By substituting (3.67) into (3.66) and simplifying, we get

$$Q(\tilde{\theta}) = Q(\hat{\theta}) + (H\hat{\theta} - h)'[H(X'X)^{-1}H']^{-1}(H\hat{\theta} - h). \tag{3.69}$$

We thus define the *numerator sum of squares* associated with the hypothesis, H, as

$$\begin{aligned} N_H &= Q(\tilde{\theta}) - Q(\hat{\theta}) \\ &= (H\hat{\theta} - h)'[H(X'X)^{-1}H']^{-1}(H\hat{\theta} - h). \end{aligned} \tag{3.70}$$

The distribution of N_H follows by noting that

$$H\hat{\theta} - h \sim N[H\theta - h, \sigma^2 H(X'X)^{-1}H'] \tag{3.71}$$

and hence, from Corollary 2.1.2,

$$N_H \sim \sigma^2\chi^2(\nu, \lambda) \tag{3.72}$$

where v, the rank of H, is the rank of $H(X'X)^{-1}H'$. The noncentrality parameter is given by

$$\lambda = (H\theta - h)'[H(X'X)^{-1}H']^{-1}(H\theta - h)/(2\sigma^2). \tag{3.73}$$

For later reference, we note that the expected value of N_H is given by Corollary 2.1.1 as

$$E(N_H) = v\sigma^2 + 2\sigma^2\lambda. \tag{3.74}$$

We recall from (3.12) that

$$Q(\hat{\theta}) \sim \sigma^2\chi^2(N - p). \tag{3.75}$$

It remains for us to establish the independence of N_H and $Q(\hat{\theta})$. To do so, let θ^* be any solution to the constraint equation $H\theta = h$. Then we may write

$$Y - X\hat{\theta} = [I - X(X'X)^{-1}X'](Y - X\theta^*) \tag{3.76}$$

and

$$H\hat{\theta} - h = H(X'X)^{-1}X'(Y - X\theta^*). \tag{3.77}$$

Thus,

$$Q(\hat{\theta}) = (Y - X\theta^*)'[I - X(X'X)^{-1}X'](Y - X\theta^*) \tag{3.78}$$

and

$$N_H = (Y - X\theta^*)'XB[H(X'X)^{-1}H']B'X'(Y - X\theta^*). \tag{3.79}$$

Noting that

$$Y - X\theta^* \sim N[X(\theta - \theta^*), \sigma^2 I] \tag{3.80}$$

and applying Theorem 2.4 establishes the independence of N_H and $Q(\hat{\theta})$.

By returning to the test statistic (3.63) and using (3.70), we get

$$\gamma^{2/N} = [1 + N_H/Q(\hat{\theta})]^{-1}. \tag{3.81}$$

Thus, small values of γ correspond to large values of the ratio $N_H/Q(\hat{\theta})$, a ratio of independent χ^2 variables. From Section 2.3.3, it follows that

$$F = \frac{N_H/v}{Q(\hat{\theta})/(N - p)} \sim F(v, N - p, \lambda). \tag{3.82}$$

These results are summarized in the following theorem.

THEOREM 3.4
A test of size α for the hypothesis

$$H_0: H\theta = h$$

against the alternative

$$H_A: H\theta \neq h$$

in the unconstrained model is given by rejecting H_0 when

$$N_H/(\nu s^2) \geq F(\alpha; \nu, N - p)$$

where, $s^2 = Q(\hat{\theta})/(N - p)$.

■

The proof of this theorem follows by noting that, under H_0, $\lambda = 0$ and (3.82) is distributed as $F(\nu, N - p)$. Note that the size of the test is not dependent on σ^2, and, further, that the power of the test depends on the parameters only through the noncentrality parameter, λ. It also follows, from Section 2.3.3, that the power of the test is monotonically increasing in λ, and it can be shown that this test is uniformly most powerful among all unbiased tests whose power depends on the parameters only through λ. The power is determined by (2.27).

The expression for N_H in (3.70) was convenient for developing the distribution of the test statistic, but in practice it is computationally easier to compute $Q(\tilde{\theta})$ from the model reduced by the constraints $H\theta = h$. Thus, $Q(\tilde{\theta}) = Q_R(\hat{\theta}_R)$, where Q_R and $\hat{\theta}_R$ are given by (3.23) and (3.21) with G replaced by H in the definition of the reduced model.

3.3.2. The Constrained Model

For testing the hypothesis (3.61) in the constrained model, the likelihood ratio statistic is given by

$$\gamma = \frac{\begin{array}{l}\max \mathrm{L}(\theta, \sigma^2) \\ \text{subject to} \quad G\theta = g,\, H\theta = h\end{array}}{\begin{array}{l}\max \mathrm{L}(\theta, \sigma^2) \\ \text{subject to} \quad G\theta = g\end{array}}. \tag{3.83}$$

The solution for the constrained maximization in the denominator is given in Section 3.2.2. For the numerator, we write the constraints as

$$F\theta = f \tag{3.84}$$

where

$$F = \begin{bmatrix} G \\ H \end{bmatrix} \qquad f = \begin{bmatrix} g \\ h \end{bmatrix}. \tag{3.85}$$

We shall assume that the system of equations (3.84) is consistent and of rank $q + v$. The consistency implies that we do not include in our hypothesis constraints any relations that conflict with the known relations on θ. The rank condition says that we have removed any redundant hypothesis constraints that are implied by the constraints $G\theta = g$, possibly in conjunction with other hypothesis constraints.

With this notation, the numerator maximization is identical to that described for the denominator, with G replaced by F. Thus, proceeding as in Section 3.3.1 we write

$$\begin{aligned} \gamma^{2/N} &= Q(\tilde{\theta})/Q(\tilde{\tilde{\theta}}) \\ &= [1 + N_H/Q(\tilde{\theta})]^{-1} \end{aligned} \tag{3.86}$$

where

$$\begin{aligned} \hat{\theta} &= (X'X)^{-1}X'Y \\ \tilde{\theta} &= A\hat{\theta} + Bg \\ B &= (X'X)^{-1}G'[G(X'X)^{-1}G']^{-1} \\ A &= I - BG \\ \tilde{\tilde{\theta}} &= C\hat{\theta} + Df \\ D &= (X'X)^{-1}F'[F(X'X)^{-1}F']^{-1} \\ C &= I - DF \\ N_H &= Q(\tilde{\tilde{\theta}}) - Q(\tilde{\theta}). \end{aligned} \tag{3.87}$$

To develop the test statistic, we write the difference $Q(\tilde{\tilde{\theta}}) - Q(\tilde{\theta})$ as a quadratic form. To do this, write

$$\begin{aligned} Y - X\tilde{\tilde{\theta}} &= Y - X\tilde{\theta} - X(\tilde{\tilde{\theta}} - \tilde{\theta}) \\ &= Y - X\tilde{\theta} + X[(DF - BG)\hat{\theta} - (Df - Bg)]. \end{aligned} \tag{3.88}$$

Using the expression for the partitioned inverse of a matrix—Appendix A.I.10—as applied to $F(X'X)^{-1}F'$, we show that

$$DF = BG + (X'X)^{-1}(HA)'[HA(X'X)^{-1}H']^{-1}HA$$

and (3.89)

$$Df = Bg + (X'X)^{-1}(HA)'[HA(X'X)^{-1}H']^{-1}(h - HBg).$$

Thus, we may write

$$-X(\tilde{\tilde{\theta}} - \tilde{\theta}) = X(X'X)^{-1}(HA)'[HA(X'X)^{-1}H']^{-1}(H\tilde{\theta} - h). \tag{3.90}$$

Further, by noting that $A\tilde{\theta} = A\hat{\theta}$ and recalling Lemma 3.1, we obtain

$$Q(\tilde{\tilde{\theta}}) = Q(\tilde{\theta}) + (H\tilde{\theta} - h)'[HA(X'X)^{-1}H']^{-1}(H\tilde{\theta} - h). \tag{3.91}$$

The numerator sum of squares associated with the hypothesis matrix, H, in the constrained model is then given by

$$\begin{aligned} N_H &= Q(\tilde{\tilde{\theta}}) - Q(\tilde{\theta}) \\ &= (H\tilde{\theta} - h)'[HA(X'X)^{-1}H']^{-1}(H\tilde{\theta} - h). \end{aligned} \tag{3.92}$$

From Theorem 3.2, we see that

$$H\tilde{\theta} - h \sim N[H\theta - h, \sigma^2 HA(X'X)^{-1}H'] \tag{3.93}$$

where the matrix, $HA(X'X)^{-1}H'$, has rank v. It thus follows from Corollary 2.1.2 that

$$N_H \sim \sigma^2\chi^2(v, \lambda) \tag{3.94}$$

where

$$\lambda = \{(H\theta - h)'[HA(X'X)^{-1}H']^{-1}(H\theta - h)\}/(2\sigma^2). \tag{3.95}$$

From Theorem 3.2, recall that $Q(\tilde{\theta}) \sim \sigma^2\chi^2(N - r)$. The independence of N_H and $Q(\tilde{\theta})$ is established by letting θ^* be any solution to the equation $F\theta = f$ and writing

$$H\tilde{\theta} - h = HA(X'X)^{-1}X'(Y - X\theta^*) \tag{3.96}$$

and

$$(Y - X\tilde{\theta}) = [I - XA(X'X)^{-1}X'](Y - X\theta^*). \tag{3.97}$$

Using Lemma 3.1, we may write

$$N_H = (Y - X\theta^*)'X(X'X)^{-1}(HA)'[HA(X'X)^{-1}H']^{-1}HA(X'X)^{-1}X'(Y - X\theta^*) \tag{3.98}$$

and

$$Q(\tilde{\theta}) = (Y - X\theta^*)'[I - XA(X'X)^{-1}X'](Y - X\theta^*). \tag{3.99}$$

Since $(Y - X\theta^*) \sim N[X(\theta - \theta^*), \sigma^2 I]$, the independence of N_H and $Q(\tilde{\theta})$ follows from Theorem 2.4. It then follows that the ratio

$$F = \frac{N_H/v}{Q(\tilde{\theta})/(N - r)} \sim F(v, N - r, \lambda). \tag{3.100}$$

The test procedure is summarized in the following theorem.

THEOREM 3.5
A test of size α for the hypothesis

$$H_0\colon H\theta = h$$

against the alternative

$$H_A\colon H\theta \neq h$$

in the constrained model is given by rejecting H_0 when

$$N_H/(\nu s^2) \geq F(\alpha; \nu, N - r)$$

where $s^2 = Q(\tilde{\theta})/(N - r)$.

■

The optimality properties for this test are the same as those for the unconstrained model, stated after Theorem 3.4. The power is given by (2.27).

The expressions for $Q(\tilde{\theta})$ and N_H as developed in this section are notationally convenient for determining the distribution of the test statistic but are cumbersome in practice. It is computationally more efficient to determine $Q(\tilde{\theta}) = Q_R(\hat{\theta}_R)$ from (3.21) and (3.23). Similarly, we may obtain $Q(\tilde{\tilde{\theta}})$ from the analogous expressions with F replacing G in the development of (3.21) and (3.23). Typically, this is done in two stages. First, the model is reduced by G, and then it is further reduced by H to determine $Q(\tilde{\tilde{\theta}})$. We shall see that this apparently cumbersome procedure is really quite simple for the models to be considered in Chapters 6 and 7.

3.3.3. Subhypotheses and Significance Levels

The test of $H\theta = h$ provides general information about the model, but we often wish to ask for more specific information. For example, in the extreme case with regression models, we may wish to test each of the m hypotheses, H: $\theta_i = 0$, $i = 1, \ldots, m$. Indeed, this seems quite natural if the initial test of the m-degree of freedom hypothesis, H: $\theta_i = 0$, $i = 1, \ldots, m$, was rejected. More generally, we may wish to test several hypotheses, $H_i\theta = h_i$, $i = 1, \ldots, k$. We shall refer to these as subhypotheses.

The testing of several subhypotheses raises an important question. That is, if we test k hypotheses, each at level α, it seems intuitive that the probability of rejecting at least one of them is greater than α, even if all hypotheses are true. To demonstrate this point, suppose the test statistics for each of the k hypotheses are independent. Then, if each of the hypotheses is tested at level α, the probability of rejecting at least one of them, even if all are true, is

$$1 - (1 - \alpha)^k. \tag{3.101}$$

For example, with $\alpha = 0.05$ and $k = 5$, the value of (3.101) is 0.23, substantially greater than the usual 1 of 20 risk.

The test statistics for such hypotheses are rarely independent. Even if the numerator chi-square statistics are independent, the denominators are all based on the residual sum of squares, $Q(\tilde{\theta})$. Thus, the ratios are not independent. The relation (3.101) is not exact in this more typical situation, but it does illustrate the cause for concern. That is, if we wish to ensure an overall (experimentwise) significance level of α when many hypotheses are to be tested, then some thought must be given to reducing the size of the individual tests.

When k dependent F-ratios are used in the analysis, the probability that at least one will exceed the critical value, even when all hypotheses are true, may be as large as $k\alpha$. To see this, we apply the Bonferroni inequality, which says that, if E_i, $i = 1, \ldots, k$ are events with complements $\bar{E}_i$, then, if we consider the intersection of these events, we have

$$\text{Prob}\left(\prod_{i=1}^{k} E_i\right) \geq 1 - \sum_{i} \text{Prob}(\bar{E}_i). \tag{3.102}$$

To apply this result to our problem, let E_i denote the event that the ith F-ratio does not exceed its critical value. Thus, $\text{Prob}(\bar{E}_i) = \alpha$, and the right side of (3.102) is equal to $1 - k\alpha$. The left side is the probability that all F-ratios are less than their critical value. It follows that the complementary event—that at least one of the F-ratios is significant—has probability less than or equal to $k\alpha$. As a conservative approach, if k tests are to be performed, then we might set the size of the individual tests at α/k to ensure the overall significance level of no more than α. Of course, some other allocation of α to the k tests might be considered.

It is easy to see that the value of k in (3.102) might be quite large for experiments of rather modest dimensions. If the experimentwise significance level is felt to be appropriate, some thought should be given to a reduced size for each test. Of course, it should be emphasized that such an approach makes it more difficult to declare significant departures from the null hypothesis.

3.4. CONFIDENCE REGIONS, CONFIDENCE INTERVALS, AND PREDICTION INTERVALS

3.4.1. Confidence Regions and Intervals

In the introduction to Section 3.3, we noted that regional estimates on the vector θ, or on linear functions of θ, provide an alternative approach to inference. The concept of a confidence region on a vector of parameters is the natural extension of a confidence interval on a single parameter. That is, we seek a region in parameter space determined by the data, say $R(Y)$, with the property,

$$\text{Prob}[\theta \in R(Y) \,|\, \theta] = 1 - \alpha. \tag{3.103}$$

While any region satisfying (3.103) will qualify as a $100(1 - \alpha)$ percent confidence region, we naturally seek a region that is, in some sense, small. In particular, we would prefer a region that has low probability of including θ^* if θ^* is not the true parameter vector. A ready source of good confidence regions is found in the "inversion" of good test statistics. In this case, the pivotal quantity used to define the confidence region is the test statistic given in Theorem 3.4.

To develop a confidence region for θ, in the unconstrained model, let $h = H\theta$ in Theorem 3.4 and consider the special case $H = I_p$. Then, we have

$$\mathrm{Prob}[N_I \leq ps^2F(\alpha; p, N - p) \,|\, \theta] = 1 - \alpha \tag{3.104}$$

where

$$N_I = (\hat{\theta} - \theta)'(X'X)(\hat{\theta} - \theta). \tag{3.105}$$

Note that, in the constrained case, we may apply the same reasoning to θ_R in the reduced model with appropriate changes in degrees of freedom. When testing the hypothesis that the parameter vector is θ, we base our decision on the location of $\hat{\theta}$ in the sample space. Specifically, we ask if $\hat{\theta}$ is in the ellipsoidal acceptance region, centered at θ, defined by

$$N_I \leq ps^2F(\alpha; p, N - p). \tag{3.106}$$

The hypothesis is rejected if $\hat{\theta}$ does not lie in this region.

In this case, the inversion of (3.106) to form a confidence region for θ is simply a matter of viewing (3.106) as an ellipsoid in parameter space, centered at $\hat{\theta}$. From (3.104) we see that the probability that the ellipsoid contains the true parameter vector is $(1 - \alpha)$, and hence (3.106) defines a $100(1 - \alpha)$ percent confidence region for θ.

We may generalize (3.106) to a confidence region for linear functions of θ. Thus, if we are interested in the linear functions defined by $H\theta$, the confidence region is given by

$$N_H \leq vs^2F(\alpha; v, N - p) \tag{3.107}$$

where

$$N_H = (H\hat{\theta} - H\theta)'[H(X'X)^{-1}H']^{-1}(H\hat{\theta} - H\theta). \tag{3.108}$$

In particular, with $v = 1$, H is a row vector, say a', and the confidence region on $a'\theta$ is

$$(a'\hat{\theta} - a'\theta)^2 \leq F(\alpha; 1, N - p)\, \widehat{\mathrm{Var}}(a'\hat{\theta}) \tag{3.109}$$

where the estimated variance of $a'\hat{\theta}$ is

$$\widehat{\mathrm{Var}}(a'\hat{\theta}) = s^2[a'(X'X)^{-1}a]. \tag{3.110}$$

We may now write (3.109) as a confidence interval for $a'\theta$ as

$$a'\theta\colon a'\hat{\theta} \pm [F(\alpha;\, 1,\, N-p)\widehat{\text{Var}}(a'\hat{\theta})]^{1/2}. \tag{3.111}$$

Typically, this interval is written in terms of the Student's t distribution, using the relation

$$F(\alpha;\, 1,\, N-p) = t^2(\alpha/2;\, N-p). \tag{3.112}$$

Two special cases of (3.111) are of interest. The first, in which the row vector, a', has a one in the ith position with zeros elsewhere, gives a confidence interval for the single component θ_i. The second, commonly considered in the means model, has a plus one in the ith position, a minus one in the jth position, and zeros elsewhere, giving a confidence interval for the difference $\theta_i - \theta_j$. Such *paired comparisons* are frequently of interest and will be examined in detail in the next chapter.

Equations (3.106) and (3.107) give us the desired confidence regions for θ or linear functions of θ. For a given vector θ, it is easy enough to determine whether or not θ or $H\theta$ is in the region. However, except for the case $p = 2$ when we can graph the ellipsoidal boundary, it is difficult to visualize the limits on the parameters. Typically, the researcher would prefer a confidence interval on each parameter function such as (3.111). Indeed, we could write individual intervals for each component of θ using (3.111), defining a rectangular confidence region, but the associated confidence coefficient is no longer $1 - \alpha$. In fact, this problem is equivalent to the one discussed in Section 3.3.3, where we considered the consequences of many tests on the same set of data. The probability that this rectangular region contains the true parameter vector, assuming independence of the test statistics, is $(1 - \alpha)^p$. That is, the probability that at least one of the intervals does not contain the true parameter is given by $1 - (1 - \alpha)^p$. Thus, the confidence coefficient may be substantially smaller than $1 - \alpha$. This problem is magnified when we include confidence intervals on linear functions of θ, such as all differences $\theta_i - \theta_j$ in the means model.

The problem, then, is to develop methods for constructing confidence intervals which will protect the overall confidence coefficient. In the following section, we describe two of the more common solutions to this problem. In Chapter 4, methods designed especially for the means model are described. For a detailed discussion of the concepts and methods, see Miller (1981).

3.4.2. Simultaneous Confidence Intervals

Our objective in this section is to develop methods for constructing confidence intervals such that the region defined by the intersection of the intervals in parameter space will have a confidence coefficient of at least $1 - \alpha$. In this way each of the intervals will have a confidence coefficient at least as large as $1 - \alpha$. The resulting intervals are known as *simultaneous confidence intervals*. In particular, we shall discuss one exact procedure, described by Scheffé, and one approximate procedure based on the Bonferroni inequality.

(a) Scheffé's Simultaneous Confidence Intervals

Scheffé (1953) [or see Miller (1981)] described an interesting method for developing simultaneous confidence intervals based on tangent planes to the ellipsoidal confidence region defined by (3.107). His procedure allows us to write as many interval statements as we wish, while ensuring that the simultaneous confidence coefficient is at least $1 - \alpha$ and the confidence coefficient on each interval is at least $1 - \alpha$. This is particularly useful in exploratory studies, as we may write intervals suggested by the data without reducing the experimentwise confidence coefficient. Of course, we anticipate that the price paid for this convenience is conservative (that is, wider) intervals.

Scheffé's method depends on a convexity property of the region defined by an inequality on a positive definite quadratic form. It is instructive to sketch the development.

Consider a general ellipsoidal region, centered at x_0, defined by

$$(x - x_0)'M(x - x_0) \leq 1 \tag{3.113}$$

where M is a positive definite symmetric matrix. Let x^* be a point on the boundary of this region. Then the tangent plane to the ellipsoid at x^* is given by

$$t(x; x^*) = (x^* - x_0)'M(x - x_0) = 1. \tag{3.114}$$

The convexity of the region (3.113) ensures that this plane does not intersect the ellipsoid. Equivalently, the half-space defined by

$$t(x; x^*) \leq 1 \tag{3.115}$$

contains the ellipsoidal region, (3.113). It follows that we may generate as many planes as we wish, for different choices of x^*, and be assured that the region defined by the intersection of the half-spaces contains the ellipsoidal region. The basis for the Scheffé method is that the confidence coefficient for the region defined by the intersection of any number of half-spaces will be at least as large as that for the ellipsoid.

In practice, we will not know the point, x^*, at which we wish to construct the tangent plane, but, rather, we will know the vector of the slopes—say, a—for the plane. Thus, we wish to determine the scalar, b, for the tangent plane

$$a'(x - x_0) = b. \tag{3.116}$$

Since the plane is determined by (3.114), we have

$$a' = b(x^* - x_0)'M \tag{3.117}$$

or, equivalently,

$$(x^* - x_0) = M^{-1}a/b. \tag{3.118}$$

Multiplying (3.118) on the left by a' and recognizing that x^* satisfies (3.116), we have

$$a'M^{-1}a = b^2. \tag{3.119}$$

We thus obtain the pair of tangent planes, with slope a', given by

$$a'(x - x_0) = \pm(a'M^{-1}a)^{1/2} \tag{3.120}$$

and the ellipsoid is bounded by the region between the two planes given by

$$|a'(x - x_0)| \leq (a'M^{-1}a)^{1/2}. \tag{3.121}$$

Applying this result to the ellipsoid defined by (3.107), with $x = H\theta$, $x_0 = H\hat{\theta}$ and

$$M = [H(X'X)^{-1}H']^{-1}[vs^2F(\alpha; v, N - p)], \tag{3.122}$$

we obtain the confidence interval on linear functions of $H\theta$ given by

$$a'H\theta\text{: } a'H\hat{\theta} \pm [vF(\alpha; v, N - p)\widehat{\text{Var}}(a'H\hat{\theta})]^{1/2} \tag{3.123}$$

where

$$\widehat{\text{Var}}(a'H\hat{\theta}) = s^2a'H(X'X)^{-1}H'a. \tag{3.124}$$

As noted above, we may write as many intervals as we wish of the form (3.123) and be assured that the simultaneous confidence coefficient is at least $1 - \alpha$. We say that the set of confidence intervals (3.123) has simultaneous confidence coefficient $1 - \alpha$.

Some comments are in order. The generality of the intervals (3.123) is a function of the initial choice of the matrix H. For example, with $H = I_p$, (3.123) becomes

$$a'\theta\text{: } a'\hat{\theta} \pm [pF(\alpha; p, N - p)s^2a'(X'X)^{-1}a]^{1/2}. \tag{3.125}$$

In this case, any linear function of θ, including the individual components of θ, may be considered. On the other hand, in the means model, $H\theta$ may correspond to, say, the differences $\theta_i - \theta_j$. In this case, $v = p - 1$, and only linear functions of these differences may be considered. In particular, (3.123) will not yield intervals for the individual θ_i. As one might suspect, the generality of (3.125) must be paid for in wider intervals than might be obtained for specific functions using (3.111). This point will be illustrated in Chapters 5 and 6.

The intervals constructed according to (3.123) have an interesting property. If the hypothesis $H\theta = h$ is rejected, then at least one interval of the form (3.123) will fail to cover the parameter, $a'H\theta$. Conversely, if $H\theta = h$ is accepted, all intervals will include the parameter. The reason for this property is that this hypothesis is rejected if h is not in the ellipsoid given by (3.107), in which case at least one pair of planes will exclude this point. Conversely, if h is in the region (3.107), then no pair of planes can exclude it. In the case in which $H\theta = h$ is rejected, it may not be a simple matter to identify a linear function of $H\theta$ which lies outside its interval.

It should be noted that, in regression models, we normally use (3.125) and are concerned with confidence intervals on individual parameters. The more general situation is encountered in the means model.

(b) Bonferroni Intervals

The Bonferroni inequality, (3.102), may be used to construct simultaneous confidence intervals on a fixed set of, say, k linear functions of θ, with confidence coefficient of at least $1 - \alpha$. Let $a_i'\theta$, $i = 1, \ldots, k$ denote the k linear functions of interest, and let E_i denote the event that the parameter $a_i'\theta$ is contained in the interval defined by

$$a_i'\theta\colon\ a_i'\hat{\theta} \pm [F(\alpha/k;\ 1,\ N - p)\widehat{\mathrm{Var}}(a_i'\hat{\theta})]^{1/2}. \tag{3.126}$$

Note that $\mathrm{Prob}(E_i) = 1 - \alpha/k$ and $\mathrm{Prob}(\bar{E}_i) = \alpha/k$. It then follows from (3.102) that the probability that all k intervals contain the corresponding parameter, $a_i'\theta$, is at least $1 - \alpha$.

A nice feature of this method is that the expressions for the confidence intervals have the same form as the individual intervals (3.111). The only difference is in the use of α/k rather than α for the percentage point of the F distribution. A possible disadvantage is that the number of functions, k, must be specified in advance, so we lose the flexibility of the Scheffé method. On the other hand, we would expect that the Bonferroni intervals would be somewhat narrower than the Scheffé intervals.

3.4.3. Prediction Intervals

A common use of regression models is that of predicting the value of a future response. To be specific, suppose we have estimated the parameters of the linear model based on N observations. Now, under the same conditions for which the estimation data were taken, we observe the p-vector of inputs, say, x_0, and ask for a prediction of what the response will be. More generally, we may wish to specify an interval in which the response is likely to lie.

This problem is most frequently associated with regression models, but we shall describe the solution in the general setting of this chapter. Thus, we will assume the p-parameter linear model

$$y = \theta' x + e. \tag{3.127}$$

To motivate the development of prediction intervals, consider first the case in which the parameters θ and σ^2 are assumed known. The prediction problem then, is just that of predicting the response based on a sample of size one from a normal population with known mean and variance. The natural point predictor is the population mean, $\theta' x_0$, and any interval predictor would most likely be centered at the mean. The width of the interval is naturally a function of how certain you wish to be that the response will fall in that interval. To describe this interval, let x_0 denote the vector of inputs and y_0 the unknown response. Then

$$z = (y_0 - \theta' x_0)/\sigma \sim N(0, 1). \tag{3.128}$$

Thus, an interval into which y_0 will fall with probability $(1 - \alpha)$—that is, a *prediction interval*—is given by

$$y_{\text{PRED}}\colon \theta' x_0 \pm z(\alpha/2)\sigma. \tag{3.129}$$

Here, $z(\alpha/2)$ denotes the upper $100(1 - \alpha/2)$ percentage point of $N(0, 1)$. In practice, the parameters are rarely known but are estimated from N independent observations on the model. The estimates of θ and σ^2 are given by Theorem 3.1 based on the data matrix X. (Here, we assume the unconstrained model without loss of generality.) In this case, it follows that

$$y_0 - \hat{\theta}' x_0 \sim N\{0, \sigma^2[1 + x_0'(X'X)^{-1}x_0]\} \tag{3.130}$$

where X is the $N \times p$ matrix of estimation data. It then follows that

$$t = (y_0 - \hat{\theta}' x_0)/\{s^2[1 + x_0'(X'X)^{-1}x_0]\}^{1/2} \tag{3.131}$$

is distributed as Student's t with $N - p$ degrees of freedom. Thus, the point predictor is given by $y_{\text{PRED}} = \hat{\theta}' x_0$ and the prediction interval by

$$y_{\text{PRED}}\colon \hat{\theta}' x_0 \pm t(\alpha/2; N - p)\{s^2[1 + x_0'(X'X)^{-1}x_0]\}^{1/2} \tag{3.132}$$

where $t(\alpha/2; N - p)$ denotes the upper $(1 - \alpha/2)$ percentage point of $t(N - p)$. The square of the denominator in (3.131), with s^2 replaced by σ^2, is called the prediction variance. We write

$$\text{Var}_{\text{PRED}}(y_0) = \sigma^2[1 + x_0'(X'X)^{-1}x_0]. \tag{3.133}$$

The prediction interval is sometimes confused with a confidence interval for the mean response. A confidence interval for the mean response at x_0 is determined from (3.111) with $a = x_0$ and given by

$$E(y_0)\colon \hat{\theta}' x_0 \pm t(\alpha/2; N - p)[s^2 x_0'(X'X)^{-1}x_0]^{1/2}. \tag{3.134}$$

The difference between (3.132) and (3.134) lies in the use of the prediction variance in the former and the variance of $\hat{\theta}' x_0$ in the latter. The developments leading to (3.129) and (3.132) emphasize the difference in the two concepts. In the case where the parameters are known and hence the mean response is known, we still cannot make an exact prediction because of the inherent variability in the model. When θ is not known, we have an added component of variability since θ must be estimated. These two components are apparent in the expression for the prediction variance.

As an extension to the concept of predicting a single response at x_0, consider predicting the mean of k responses for that same input. Letting $\bar{y}_0$ denote the average response, the results follow as above by noting that $\bar{y}_0 \sim N(\theta' x_0, \sigma^2/k)$. The prediction interval is then given by

$$\bar{y}_{\text{PRED}}\colon \hat{\theta}' x_0 \pm t(\alpha/2; N - p)\{s^2[1/k + x_0'(X'X)^{-1}x_0]\}^{1/2}. \tag{3.135}$$

This result allows us to relate the concepts of prediction and confidence intervals in another way. The relation follows by letting k approach infinity in (3.135). We note that the sample average converges to the mean response and that the prediction interval converges to the confidence interval.

It is important to distinguish between these two types of intervals, since it is often true that $1 \gg x_0'(X'X)^{-1}x_0$ in (3.133). Using the confidence interval (3.134) as a prediction interval could give prediction limits which are much too narrow.

3.5. ADDITIONAL TOPICS ON REGRESSION MODELS

Implicit in the theory developed in this chapter are two very strong assumptions. These are that (a) we have the correct mathematical expression for the mean response and (b) we have good data in the sense that the distributional assumptions are satisfied and that it is free from recording or other errors. In many situations, especially with regression models, these assumptions may be violated. The assumption $E(y) = \theta' x$ implies that the model is linear in θ and that all relevant inputs, in proper functional form, have been included. With regard to the assumption of normality, it may be more appropriate to consider a more heavily tailed distribution. Often data problems are caused by simple recording or measurement errors on either the response or the input variables. Also, discrepancies can be caused by the fact that some condition external to the model has been changed for some of the responses.

Frequently, the functional form for the mean response is assumed to be valid only over a restricted range of the inputs. On occasions, one or more input vectors may lie outside of this region. Such observations can have an unwarranted influence on the estimates. It is also important that the choices of the inputs in the sample are representative of the range of interest.

Another concept which often arises when the regression model is to be used for prediction is that of eliminating some of the terms in the model. It may be that, even though the initial model is correct, we can predict almost as well with fewer terms in the equation.

A thorough discussion of all of these topics is beyond the scope or intent of this book. In addition, there are now many excellent books on these topics, including those mentioned in Section 1.1.1. However, some of the topics are relevant for our purposes, so we provide a brief review of the essential ideas.

3.5.1. Detection of Problems with the Data

Problems with the data include (a) abnormal values for the response (outliers), (b) abnormal values of the input values (extreme points), and (c) violation of the assumption of normality. A natural approach for detecting such problems is to examine the vector of residuals defined as

$$r = Y - \hat{Y} \tag{3.136}$$

where $\hat{Y}$ is the vector of *predicted values*

$$\hat{Y} = X\hat{\theta}. \tag{3.137}$$

Plots of the components of r versus those of $\hat{Y}$ are helpful in detecting outliers, plots of these residuals versus the input variables may reveal extreme points or model deficiencies, and a histogram or sample distribution function of r may be examined for normality.

A criticism of the use of r is that $\hat{Y}$ has already been influenced by any unusual data points. This suggests that, to measure the influence of a particular observation (case), we might fit the model with and without that case. We may then note the effect of that single case on the estimates of θ, on the vector of predicted values, and on the estimate of σ^2. Letting a subscript (i) denote a quantity computed without the ith case, we examine the effect on the coefficient estimates by evaluating the differences

$$\hat{\theta}_j - \hat{\theta}_{j(i)} \tag{3.138}$$

for $j = 1, \ldots, p$ and for each $i = 1, \ldots, N$. The effect on prediction is measured by the components of the vector of differences

$$\hat{Y} - \hat{Y}_{(i)}. \tag{3.139}$$

The estimate of σ^2, without the ith case, is denoted by $s^2_{(i)}$. The ratio of this with s^2 would indicate the influence on the estimate of σ^2.

The books by Belsley, Kuh, and Welsch (1980) and Cook and Weisberg (1982) provide detailed accounts of how these measures may be examined. The review paper by Hocking (1983b) provides a summary of these ideas and gives several additional references.

An extreme vector of inputs is easily detected in cases where $m = 1$ or 2 in the regression model, but multidimensional extremes may be less evident. A simple yet effective way to identify such cases is to examine the diagonal elements of the matrix

$$H = X(X'X)^{-1}X'. \tag{3.140}$$

Note that the vector of predicted values (3.137) is given by $\hat{Y} = HY$. Thus, the jth component of the ith row of H measures the effect of the jth response on the ith predicted value. In particular, the ith diagonal element, h_i, measures the effect of the ith response on the ith predicted value. Hoaglin and Welsch (1978) discuss this "hat" matrix and suggest that diagonal elements exceeding $2p/N$ indicate extreme cases.

3.5.2. Problems with the Model

Generally, the use of various types of residual plots has been recommended to suggest additional terms that may be added to the model or to suggest transformations on either the input or the response variables to yield a better model. References to papers on residual plotting techniques and methods for determining transformations are found in Hocking (1983b).

In this section, we shall restrict our attention to a particular problem known as *collinearity* among the input variables. This term is used to describe situations in which the input variables are approximately linearly related. That is, there exist constants a_j such that

$$\sum_{j=1}^{p} a_j x_{ij} \doteq 0 \tag{3.141}$$

for the estimation data. The effect of such a relation is that the coefficient matrix $X'X$ in the normal equations is nearly singular. Even if the inversion of $X'X$ is accurately computed, the effect on the estimates may be dramatic. The coefficient estimates may be larger in absolute value than anticipated, or the sign may be the opposite of what was expected.

The papers by Hoerl and Kennard (1970a, b) discuss and illustrate this problem. The reasons for the unusual results are identified, and a proposed remedy called *ridge regression* is described. If we let β denote the m-vector of slopes after having standardized the input vectors to have zero sum and unit sum of squares, the ridge estimator is given by the solution of the equations

$$(X_s'X_s + kI)\beta = X_s'Y. \tag{3.142}$$

Here, X_s denotes the $N \times m$ standardized data matrix. The scalar k is a suitably chosen positive constant.

The *ridge estimator* is one of a class of *biased estimators* based on modifications of the *least squares estimator*. A discussion of such estimators is given by Hocking, Speed, and Lynn (1976).

We have described collinearity as a model problem, but it may also be a data problem. This can occur if the sampling is restricted so that the input vectors lie on a linear subspace of the p-dimensional space. It is important to distinguish between collinearities that are inherent in the model and those that are artifacts of the data collection.

3.5.3. The Assumption of Normality

We have indicated that the outliers (unusually large or small values of the response) can greatly influence the parameter estimates. In practice, it is not always clear whether such a response is an erroneous observation or the result of a nonnormal distribution of errors. The least squares estimator reacts strongly to such observations. It has been suggested that functions other than the squared deviations be considered for estimation. For example, we might consider minimizing the sum of the absolute values, rather than the sum of squares of the deviations. This procedure corresponds to assuming a double exponential model for the errors and has the effect of giving less weight to outlying cases.

The basic objective is to develop estimators which are robust—that is, less sensitive—to the assumption of normality. Several estimators have been proposed, some motivated by specific error models and others based simply on the desire to downweight cases with large residuals. For further reading on this topic, see Andrews (1974) and Huber (1981). Additional references are found in Hocking (1983b).

3.5.4. Variable Selection

When the objective of the modeling is to produce an equation for prediction, we may wish to reduce the number of terms in the model. Often, this can be done without substantial loss in the predictive capability of the equation. It may be argued that variable elimination should be considered in any regression model. The reason for this is that we often include far too many variables in the model to ensure that we have all of the important variables.

One issue to be considered is the objective of the variable selection. If it is simply to remove extraneous variables, our criteria may be different than if we are striving for an adequate prediction equation. The selection problem is complicated by the fact that we usually cannot isolate the effect of one variable from that of the others. Models with collinearities and samples with extreme cases make the selection problem difficult. Computing statistics with which we make the decisions can be expensive, but this depends on the technique used and the size of the model. Complete analyses are reasonable for $p < 30$, and this limit will be extended as computer speeds increase.

With the recent advances in data and model inspection, many of the problems with variable selection have been removed. By the time the analyst has completed the thorough examination of the data and the model indicated in Sections 3.5.1 and 3.5.2, there should be few real surprises in the choice of variables to retain in the model.

Most texts on regression contain extensive discussions of the variable selection problem. The review papers by Hocking (1976) and (1983b) contain most of the ideas and many references on the topic.

3.6. A NUMERICAL EXAMPLE

In this section, we provide a simple numerical example which will illustrate several of the ideas introduced in this chapter. Since Chapters 4 through 7 will be devoted to means models, we have chosen Example 1.2, a regression problem, for this illustration. Also, since constrained models are examined in detail in later chapters, we consider only the unconstrained case.

The model in Example 1.2 is

$$y = \theta_0 + \theta_1 x_1 + \theta_2 x_2 + \theta_3 x_3 + e \tag{3.143}$$

where the inputs are the width (WID), density (DENS), and strength (STR) of stainless steel sheets used to produce the number of units, y. The normal equations may be written directly in the form (3.5), using the data from Table 1.2. In practice, it is convenient to rewrite (3.143), expressing the input variables as the difference between the observed values and their means. Letting $\bar{x}_j$, $j = 1, 2, 3$ denote the means for the 20 observations on the three input variables, the model is written as

$$y = \beta_0 + \sum_{j=1}^{3} \theta_j(x_j - \bar{x}_j) + e. \tag{3.144}$$

The simple relation between the two models is given by

$$\beta_0 = \theta_0 + \sum_{j=1}^{3} \theta_j \bar{x}_j. \tag{3.145}$$

In this form, the estimate of β_0 is $\hat{\beta}_0 = \bar{y}$, the average response, and the equations for estimating the slopes are given by

$$\begin{bmatrix} 250.0 & 248.6 & 96.2 \\ 248.6 & 4093.4 & 2537.3 \\ 96.2 & 2537.3 & 1697.1 \end{bmatrix} \begin{bmatrix} \hat{\theta}_1 \\ \hat{\theta}_2 \\ \hat{\theta}_3 \end{bmatrix} = \begin{bmatrix} 1326.7 \\ 24459.0 \\ 15941.9 \end{bmatrix}. \tag{3.146}$$

The estimation results are summarized in Table 3.1. The estimate of σ^2 is computed from (3.14). The variances of the estimates of the slopes are given by the diagonal elements of (3.11) and are estimated by replacing σ^2 by s^2. In this case, the diagonal elements of the inverse of the coefficient matrix in (3.146), multiplied by s^2, gives the estimated variances of the $\hat{\theta}_j$. The variance of $\hat{\beta}_0$ is σ^2/N, and the variance of $\hat{\theta}_0$ is computed using the expression for the variance of a linear combination of random variables in (2.4).

TABLE 3.1 Estimation summary.

Parameter	θ_0	θ_1	θ_2	θ_3	σ^2
Estimate	6.38	−0.92	5.41	1.56	365.6
Estimated Variance	1656.5	1.54	0.35	0.83	—

A common hypothesis to consider for regression models is that all of the parameters other than θ_0 are zero. The motivation for this is to see whether this set of inputs is useful in describing the response. In this case, the hypothesis is given by

$$H_0\colon \theta_j = 0 \qquad j = 1, 2, 3. \tag{3.147}$$

Clearly, the model reduction approach is most efficient for determining the sum of squares for testing this hypothesis. The reduced model is

$$y = \theta_0 + e. \tag{3.148}$$

In this model, $\hat{\theta}_0 = \bar{y}$, and it is a simple matter to compute the numerator sum of squares, N_H, as the difference in the residual sums of squares for the two models (3.143) and (3.148). The statistics for this are summarized in Table 3.2. [This display is referred to as an *Analysis of Variance (ANOVA)* table. While that name is inappropriate since we are not analyzing variance, it is commonly used. In Chapters 4 through 7, where such tables are used specifically to test hypotheses on means, we shall use the term *Analysis of Means (AOM)* table.]

TABLE 3.2 Analysis of the regression model.

Description	*df*	*SS*	*MS*	*F*
N_H	3	146684	48895	133.8
$Q(\hat{\theta})$	16	5849	366	—

The first column describes the quadratic forms to be found in that row. The next three columns give the degrees of freedom (df) associated with the quadratic forms, the values of the quadratic forms, or sums of squares (SS), and the ratios, or mean squares, MS = SS/df. The last column gives the value of the F statistic computed as the ratio of these two mean squares. In this case, $F \sim F(3, 16, \lambda)$ where $\lambda = \theta' C\theta/(2\sigma^2)$, and C is the coefficient matrix in (3.146). With $F(0.0001; 3, 16) = 9.1$, the hypothesis is soundly rejected, indicating that at least one of the slopes is not zero.

Continuing with our examination of the model, we consider 95 percent confidence intervals on each of the slopes. For comparison, we evaluate the individual intervals (3.111), the Scheffé intervals (3.123), with $H = (0\ I_3)$ corresponding to (3.147), and the Bonferroni intervals (3.126) for $k = 3$. The results are summarized in Table 3.3, where we show the half-widths for each of the intervals. The relative widths of the intervals in this table are typical. That is, the individual intervals are shortest, but we recall that this may be misleading since the probability that at least one of these intervals does not contain the true parameter is approximately $1 - .95^3 = 0.14$. The Scheffé intervals ensure that this probability is not greater than 0.05, and this is approximately true for the Bonferroni intervals. The penalty paid for this simultaneous protection is evidenced by the increased widths in the latter two intervals. The Scheffé intervals are the most conservative (widest).

TABLE 3.3 Half-widths of confidence intervals.

Parameter	*Individual*	*Scheffé*	*Bonferroni*
θ_1	2.63	3.87	3.31
θ_2	1.26	1.85	1.59
θ_3	1.93	2.84	2.43

The analysis described to this point is classical and illustrates most of the concepts in Sections 3.2 through 3.4. However, it may be premature in the sense that we should have first expended some effort to examine the data. To illustrate some of the ideas suggested in Section 3.5, we first examine the diagonal elements of the hat matrix H as defined by (3.140). In this case, $h_8 = 0.8$, which exceeds the suggested cutoff of $2p/N = 0.4$. We thus single out this observation for further investigation. Fitting the model with the remaining 19 observations, we obtain the new estimates shown in Table 3.4.

TABLE 3.4 Estimation summary with case eight deleted.

Parameter	θ_0	θ_1	θ_2	θ_3	σ^2
Estimate	−42.3	0.98	1.74	6.74	106.6
Estimated Variance	542.4	0.54	0.44	1.02	—

Comparing Tables 3.1 and 3.4 reveals the dramatic changes in these parameter estimates caused by the deletion of this single case. A comparison of predicted values also would reflect the influence of this observation. In this example, an examination of the plot of PROD versus STR and of DENS versus STR reveals the sense in which case eight is influential. From Figure 3.1, we see that case eight is an extreme point in the DENS × STR plot in the sense that it is removed from the bulk of the data. From the PROD × STR plot, we see that the response appears to be unusually low. These two factors account for the sizeable changes in the parameter estimates.

The influence of this single case warrants further study. In this case, it was determined that a recording error was made, and the case was deleted from the study. In other situations, the explanation may not be so simple, but it is clear that we cannot ignore this influence and accept the original estimates in Table 3.1.

Fortunately, we do not have to refit the model with each of the N cases deleted in order to study their influence. It is a relatively simple matter to compute $\hat{\theta}_{j(i)}$ from the initial regression computations. Cook (1977) described a simple measure of influence based on the squared length of the vector $\hat{Y} - \hat{Y}_{(i)}$. The proposed statistic is

$$\begin{aligned} D_i &= (\hat{Y} - \hat{Y}_{(i)})'(\hat{Y} - \hat{Y}_{(i)})/(ps^2) \\ &= r_i^2 h_i/[(1 - h_i^2)ps^2]. \end{aligned} \tag{3.149}$$

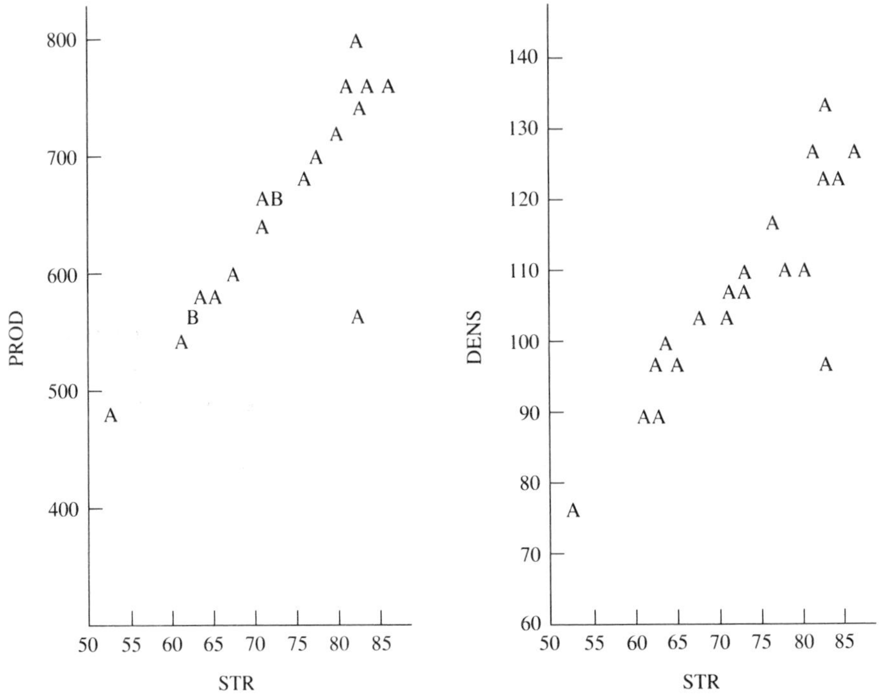

FIGURE 3.1
Scatter plots for steel-production data

This statistic can be computed for each observation using the information from the initial regression computation, and it does not require refitting the equation. Cook (1977) noted the D_i is approximately distributed as $F(p, N - p)$ and suggested that values of D_i exceeding $F(0.1; p, N - p)$ indicate influential observations. Belsley, Kuh, and Welsch (1980) described a similar measure of the influence on prediction, differing essentially by the use of $s_{(i)}^2$ rather than s^2 in (3.149). They also proposed measures of the influence on the estimates of the θ_j. A summary of many of the proposed influence measures and related references is found in Hocking (1983b).

In our example, we obtain $D_8 = 11.44$. When compared with $F(0.1; 4, 16) = 0.26$, this case is clearly indicated as influential. The advantage of these numerical diagnostic measures is not evident from this example, however, since the plots in Figure 3.1 clearly identified case eight as being unusual. In higher dimensional problems, extreme points and outliers may not be obvious in the two-dimensional plots.

Continuing with the data examination, we see from Figure 3.1 that the deletion of case eight suggests a near linear relation between DENS and STR. The standard collinearity measures as described in Hocking and Pendleton (1983) indicate that this collinearity is not extremely strong, but it merits some attention. If this near linear relation is typical of the relation between strength and density, then one of these variables might be eliminated from the prediction equation which was the ultimate objective of the study. On the other hand, if this relation is not typical, we should not consider elimination of either variable, and, further, we should be warned against using the fitted equation for prediction when the values of STR and DENS lie away from the region shown in Figure 3.1 even though each variable lies within the observed range for that variable.

We shall not pursue the example further since most of its points have been illustrated, but the reader is encouraged to do so. In practice, we would certainly recompute the confidence intervals and the test statistic in Table 3.2. We may be inclined to consider eliminating the variable WID from the prediction equation. We might also consider the use of a biased estimator such as ridge regression to dampen the effects of the collinearity. The fine tuning of a regression model is clearly an iterative process, which certainly leads us to question the confidence coefficients and significance levels as applied to the final equation. The precise application to regression models of the theory developed in this chapter should be viewed rather casually, but the ideas are certainly an aid to finding a good model.

CHAPTER 3 EXERCISES

1. **(a)** Develop the expressions for the normal equations and $Q(\hat{\theta})$ for the simple regression model

$$y = \theta_0 + \theta_1 x + e.$$

(b) Determine the estimates of the slope and intercept for both the male and female test score data from Table 1.1.

(c) Develop the normal equations for the two models (1.39) and (1.40), and estimate all parameters.

(d) Test the hypothesis, with $\alpha = 0.05$, that the two slopes are equal against the two-sided alternative.

2. Repeat the analysis of the example in Section 3.6 with observation eight deleted from the data. Compute the ridge regression estimator for $k = 0.01$ and compare with the least squares estimator.

3. Estimate the parameters for the model in Example 1.3. Estimate the values of x_1 and x_2 that will maximize the quality of the product.

4. Verify the properties of the A matrix in Lemma 3.1.

5. Fill in the details for the development leading to the expression for N_H in (3.92).

6. Verify the Bonferroni inequality in (3.102).

7. **(a)** For the special case $A = I$, $X'X = I$, $H = I_p$ and $p = 2$, $\sigma^2 = 1$, $\alpha = 0.05$, write the Scheffé simultaneous confidence intervals for

$$a_1' = (1, 0) \qquad a_2' = (0, 1)$$
$$a_3' = (1, 1) \qquad a_4' = (1, -1).$$

(b) Sketch the confidence ellipsoid and the tangent planes associated with these confidence intervals.

(c) Sketch the lines corresponding to the analogous Bonferroni intervals on the same figure.

8. Let X_s denote the standardized data matrix used in (3.142). Let $R = X_s'X_s$, and let λ_i and v_i, $i = 1, \ldots, m$ denote the eigenvalues and eigenvectors of R. Show that, if $\lambda_i = 0$, then v_i identifies an exact collinearity in X_s and hence in X. If λ_i is "near" zero, a near collinearity is indicated.

9. **(a)** Develop an expression for $\hat{\theta}_{(i)}$ from the information obtained for the regression on all the data.

(b) Use the result in (a) to verify the expression for Cook's D_i in (3.149).

(c) Determine an expression for $s^2_{(i)}$ in terms of s^2, r_i, and h_i.

CHAPTER FOUR
THE ANALYSIS OF MEANS

4.1 INTRODUCTION

The theories for estimation and hypothesis testing in linear models are applicable to either regression or means models. The special characteristics of means models allow us to examine more specific questions and to introduce problems that are unique to such models. In this chapter, we review the general results developed in Chapter 3, using the notation for the means model and noting special features of the results. In addition, we consider problems that arise when the populations in the means model are not all sampled with the same frequency or when some populations are not sampled.

It will be sufficient for the general discussion to assume the means model described for Example 1.4 and given in (1.19). That is, it is assumed that we have sampled p normal populations with distinct means, μ_i, $i = 1, \ldots, p$, and that we have n_i observations on the ith population. Letting y_{ij}, $i = 1, \ldots, p; j = 1, \ldots, n_i$ denote the jth observation on the ith population, the unconstrained model is written algebraically as

$$y_{ij} = \mu_i + e_{ij} \qquad \begin{array}{l} i = 1, \ldots, p \\ j = 1, \ldots, n_i. \end{array} \tag{4.1}$$

On occasion, it may be reasonable to assume that these means satisfy certain linear equalities. If so, these constraints are adjoined to the model. We shall consider only the fixed effects model in this chapter; hence we assume that the e_{ij} are independent, $N(0, \sigma^2)$. For our work it will be convenient to describe the model in matrix notation, as in (1.24) and (1.35). For reference, the model is defined formally as follows.

DEFINITION 4.1 THE MEANS MODEL
Letting Y denote the N-vector of responses in (4.1), where $N = \Sigma_i n_i$, the model is given by

$$\begin{array}{l} Y = W\mu + e \\ \text{subject to} \quad G\mu = g. \end{array} \tag{4.2}$$

Here, $W = D_p(J_{n_i})$ is the $N \times p$ matrix that indicates the mean associated with an observation. We shall refer to W as the *counting matrix,* recognizing that W also indicates the number of observations on each population. The p-vector of means in (4.1) is denoted by μ, and G and g are assumed known, with G being $q \times p$ of rank q. The fixed effects model is characterized by assuming $e \sim N(0, \sigma^2 I)$.

■

The analysis of this model will consist primarily of making inferences on the population means; hence, we shall refer to the analysis collectively as the *Analysis of Means* (*AOM*). This will include the tabular display of quadratic forms, illustrated in Chapters 5 through 7, for testing certain hypotheses. This display is commonly known as an *Analysis of Variance* (*ANOVA*) *table*. We shall reserve the term *ANOVA,* however, for the model in which the covariance matrix is more general.

To relate to other writings on linear model analysis, we shall see that most references would use a model of the form

$$Y = X\theta + e \tag{4.3}$$

where X is $N \times (m + 1)$ of rank $p < m + 1$ in place of the means model. Here, the vector θ is unconstrained and can be related to μ. In particular, we refer to the excellent texts by Graybill (1976), Scheffé (1959), and Searle (1971a). The complexities and confusion caused by the fact that X does not have full column rank and that the components of θ in this model are not uniquely defined prompted the use of the means model. The analysis of the means model is generally simpler, and the parameters, being means, are easy to interpret. In later chapters, we shall find it convenient to work with linear functions of μ as our basic parameters. That discussion will provide the link between the means model and the overparameterized or nonfull rank model, (4.3).

Our use of the term *model* will refer collectively to (a) the assumed structure on the mean vector, μ, including constraints, (b) the available data given by the response vector, Y, and related to μ by the counting matrix, W, and (c) the distributional assumptions made on the response vector Y. This general definition of the term *model* will be important when we discuss problems with missing cells where both the design of the experiment and the structure on the mean vector will be necessary to determine whether some or all of the means are estimable and whether or not desired hypotheses can be tested.

Examples illustrating the means model were given in Chapter 1. Since this model formulation may be different from that previously encountered by the reader, it will be instructive to describe additional situations for which this model is appropriate. The following sequence of increasingly complex problems should illustrate the flexibility and clarity of the means model formulation. For reference, we shall associate the models for these examples with the terms commonly used to describe them in the literature.

EXAMPLE 4.1

Suppose the data vector, Y, consists of test scores of a random sample of incoming freshmen at a particular university. We assume that the scores, y_i, $i = 1, \ldots, N$, are independently normally distributed, $N(\mu, \sigma^2)$. In the notation of the means model, we may write

$$
\begin{aligned}
Y &= \mu J + e, \\
e &\sim N(0, \sigma^2 I)
\end{aligned}
\tag{4.4}
$$

where μ is a scalar and $W = J$ is an N-vector of ones. [Note that this model is a special case of (4.1) or (4.2) in which we observe a single population.]

The parameters, μ and σ^2, in this model are easily estimated by

$$
\begin{aligned}
\hat{\mu} &= \bar{y}. = \textstyle\sum y_i/N, \\
s^2 &= \textstyle\sum (y_i - \bar{y}.)^2/(N - 1).
\end{aligned}
\tag{4.5}
$$

(Here we have used the convention that a "." replacing a subscript implies summation over the range on that subscript.) These estimates will be seen to follow from the general theory developed in Section 3.2, but they are no doubt familiar to the reader.

The analysis of μ may be in the form of a hypothesis such as

$$
\begin{aligned}
&H_0\colon \mu = \mu_0 \\
&\text{versus} \\
&H_A\colon \mu \neq \mu_0.
\end{aligned}
\tag{4.6}
$$

Alternatively, we may wish an interval estimate (confidence interval) for μ of the form

$$
\mu_L \leq \mu \leq \mu_U. \tag{4.7}
$$

The parameter, σ^2, is usually estimated only as needed for the test statistics or for the interval estimates. On occasion, inferences on σ^2 may be of interest. Here, for example, the variability of the test scores may be of interest.

This problem is generally referred to as the *one-sample problem,* emphasizing the fact that only one population is observed.

■

EXAMPLE 4.2

In the same setting as in Example 4.1, suppose the N students sampled consist of n_1 males and $n_2 = N - n_1$ females. Denote the means for these two groups by μ_1 and μ_2, respectively, and assume that the variance is the same for each group. Thus, we can view this data as a random sample of size n_1 from the male population whose scores are distributed as $N(\mu_1, \sigma^2)$ and an independent random sample of size n_2 from the female population whose scores are distributed as $N(\mu_2, \sigma^2)$. Letting y_{ij},

$i = 1, 2; j = 1, \ldots, n_i$ denote the jth response on the ith group, we may write the model as

$$y_{ij} = \mu_i + e_{ij}. \tag{4.8}$$

Alternatively, with Y denoting the N-vector of scores ordered with group one first, the model may be written, as in Definition 4.1, by defining

$$W = \begin{bmatrix} J_{n_1} & 0 \\ 0 & J_{n_2} \end{bmatrix} \qquad \mu = \begin{bmatrix} \mu_1 \\ \mu_2 \end{bmatrix}. \tag{4.9}$$

In this case, estimates of the μ_i are given by

$$\hat{\mu}_i = \bar{y}_{i\cdot} = \sum_j y_{ij}/n_i \qquad i = 1, 2. \tag{4.10}$$

The analysis may consist of interval estimates on the μ_i or of tests of hypotheses. For example, a natural question is whether the two groups have the same mean. Thus, we might consider the hypothesis

$$H_0\colon \mu_1 = \mu_2$$

$$\text{versus} \tag{4.11}$$

$$H_A\colon \mu_1 \neq \mu_2.$$

An estimate of σ^2 may be obtained for each group separately as

$$s_i^2 = \sum_j (y_{ij} - \bar{y}_{i\cdot})^2/(n_i - 1) \qquad i = 1, 2. \tag{4.12}$$

These two estimates may be combined into a single estimate as

$$s^2 = [(n_1 - 1)s_1^2 + (n_2 - 1)s_2^2]/(n_1 + n_2 - 2). \tag{4.13}$$

Again, these intuitive estimates will be seen to follow from the general theory. This will be demonstrated in Section 4.2, but the reader will note that this is the classical *two-sample problem*. We also note that this model is mathematically equivalent to the model for the bread-dough experiment given in Example 1.2 except that, in Example 1.2, more than two populations were observed. The model for $p \geq 2$ is called the *one-way classification model* and is discussed in detail in Chapter 5. We also observe that, for this model, the notation is exactly that used to describe the general means model. This is an important point. We shall see in the next example that more complex subscripting may be notationally more convenient for some models, but this does not alter the generality of the model in Definition 4.1.

■

EXAMPLE 4.3

Continuing with Example 4.2, suppose the students are also classified by race—say, Caucasian, Negroid, and Oriental. In this case, we may view the data as consisting of six independent random samples from the six populations defined by all combinations of sex and race. No new concepts are involved here, but some notation is introduced for convenience. Rather than using μ_i, $i = 1, \ldots, 6$, to denote the population means, we shall use μ_{ij}, $i = 1, 2; j = 1, 2, 3$. Thus, μ_{11} might correspond to Caucasian males, and so on. Similarly, the data are described by y_{ijk}, $i = 1, 2; j = 1, 2, 3; k = 0, 1, 2, \ldots, n_{ij}$ where, for example, n_{11} is the number of Caucasian males in the sample. The range on k allows for the possibility that there may be zero observations on a particular population. We discuss such a situation in Example 4.5.

In the notation of Definition 4.1 we have

$$W = \begin{bmatrix} J_{n_{11}} & 0 & 0 & 0 & 0 & 0 \\ 0 & J_{n_{12}} & 0 & 0 & 0 & 0 \\ 0 & 0 & J_{n_{13}} & 0 & 0 & 0 \\ 0 & 0 & 0 & J_{n_{21}} & 0 & 0 \\ 0 & 0 & 0 & 0 & J_{n_{22}} & 0 \\ 0 & 0 & 0 & 0 & 0 & J_{n_{23}} \end{bmatrix} \qquad \mu = \begin{bmatrix} \mu_{11} \\ \mu_{12} \\ \mu_{13} \\ \mu_{21} \\ \mu_{22} \\ \mu_{23} \end{bmatrix}. \tag{4.14}$$

Alternatively, we may write the model algebraically as

$$y_{ijk} = \mu_{ij} + e_{ijk} \qquad \begin{aligned} i &= 1, 2 \\ j &= 1, 2, 3 \\ k &= 0, 1, \ldots, n_{ij}. \end{aligned} \tag{4.15}$$

The data for this model may be displayed in a two-way table with the cells of the table being indicated by the subscript pair (i, j). These cells may be associated with the corresponding populations, and the μ_{ij} are often referred to as *cell means*. For this reason, the means model is often called the *cell means model*. We shall use either term in the sequel, with preference for the latter since it is in common usage.

Although the only mathematical difference between the one-way model of Example 4.2 and this *two-way classification model* is in the use of two subscripts to describe the means, the practical differences are substantial. This example suggests many questions that might potentially be raised. For example, we may be interested in comparison of performance by race or by sex. A more detailed examination might consider the difference in performance by sex within a particular race, and so on.

Reference to Example 1.6—where we considered the effect of the two factors, freezing time and type of dough—reveals that this two-factor model applies equally well to that situation. A thorough discussion of this model is given in Chapter 6.

■

EXAMPLE 4.4

A hypothesis that might be of interest in Example 4.3 is that the difference in the performance of males and females is the same for each race. Stated in terms of the cell means, this relation is written as

$$\begin{aligned} \mu_{11} - \mu_{21} &= \mu_{12} - \mu_{22} \\ \mu_{11} - \mu_{21} &= \mu_{13} - \mu_{23}. \end{aligned} \tag{4.16}$$

In some special situations, the researcher may be willing to assume that this relation is inherently true for the population means. Under this assumption, the model is given in Example 4.3 subject to these constraints on the means. We write this constrained model as

$$\begin{aligned} y_{ijk} = \mu_{ij} + e_{ijk} \qquad & i = 1, 2 \\ & j = 1, 2, 3 \\ & k = 0, 1, \ldots, n_{ij} \\ \text{subject to} \qquad \mu_{11} - \mu_{21} &= \mu_{12} - \mu_{22} \\ \mu_{11} - \mu_{21} &= \mu_{13} - \mu_{23}. \end{aligned} \tag{4.17}$$

In the usual terminology, (4.17) is known as the *two-way classification* (or *two-factor*) *model without interaction*. We note the equivalence of this model and the one used in the bread-dough experiment when it was assumed that the factors in that study did not interact in the sense of (4.16). These relations are known as the *constraints of no interaction.*

The introduction of constraints into the model will introduce a new level of complexity into the analysis, but the concepts remain quite simple. Notationally, it will usually be convenient to retain the simple form of the model, (4.15), and adjoin the constraints as in (4.17). Computationally, it is better to use the constraints to reduce the dimensions of the model. For example, we may solve the constraints for μ_{11} and μ_{12} and substitute into (4.15). The resulting unconstrained model is given by noting that, after this substitution,

$$E(Y) = \begin{bmatrix} J_{n_{11}} & J_{n_{11}} & 0 & -J_{n_{11}} \\ J_{n_{12}} & 0 & J_{n_{12}} & -J_{n_{12}} \\ J_{n_{13}} & 0 & 0 & 0 \\ 0 & J_{n_{21}} & 0 & 0 \\ 0 & 0 & J_{n_{22}} & 0 \\ 0 & 0 & 0 & J_{n_{23}} \end{bmatrix} \begin{bmatrix} \mu_{13} \\ \mu_{21} \\ \mu_{22} \\ \mu_{23} \end{bmatrix}. \tag{4.18}$$

The coefficient matrix for the reduced model in (4.18) is denoted by W_R. We see that this matrix no longer has the simple structure of the counting matrix, W, but the general results of Chapter 3 may be applied.

■

EXAMPLE 4.5

Consider the same situation as in Example 4.3, but suppose $n_{11} = 0$. That is, for some reason there are no Caucasian males included in the study. This phenomenon, referred to as the *empty*, or *missing, cell problem,* frequently occurs in sociological data, such as this example, when the number of classification criteria is increased. It may also occur in carefully designed experiments when some of the units being examined fail or are lost. On other occasions, the experiment may be intentionally designed with missing cells when the experimental units or procedures are costly.

With regard to the analysis, the unconstrained case in Example 4.3 and the constrained case in Example 4.4 present interesting problems. In Example 4.3, the estimation problem is simple since only those means μ_{ij} with $n_{ij} > 0$ can be estimated. On the other hand, formulating hypotheses of interest is a problem, since hypotheses involving μ_{11} cannot be tested. For example, with $n_{11} = 0$, how should we compare males against females in this study? Should we exclude Caucasian females as well, or should a compromise hypothesis be considered?

In Example 4.4, it is possible to estimate μ_{11}, even though $n_{11} = 0$, by using the constraint

$$\mu_{11} = \mu_{21} + \mu_{12} - \mu_{22} \tag{4.19}$$

since μ_{21}, μ_{12}, and μ_{22} are estimable. In other, more complex experiments with several missing cells, we may find that it is not possible to estimate all cell means.

In this simple example, we shall see that any hypothesis that could have been tested with all $n_{ij} > 0$ can still be tested. With more missing cells, this may not be true.

■

These examples should indicate the generality of the cell means model. In the special case to be considered in this chapter, where we assume $\text{Var}(Y) = \sigma^2 I$, the data is viewed as arising by sampling populations that may differ only in their means. When we refer to the structure on the means, we include any *a priori* constraints on μ.

The remaining sections of this chapter will be devoted to spelling out the results from Chapter 3 as applied to the cell means model. Additional results on simultaneous confidence intervals and estimation and hypothesis testing with missing cells also will be developed. In Chapters 5, 6, and 7 specific models will be examined and illustrated.

4.2. ESTIMATION OF PARAMETERS

The results in this section follow directly from those in Section 3.2. Again, it is useful to describe separately the unconstrained and the constrained models. An additional complexity arises for the cell means model when some of the cell frequencies are zero—that is, when some of the populations are not observed. For this reason, we initially assume nonzero cell frequencies ($n_i > 0$) and then treat the missing cell case in Section 4.2.3.

4.2.1. The Unconstrained Model: $n_i > 0$

The likelihood equations are given by (3.5) and (3.6), with X and θ replaced by W and μ. The unbiased estimators of μ and σ^2 are given as

$$\hat{\mu} = (W'W)^{-1}W'Y \tag{4.20}$$

$$\begin{aligned} s^2 &= Q(\hat{\mu})/(N - p) \\ &= Y'[I - W(W'W)^{-1}W']Y/(N - p). \end{aligned} \tag{4.21}$$

The structure of W allows us to write

$$W'W = D_p(n_i). \tag{4.22}$$

Hence, $(W'W)^{-1}$ is just a diagonal matrix with elements $1/n_i$. Also,

$$W'Y = C_p(y_{i}.). \tag{4.23}$$

That is, $W'Y$ is a p-vector with elements $y_{i}.$. It follows from (4.20) that

$$\hat{\mu} = C_p(\bar{y}_{i}.). \tag{4.24}$$

That is, the components of $\hat{\mu}$ are the sample means $\bar{y}_{i}.$ for the ith population.
The estimate of σ^2 is simplified by noting that

$$\begin{aligned} Q(\hat{\mu}) &= \sum_{i=1}^{p} \sum_{j=1}^{n_i} (y_{ij} - \bar{y}_{i}.)^2 \\ &= \sum_{i} (n_i - 1)s_i^2 \end{aligned} \tag{4.25}$$

where

$$s_i^2 = \sum_{j} (y_{ij} - \bar{y}_{i}.)^2/(n_i - 1) \tag{4.26}$$

denotes the unbiased estimate of σ^2 based on the data from the ith population. The "pooled" estimator of σ^2 is thus given by the weighted average of these individual estimates as

$$s^2 = \sum_{i} (n_i - 1)s_i^2/(N - p). \tag{4.27}$$

The properties of these estimators follow from Theorem 3.1 and are summarized in the following theorem.

THEOREM 4.1
For the unconstrained cell means model, $\hat{\mu}$ and s^2 are independent and are minimum variance, unbiased estimators. The distributions of these estimators are determined by

$$\hat{\mu} \sim N[\mu, \sigma^2(W'W)^{-1}] \tag{4.28}$$

and

$$Q(\hat{\mu})/\sigma^2 \sim \chi^2(N - p). \tag{4.29}$$

Note that the estimates of the components of μ are independent, $N(\mu_i, \sigma^2/n_i)$.

■

4.2.2. The Constrained Model: $n_i > 0$

Explicit expressions for the constrained estimators depend on the particular constraints assumed on the model. We shall defer a discussion of special constraints to later chapters. The expressions for the estimators follow from Theorem 3.2, which we restate here for reference in the following two theorems.

THEOREM 4.2
For the constrained cell means model, $\tilde{\mu}$ and s^2 are independent, minimum variance, unbiased estimators. The estimate of μ is given by

$$\tilde{\mu} = A\hat{\mu} + Bg \tag{4.30}$$

where $\hat{\mu}$ is the unconstrained estimator from Theorem 4.1. Also,

$$\begin{aligned} B &= (W'W)^{-1}G'[G(W'W)^{-1}G']^{-1} \\ A &= I - BG. \end{aligned} \tag{4.31}$$

The estimate of σ^2 is given by

$$s^2 = Q(\tilde{\mu})/(N - r) \tag{4.32}$$

where $r = p - q$ and

$$Q(\tilde{\mu}) = (Y - W\tilde{\mu})'(Y - W\tilde{\mu}). \tag{4.33}$$

The distributions of these estimators are determined by noting that

$$\tilde{\mu} \sim N[\mu, \sigma^2 A(W'W)^{-1}] \tag{4.34}$$

and

$$Q(\tilde{\mu})/\sigma^2 \sim \chi^2(N - r). \tag{4.35}$$

■

Alternative expressions for the estimators are obtained by using the constraints to reduce the number of parameters in the model. Thus, if we write the model as

$$\begin{aligned} Y &= W_1\mu_1 + W_2\mu_2 + e \\ \text{subject to} \quad & G_1\mu_1 + G_2\mu_2 = g \end{aligned} \tag{4.36}$$

where G_1 is nonsingular, the reduced model is

$$Y_R = W_R\mu_R + e \tag{4.37}$$

where

$$\begin{aligned} Y_R &= Y - W_1G_1^{-1}g \\ W_R &= W_2 - W_1G_1^{-1}G_2 \\ \mu_R &= \mu_2. \end{aligned} \tag{4.38}$$

It is emphasized that W_R no longer has the simple structure of W; hence, the expression for the estimate of μ_R is more complex than $\hat{\mu}$. The following theorem presents the results for the constrained model in terms of the reduced model.

THEOREM 4.3

For the constrained cell means model, the estimators for μ and σ^2 in Theorem 4.2 may be written as

$$\begin{aligned} \tilde{\mu} &= \begin{bmatrix} \tilde{\mu}_1 \\ \tilde{\mu}_2 \end{bmatrix} \\ s^2 &= s_R^2 \end{aligned} \tag{4.39}$$

where

$$\begin{aligned} \tilde{\mu}_2 &= \hat{\mu}_R = (W_R'W_R)^{-1}W_R'Y_R \\ \tilde{\mu}_1 &= G_1^{-1}g - G_1^{-1}G_2\tilde{\mu}_2 \end{aligned} \tag{4.40}$$

and

$$s_R^2 = Q_R(\hat{\mu}_R)/(N - r) \tag{4.41}$$

with

$$Q_R(\mu_R) = (Y_R - W_R\mu_R)'(Y_R - W_R\mu_R). \tag{4.42}$$

The distribution of $\tilde{\mu}$ is determined by noting that

$$\hat{\mu}_R \sim N[\mu_R, \sigma^2(W_R'W_R)^{-1}] \tag{4.43}$$

and recognizing that $\tilde{\mu}$ is a linear function of $\hat{\mu}_R$. The covariance matrix of $\tilde{\mu}$ is given either by (4.34) or by (3.29), with X and θ replaced by W and μ. Using the results of Theorem 3.1, it follows that

$$s_R^2 \sim [\sigma^2/(N - r)]\chi^2(N - r). \tag{4.44}$$

The equivalence of $\tilde{\mu}$ as given by (4.30) and (4.39) and of s^2 as given by (4.32) and (4.41) is easily established.

■

4.2.3. Missing Cells, Effective Models, and Connectedness

Point estimation for both the unconstrained and the constrained model has been developed under the assumption that $n_i > 0$. In the unconstrained case, there is no problem with missing cells since $n_i = 0$ means that the ith population was not observed, and hence its mean, μ_i, cannot be estimated. The estimates for the remaining population means are unchanged.

For the constrained model, it may be possible to estimate μ_i when $n_i = 0$, since the constraints may relate μ_i to other means whose populations are observed. The following development will describe situations in which this is possible, and will reveal some interesting aspects of the problem.

In the constrained model given by Definition 4.1, suppose $n_i = 0$ for one or more of the indices $i = 1, \ldots, p$. Assuming, for simplicity, that these are the first m indices—that is, $n_i = 0$ for $i = 1, \ldots, m$—we may write the matrix, W, and the mean vector, μ, in partitioned form as

$$W = (W_m | W_o)$$

and (4.45)

$$\mu = \begin{bmatrix} \mu_m \\ \mu_o \end{bmatrix}.$$

Here W_m is an $N \times m$ matrix of zeros, $W_o = D_{p-m}(J_{n_i})$, and μ_m and μ_o are the vectors of means corresponding to the missing and observed cells, respectively. We also partition G conformably as

$$G = (G_m | G_o). \tag{4.46}$$

Our objective is to remove μ_m from the model by expressing μ_m in terms of μ_o, using the constraints. Formally, we may do this by row-reducing G and g in (4.46), obtaining

$$G \rightarrow \left[\begin{array}{c|c} G_{mm} & G_{mo} \\ 0 & G_E \end{array} \right] \qquad g \rightarrow \begin{bmatrix} g_m \\ g_E \end{bmatrix} \tag{4.47}$$

where G_{mm} has full row rank, say t. The elimination of μ_m from the model then leaves the constraints defined by the lower partition in (4.47). The resulting model will be called the *effective model* and is defined as follows.

DEFINITION 4.2
In the case of missing cells, we define the effective model as

$$Y = W_o\mu_o + e \tag{4.48}$$

$$\text{subject to} \quad G_E\mu_o = g_E. \tag{4.49}$$

Here W_o and μ_o are defined by (4.45), while G_E and g_E are defined by (4.47). [Note that the *effective model* for the unconstrained case is given by (4.48) without the constraints.] The constraints in (4.49) will be called the *effective constraints*.

■

The results of Section 4.2.2 may now be applied to the effective model to estimate μ_o. The estimability of μ_m rests on the structure of the equations

$$G_{mm}\mu_m + G_{mo}\mu_o = g_m. \tag{4.50}$$

If G_{mm} is nonsingular, μ_m is uniquely defined in terms of μ_o and, hence, μ_m is estimated by

$$\tilde{\mu}_m = G_{mm}^{-1}(g_m - G_{mo}\tilde{\mu}_o). \tag{4.51}$$

(Without loss of generality, we can assume that $G_{mm} = I_m$ in this case.)
If G_{mm} is singular, we can write it as

$$G_{mm} = (I_t|C) \tag{4.52}$$

where C is a $t \times (m - t)$ matrix. In this case, it is not possible to solve uniquely for μ_m, but we can solve for linear functions of μ_m. Of course, some components of μ_m may have unique expressions in terms of μ_o.

For reasons that will be made clear in Chapter 6, we shall refer to the case $G_{mm} = I_m$ as the *connected case*. The term *connected* is used for historical reasons and is synonymous with estimable; that is, it refers to the case in which the model and data are such that all cell means are linearly estimable. Linear estimability of μ simply means that there is a linear function of the data, say $AY + a$, such that $E(AY + a) = \mu$. Clearly, μ_o is linearly estimable by Theorem 4.2. The (linear) estimability of μ_m depends on the nonsingularity of G_{mm} in (4.50). To summarize, we make the following definition.

DEFINITION 4.3
A model is said to be connected if all components of μ are linearly estimable—that is, $G_{mm} = I_m$ in (4.50).

■

Note that, in the unconstrained case, connectedness requires that all $n_i > 0$. In the constrained case, the requirement is that G_{mm} be nonsingular. This emphasizes the fact that connectedness is a function of all aspects of the model. The available data and the constraints must both be examined.

The term *connected* is used in the literature [see Graybill (1976) or John (1971)] with reference to the data for a special case of constrained models. The relation between that usage and our general definition will be developed in Chapter 6.

The reduction of the original model to the effective model reveals the situation with regard to estimability. In practice, we rarely work with the effective model but, rather, with the reduced model described in Section 4.2.2 under the assumption that $n_i > 0$. Recalling that development, we see that the formulation of the reduced model did not depend on this assumption. The critical step occurs when trying to estimate μ_R, which requires W_R to have full column rank, r. The following theorem establishes the condition for this to hold.

THEOREM 4.4

The cell means model is connected if and only if the matrix W_R in the reduced model has full column rank.

Proof: Recall from Section 4.2.2 that, in the reduced model, W_R was given by

$$W_R = W_2 - W_1 G_1^{-1} G_2 \tag{4.53}$$

where μ_1 was related to $\mu_2 = \mu_R$ by

$$\mu_1 = G_1^{-1} g - G_1^{-1} G_2 \mu_R. \tag{4.54}$$

If W_R has full column rank, then $W_R' W_R$ is nonsingular and

$$\hat{\mu}_R = (W_R' W_R)^{-1} W_R' Y_R. \tag{4.55}$$

The estimate of μ_1 is then obtained from (4.54), and the model is connected. Conversely, if the model is connected, there exists a matrix T such that

$$E(T' Y_R) = T' W_R \mu_R = \mu_R. \tag{4.56}$$

Since this result is to hold for any μ_R, then

$$I_r - T' W_R = 0. \tag{4.57}$$

Thus,

$$r = r(I_r) = r(T' W_R) \leq r(W_R) \leq r \tag{4.58}$$

and, hence, W_R has rank r and the proof is complete.

■

Our development of the reduced model required the identification of a nonsingular submatrix of G, say G_1, without regard to whether μ_2 consisted of observed cells. The result of Theorem 4.4 is thus independent of whether or not the cells represented in the reduced model are observed. If the model is connected, this does not present a problem since all components of μ are estimable. The estimate of $\mu_2 = \mu_R$ is obtained from the reduced model, and the estimate of μ_1 is obtained as in (4.54) quite apart from the location of the unobserved cells.

Similarly, the estimate of σ^2 in the connected case follows directly from (4.32) and (4.41). The distribution of s^2 is still determined by (4.35) with no change in the degrees of freedom.

If W_R does not have full column rank, then we may proceed in two ways: (a) revert to the effective model or (b) delete $m - t$ columns in W_R, where t is the rank of G_{mm}, so that the remaining matrix has full column rank. If we assign the value zero to the corresponding parameters, then the remaining components of μ_R are estimated as usual and μ_1 is estimated by (4.40). This procedure, which is commonly used in practice when we work with the reduced model—or a transformed version of it—will yield correct estimates for observed cells and for unobserved cells that are uniquely estimable through (4.50). The remaining unobserved cells are not uniquely estimated, although the estimates of the linear combinations of them implied by (4.50) are uniquely estimated. Unfortunately, without the decomposition in (4.47), we cannot distinguish between these two groups of unobserved cells. For this reason, as well as problems encountered later, the unconnected case should be handled with care. The effective model analysis will avoid such difficulties.

Finally, we note that the estimate of σ^2 in the disconnected case can be obtained as usual from the reduced model with deleted columns. The degrees of freedom associated with the residual sum of squares is now given by $N - [r - (m - t)]$. This result follows easily from the effective model.

The following example will illustrate some of the concepts of this section. Additional examples will be given in Chapters 6 and 7, where we discuss the two-factor and three-factor models in detail.

EXAMPLE 4.6

Consider Example 4.5 with model given by (4.17) and $n_{11} = 0$. In this case, we write $\mu' = (\mu_{11} | \mu_{12}\mu_{13}\mu_{21}\mu_{22}\mu_{23})$, where μ_{11} is the mean of the unobserved population. From (4.17), the constraint matrix is written in the partitioned form of (4.47) as

$$G = \left[\begin{array}{c|ccccc} 1 & -1 & 0 & -1 & 1 & 0 \\ 1 & 0 & -1 & -1 & 0 & 1 \end{array}\right]. \tag{4.59}$$

Performing the reductions as indicated in (4.47), we obtain

$$\left[\begin{array}{c|ccccc} 1 & -1 & 0 & -1 & 1 & 0 \\ 0 & 1 & -1 & 0 & -1 & 1 \end{array}\right]. \tag{4.60}$$

From (4.60) we see that $G_{mm} = 1$, and the model is connected. The single effective constraint is

$$\mu_{12} - \mu_{13} - \mu_{22} + \mu_{23} = 0 \tag{4.61}$$

and the estimate of μ_{11} is obtained from the relation

$$\mu_{11} - \mu_{12} - \mu_{21} + \mu_{22} = 0. \tag{4.62}$$

The effective model is thus given by

$$\begin{aligned} y_{ijk} = \mu_{ij} + e_{ijk} \qquad & (i, j) \neq (1, 1) \\ & i = 1, 2 \\ & j = 1, 2, 3 \\ & k = 1, 2, \ldots, n_{ij} \end{aligned} \tag{4.63}$$

subject to the constraint (4.61).

The analysis of this effective model may be accomplished as usual using the results of either Theorem 4.2 or Theorem 4.3. Note that the dimension of the problem is $r = 4$ and that the residual degrees of freedom is $n_{..} - r$.

In practice, we may not work with the effective model but simply use the reduced model as described in Example 4.5. Referencing (4.18), with $n_{11} = 0$, we immediately see that the model is connected, by application of Theorem 4.4, since W_R has rank 4. To further illustrate the importance of the connectedness of the model, suppose we construct the reduced model by solving the constraints for μ_{22} and μ_{23}. The mean structure is then given by

$$E(Y) = \begin{bmatrix} 0 & J_{n_{12}} & 0 & 0 \\ 0 & 0 & J_{n_{13}} & 0 \\ 0 & 0 & 0 & J_{n_{21}} \\ -J_{n_{22}} & J_{n_{22}} & 0 & J_{n_{22}} \\ -J_{n_{23}} & 0 & J_{n_{23}} & J_{n_{23}} \end{bmatrix} \begin{bmatrix} \mu_{11} \\ \mu_{12} \\ \mu_{13} \\ \mu_{21} \end{bmatrix}. \tag{4.64}$$

Note that even though this form of the reduced model involves the mean of the unobserved population, μ_{11}, W_R still has full column rank, and, hence, the model is connected. The impact of Theorem 4.4 is that we need not concern ourselves with missing data in the connected case.

■

4.3. TESTS OF LINEAR HYPOTHESES ON CELL MEANS

The likelihood ratio tests for linear hypotheses on μ for the unconstrained and the constrained cases follow directly from Section 3.3 and are repeated here for reference. Special results are developed for the missing cell problems, and we also discuss the concept of orthogonal contrasts.

4.3.1. The Unconstrained Model: $n_i > 0$

As in Section 3.3.1, we wish to test the linear hypothesis, $H\mu = h$, against the two-sided alternative that at least one component of the vector of equalities fails to hold. We assume that H is $v \times p$ of rank v. The results are summarized in the following theorem.

THEOREM 4.5
A test of size α for the hypothesis

$$H_0\colon H\mu = h \quad \text{versus} \quad H_A\colon H\mu \neq h \tag{4.65}$$

in the unconstrained model is given by rejecting H_0 when

$$F = N_H/(vs^2) > F(\alpha; v, N - p). \tag{4.66}$$

Here, the numerator sum of squares is given by

$$\begin{aligned} N_H &= (H\hat{\mu} - h)'[H(W'W)^{-1}H']^{-1}(H\hat{\mu} - h) \\ &= Q(\tilde{\mu}) - Q(\hat{\mu}) \\ &= Q_R(\hat{\mu}_R) - Q(\hat{\mu}). \end{aligned} \tag{4.67}$$

Here, $\hat{\mu}$ and s^2 are as given in Theorem 4.1. The expressions for $\tilde{\mu}$, $\hat{\mu}_R$, and Q_R are given in Theorems 4.2 and 4.3 with H and h replacing G and g.

■

4.3.2. The Constrained Model: $n_i > 0$

From Theorem 3.5, with appropriate notational changes, we obtain the following theorem.

THEOREM 4.6
A test of size α for the hypothesis (4.65) in the constrained model is given by rejecting H_0 when

$$F = N_H/(vs^2) > F(\alpha; v, N - r). \tag{4.68}$$

The statistic s^2 in the denominator is given in either Theorem 4.2 or Theorem 4.3. The numerator sum of squares is given by

$$\begin{aligned} N_H &= (H\tilde{\mu} - h)'[HA(W'W)^{-1}H']^{-1}(H\tilde{\mu} - h) \\ &= Q(\tilde{\tilde{\mu}}) - Q(\tilde{\mu}) \\ &= Q_{RR}(\tilde{\mu}_{RR}) - Q_R(\hat{\mu}_R). \end{aligned} \tag{4.69}$$

In the first expression for N_H in (4.69), $\tilde{\mu}$ and A are as defined in Theorem 4.2. In the second expression, $\tilde{\tilde{\mu}}$ is given by Theorem 4.2 with G and g replaced by F and f in (4.30) and (4.31), where F and f define the combined constraint and hypothesis equations as in (3.85). In the final expression, $\hat{\mu}_R$ and Q_R are given in Theorem 4.3, and $\tilde{\mu}_{RR}$ and Q_{RR} are as defined in Theorem 4.3 with F and f replacing G and g in the reduced model.

The test statistic, F, in (4.68) is distributed as $F(v, N - r, \lambda)$, where $r = p - q$ and

$$\lambda = (H\mu - h)'[HA(W'W)^{-1}H']^{-1}(H\mu - h)/(2\sigma^2). \qquad (4.70)$$

■

In either Theorem 4.5 or Theorem 4.6, the last expression for the numerator sum of squares is commonly used for computation. For example, in (4.69) the model would first be reduced by the constraints $G\mu = g$ to obtain $Q_R(\hat{\mu}_R)$ and then reduced further by the hypothesis equations $H\mu = h$ to obtain $Q_{RR}(\tilde{\mu}_{RR})$. We shall see, in Chapter 6, that reparameterizations of the cell means model make this approach especially simple for certain hypotheses.

4.3.3. Missing Cells and Effective Hypotheses

For the unconstrained model, the requirement $n_i > 0$ seems unnecessary since populations not observed may simply be deleted from the analysis. Hypotheses may then be tested on the means for the observed populations. The problem that arises in practice, however, is that certain hypotheses may be of interest, and these hypotheses are formulated in terms of all of the populations. Since such hypotheses can no longer be tested, the question of a compromise hypothesis is of interest—that is, a hypothesis that will in some sense provide the user with as much information as possible regarding the effects being examined.

In the constrained model, it was noted that means for unobserved populations may be estimable via the constraints. It may also be true that the hypothesis of interest is testable even though it is stated in terms of means of unobserved populations. Our task is to identify situations in which the original hypothesis is testable and, if it is not, determine a feasible alternative.

Specific answers to these questions will be developed for particular models in Chapters 6 and 7. For now, we shall suggest a general approach that is logically consistent but that may not satisfy the needs of the researcher in all cases. The idea is based on the effective model introduced in Section 4.2.3. The natural next step is to develop an *effective hypothesis* to be considered as a compromise to the desired hypothesis. The concepts will be developed for the constrained model, and we shall simply note the results for the unconstrained case.

Recall, from Definition 4.2, that the effective model is given by

$$\begin{aligned} Y &= W_o\mu_o + e \\ \text{subject to} \quad G_E\mu_o &= g_E. \end{aligned} \qquad (4.71)$$

The means corresponding to missing cells are related to μ_o by

$$G_{mm}\mu_m + G_{mo}\mu_o = g_m. \tag{4.72}$$

To define the effective hypothesis corresponding to the hypothesis $H\mu = h$, we write the hypothesis in partitioned form as

$$H_m\mu_m + H_o\mu_o = h. \tag{4.73}$$

There are two cases to consider. First, consider the connected case in which we may assume $G_{mm} = I$. Substituting (4.72) into (4.73) yields the effective hypothesis

$$(H_o - H_mG_{mo})\mu_o = h - H_mg_m. \tag{4.74}$$

In the unconnected case, we may assume that G_{mm} has the form (4.52) and write (4.72) as

$$\mu_{m1} + C\mu_{m2} + G_{mo}\mu_o = g_m \tag{4.75}$$

where μ_m has been partitioned into μ_{m1} and μ_{m2} of length t and $m - t$, respectively.
Partitioning H conformably as

$$H = (H_{m1}|H_{m2}|H_o) \tag{4.76}$$

and substituting from (4.75) yields

$$(H_{m2} - H_{m1}C)\mu_{m2} + (H_o - H_{m1}G_{mo})\mu_o = h - H_{m1}g_m. \tag{4.77}$$

Rewriting (4.77) as

$$H_m^*\mu_{m2} + H_o^*\mu_o = h^* \tag{4.78}$$

we apply row reductions as in the development of the effective constraints to obtain

$$\begin{aligned} H^* = (H_m^*|H_o^*) &\rightarrow \begin{bmatrix} H_{mm}^* & H_{mo}^* \\ 0 & H_E \end{bmatrix} \\ h^* &\rightarrow \begin{bmatrix} h_m^* \\ h_E \end{bmatrix}. \end{aligned} \tag{4.79}$$

Eliminating the rows involving unobserved populations yields the effective hypothesis. The following definition summarizes these results.

DEFINITION 4.4
For the general case of missing cells, the effective hypothesis is defined to be

$$H_E\mu_o = h_E. \tag{4.80}$$

In the connected case, the effective hypothesis is given by (4.74). In the unconstrained case, the development of H_E begins with (4.79), and the row reductions are applied to H, rather than to H^*.

■

In general, the rank of H_E will be less than that of H and, hence, the effective hypothesis is testing, in some sense, only a part of the original hypothesis. An appealing feature of the effective hypothesis, however, is that, if (4.80) is not true, then the original hypothesis, $H\mu = h$, is not true. Further, the effective hypothesis is the hypothesis of maximal rank for which this holds (Hocking, Speed, and Coleman, 1980).

In Section 4.2.3 we observed that, if G_{mm} was nonsingular, then all components of μ could be estimated. We referred to this as the *connected case*. In this case we have seen that the hypothesis is given by (4.74) and that the row reduction in (4.79) is not necessary. It is easily shown that the coefficient matrix in (4.74) has the same rank as H and, further, that (4.74) is equivalent to the original hypothesis. We thus have the following theorem.

THEOREM 4.7

If the model is connected, then any hypothesis that may have been planned for the experiment with $n_i > 0$ can be tested.

■

The importance of this simple theorem should be emphasized. It says that, if the model is connected, we may proceed with the same analyses that would have been performed if there had been no missing cells. Recalling Theorem 4.4, we see that if W_R in the reduced model has full column rank, we may estimate all parameters and test any linear hypotheses on μ. As a consequence of this we shall also be able to construct confidence intervals on any linear function of μ. The analysis of models with missing cells is thus straightforward in the connected case. The unconnected case will be examined in Chapters 6 and 7 for specific models.

If the model is not connected, then the particular hypothesis of interest may still be testable. From (4.77), we see that the condition for this to be true is

$$H_{m2} - H_{m1}C = 0. \tag{4.81}$$

In this case, the hypothesis is given by

$$(H_0 - H_{m1}G_{mo})\mu_0 = h - H_{m1}g_m. \tag{4.82}$$

If the model is not connected and (4.81) does not hold, then the researcher must decide on an appropriate compromise analysis. The effective hypothesis is a candidate, but we shall see that it must be used with caution. In general, the analyst must face up to those comparisons that can be made and try to salvage as much as possible from the experiment. This may be done in terms of a specially constructed hypothesis or possibly by using confidence intervals as described in Section 4.4.

A source of confusion in the statistical literature on the analysis of experiments with missing data is that the hypotheses tested are determined as a consequence of the peculiarities of a computational algorithm and are not specified by the user. In most cases, the user is not aware of the hypothesis associated with the analysis and is confused when different algorithms yield different results for the test statistics. Actually, these comments may be applied to the case $n_i > 0$ and even to the "balanced case" $n_i = n$, $i = 1, \ldots, p$. One of the motivations for the cell means model approach is to resolve such difficulties. This problem will be discussed in detail in Chapters 6 and 7. The following example will illustrate the essential ideas.

EXAMPLE 4.7

Consider the two-factor model from Example 4.2—that is $E(y_{ijk}) = \mu_{ij}$, with $i = 1, 2$; $j = 1, 2, 3$; and $k = 0, 1, \ldots, n_{ij}$. The hypothesis of no interaction is given by

$$\begin{aligned} H_I\colon \mu_{11} - \mu_{21} &= \mu_{12} - \mu_{22} \\ &= \mu_{13} - \mu_{23}. \end{aligned} \tag{4.83}$$

Now suppose that $n_{11} = 0$. Since there are no constraints in this model, the effective hypothesis based on (4.83) is obtained by eliminating μ_{11} as implied in (4.79), yielding the single degree of freedom hypothesis

$$H_{I_E}\colon \mu_{12} - \mu_{22} - \mu_{13} + \mu_{23} = 0.$$

A hypothesis for comparing the differences in the performances of the two sexes is given by considering the average response over all races—that is,

$$H_S\colon \bar{\mu}_{1\cdot} = \bar{\mu}_{2\cdot}. \tag{4.84}$$

With $n_{11} = 0$, it follows from (4.79) that there is no effective hypothesis for sex differences based on (4.84). Of course, the analyst may choose to formulate other hypotheses—for example, a comparison of the average of the mean responses for males in the two observed races to the average for the females over all three races. Alternatively, one could make the comparison excluding the Caucasian race. The important point to emphasize here is that the particular sex hypothesis should be spelled out explicitly.

The analogous hypothesis for comparing races is

$$H_R\colon \bar{\mu}_{\cdot 1} = \bar{\mu}_{\cdot 2} = \bar{\mu}_{\cdot 3}. \tag{4.85}$$

With $n_{11} = 0$, the reduction (4.79) yields the hypothesis

$$H_{R_E}\colon \bar{\mu}_{\cdot 2} = \bar{\mu}_{\cdot 3}.$$

This hypothesis ignores the Caucasian race, but again the analyst may choose to specify a different measure of racial differences.

For the same problem, but now considering the constraint of no interaction, the reduction implied in (4.47) yields the single effective constraint given by H_{I_E}. Under the constrained model it is seen that the hypotheses H_S and H_R are equivalent to

$$H_S^*: \mu_{1j} = \mu_{2j} \text{ for any } j,$$

and

$$H_R^*: \mu_{i1} = \mu_{i2} = \mu_{i3} \text{ for any } i.$$

These expressions follow by using the interaction constraints to simplify (4.84) and (4.85).

For this example, we see that these hypotheses may still be tested with $n_{11} = 0$. That is, we may choose $j = 2$ or 3 in H_S^* and set $i = 2$ in H_R^*.

These two situations serve to illustrate Definition 4.4 and Theorem 4.7; that is, since this no-interaction model is connected, the hypotheses that would be tested with no missing data may still be tested. In the model allowing for interaction, which is not connected, we see that we are able to test part of H_I and H_R but could not test H_S. These represent typical results for missing data problems. Additional examples are given in Chapters 6 and 7.

■

4.3.4. Subhypotheses and Significance Levels

The test of $H\mu = h$ may provide an indication of the presence of an effect as measured by these linear functions of the cell means. It is then natural to ask for more specific information on the nature of these effects. For example, in the sex-by-race study in Example 4.3, the initial hypothesis may have tested that all of the cell means, μ_{ij}, were the same. Rejection of this hypothesis leads to many questions as to the nature of these differences. For example, we may wish to test the interaction hypothesis discussed in Example 4.4—that is, the hypothesis H_I in (4.83). In addition, we may wish to test for sex differences averaged over all races or for racial differences averaged over both sexes. These are the hypotheses H_S and H_R in (4.84) and (4.85). The latter two hypotheses are often referred to as *main effect hypotheses*.

Typically, hypotheses such as H_I, H_S, and H_R would be considered *a priori*, rather than being suggested by the more general hypothesis. Note that these three hypotheses have numerator sums of squares with two, one, and two degrees of freedom (rank of the hypothesis matrix), respectively. This suggests that these hypotheses represent, in some sense, a decomposition of the original hypothesis of equal means with five degrees of freedom. Although the degrees of freedom sum to the total degrees of freedom for the general hypothesis, it is not necessarily true that the numerator sums of squares for the individual tests add up to the sum of squares for the general test. Even if they do, so that Cochran's Theorem (Theorem 2.6) may be used to establish their independence, it is natural to ask for the significance level associated with the three hypotheses. In particular, if we test each hypothesis at level α, it appears, intuitively, that the probability of rejecting at least one of the hypotheses is greater than α even if all are true.

To demonstrate this point, consider the balanced case ($n_{ij} = n$, all (i, j)), in which we shall see that the three numerator sums of squares are independent chi-squared statistics. Assuming further that σ^2 is known, the tests would be based on these numerator statistics. In this case, the probability that we reject at least one of the hypotheses, even if all are true, is obtained from (3.101) as $1 - (1 - \alpha)^3$. Since σ^2 is rarely known, the F-ratios developed in Section 4.3.1 must be used, and they are not independent even in the balanced case since they share the same denominator. In the unbalanced case, the numerators are not independent. Thus, (3.101) does not apply in these cases, but it does illustrate the cause for concern; that is, if we wish to ensure an overall (experimentwise) significance level of α when many hypotheses are to be tested, then some thought must be given to reducing the size of the individual tests. The approximate solution suggested in Section 3.3.4 using the Bonferroni inequality provides a simple and reasonable answer by performing all tests at level $\alpha/3$ in this example.

In most analyses, further decompositions may be applied to each of the hypotheses H_I, H_S, and H_R. For example, we might consider single degree of freedom hypotheses on each of the two interaction constraints in (4.83) or each of the differences $\bar{\mu}_{\cdot i} - \bar{\mu}_{\cdot t}$ implied by (4.85). It is easy to see that the value of k in (3.101) might be quite large for experiments of rather modest dimensions. If the experimentwise significance level is believed to be appropriate, the size of the individual tests must be reduced. Of course, it should be emphasized that such an approach makes it more difficult to declare significant departures from the null hypothesis.

4.3.5. Single Degree of Freedom Hypotheses, Contrasts, and Orthogonal Contrasts

Although any subhypothesis might be considered when attempting to isolate the nature of the effects implied by a general hypothesis, we frequently consider single degree of freedom hypotheses or comparisons,

$$H_i'\mu = h_i \qquad i = 1, \ldots, k \tag{4.86}$$

where H_i' is a row vector. The number, k, of comparisons depends on the interests of the researcher. The problem of the overall significance level, discussed in Section 4.3.4, should be kept in mind, but otherwise there is no restriction on the comparisons considered. The results of Theorems 4.5 and 4.6 apply immediately with numerator degrees of freedom $\nu = 1$. Thus, in the unconstrained model, the numerator sum of squares corresponding to (4.86) is given by

$$N_i = (H_i'\hat{\mu} - h)^2 / H_i' D_p^{-1}(n_i) H_i. \tag{4.87}$$

In many cases, the coefficients in (4.86) satisfy the condition $H_i'J_p = 0$ and $h_i = 0$. Such a comparison of means is called a *contrast*.

When considering more than one contrast, it is of interest to ask for conditions under which the numerator sums of squares are independent chi-square variates. Suppose two such contrasts are denoted by $H_i'\mu$ and $H_j'\mu$ and, further, assume that

$$H_i' D_p^{-1}(n_i) H_j = 0. \tag{4.88}$$

In this case, the two contrasts are said to be *orthogonal*. The motivation for this definition is that

$$\text{Cov}(H_i'\hat{\mu}, H_j'\hat{\mu}) = \sigma^2 H_i' D_p^{-1}(n_i) H_j. \tag{4.89}$$

Thus, $H_i'\hat{\mu}$ and $H_j'\hat{\mu}$ are uncorrelated for orthogonal contrasts. Further, application of Theorem 2.4 immediately shows that N_i/σ^2 and N_j/σ^2 are independently distributed as single degree of freedom chi-square variates.

As noted in Section 4.3.4, the test statistics for these hypotheses are still correlated since they share the same denominator, and, even if they were not, the problem of testing a large number of hypotheses remains.

One motivation for considering orthogonal contrasts is that they represent a decomposition of the test for equality of means. Thus, suppose in the unconstrained model we test the $p - 1$ degree of freedom hypothesis

$$H_0\colon \mu_i = \mu_j \qquad i \neq j = 1, \ldots, p. \tag{4.90}$$

Then, we may establish the following theorem.

THEOREM 4.8

Given a set of $p - 1$ orthogonal contrasts, $H_i'\mu = 0$, $i = 1, \ldots, p - 1$, then the sum of squares for testing (4.90) is given by

$$N_H^* = \sum_{i=1}^{p-1} N_i \tag{4.91}$$

with N_i given by (4.87). Further, the sums of squares, N_i/σ^2, are independent, single degree of freedom, chi-square variates.

Proof: Let H denote the matrix whose rows are H_i', $i = 1, \ldots, p - 1$. Then the sum of squares for testing the hypothesis $H\mu = 0$ is given by

$$\begin{aligned} N_H &= \hat{\mu}'H'[HD_p^{-1}(n_i)H']^{-1}H\hat{\mu} \\ &= \sum_{i=1}^{p-1} (H_i'\hat{\mu})^2/[H_i'D_p^{-1}(n_i)H_i] \\ &= \sum_{i=1}^{p-1} N_i. \end{aligned} \tag{4.92}$$

Implicit in this result is the fact that H has rank $p - 1$. This may also be seen directly by noting that if, say, H_1' is linearly dependent on H_i', $i = 2, \ldots, p - 1$, we would have a contradiction. That is,

$$H_1' = \sum_{i=2}^{p-1} a_i H_i'$$

implies that (4.93)

$$H_1'D_p^{-1}(n_i)H_1 = 0$$

which would imply that H_1 is identically zero.

Now, let the hypothesis (4.90) be written in matrix form as

$$H^*\mu = 0 \tag{4.94}$$

where $H^* = (I_{p-1} | -J_{p-1})$. We now show that there exists a nonsingular matrix, T, such that $H^* = TH$. To see this, note that, since H is $(p - 1) \times p$ of rank $p - 1$, we may write,

$$H = (H_1 | H_2) = H_1(I | H_1^{-1}H_2) = H_1H^* \tag{4.95}$$

where H_1 is any set of linearly independent columns of H. The last result in (4.95) follows since $HJ = 0$, which implies that $H_1^{-1}H_2 = -J$. Letting $T^{-1} = H_1$, the sum of squares for testing (4.90) is given by

$$\begin{aligned} N_{H^*} &= \hat{\mu}'H^{*\prime}[H^*D_p^{-1}(n_i)H^{*\prime}]^{-1}H^*\hat{\mu} \\ &= \hat{\mu}'H'T'[THD_p^{-1}(n_i)H'T']^{-1}TH\hat{\mu} \\ &= N_H. \end{aligned} \tag{4.96}$$

Reference to (4.92) establishes the first part of the theorem. The distributional properties follow from Theorem 2.4 and from the orthogonality of the contrasts defined by (4.88).

■

The decomposition of the hypothesis (4.90) into $p - 1$, single degree of freedom hypotheses whose numerator sums of squares are independent chi-square variates summing to the hypothesis sum of squares, has certain esthetic appeal. Beyond that, one should not be lulled into a false sense of security, as the problem of the overall significance level still remains.

Of more importance is the fact that the orthogonal contrasts may not be the comparisons desired by the analyst. Assuming that specific comparisons are of interest, then they should be stated and tested, with appropriately adjusted significance levels, quite apart from any desire for expressing the comparisons as contrasts that may satisfy (4.88).

Finally, these results extend directly to the constrained case by defining the orthogonality condition as

$$H_i'AD_p^{-1}(n_i)H_j = 0 \tag{4.97}$$

where A is given by (4.31). The sums of squares for the single degree of freedom hypotheses are given by Theorem 4.6 as

$$N_i = (H_i'\tilde{\mu} - h)^2/H_i'AD_p^{-1}(n_i)H_i. \tag{4.98}$$

4.4. SIMULTANEOUS CONFIDENCE INTERVALS AND MULTIPLE COMPARISONS

In Section 3.4 we described a method for constructing confidence regions for θ or linear functions of θ in the linear model, which applies to the cell means model. Confidence intervals on linear functions of μ are obtained as in Section 3.4. The problems associated with writing multiple confidence intervals are still applicable. The method of constructing simultaneous confidence intervals due to Scheffé (1953) and the approximate procedure based on the Bonferroni inequality may be used in the cell means model. For reference, we restate these results.

For the Scheffé intervals, we recall that sets of intervals based on linear functions of μ may be obtained. The generality of the set of intervals depends on the choice of the initial set of linear functions—say, $H\mu$. Intervals on linear combinations of the components of the vector $H\mu$ are given by

$$a'H\mu\text{: } a'H\hat{\mu} \pm [vF(\alpha; v, N - p)\widehat{\text{Var}}(a'H\hat{\mu})]^{1/2}$$

where (4.99)

$$\widehat{\text{Var}}(a'H\hat{\mu}) = s^2a'H(W'W)^{-1}H'a.$$

Here, v is the rank of H. The expression applies to the unconstrained model or the reduced model. As noted above, we may write as many intervals as we wish of the form (4.99) and be assured that the experimentwise confidence coefficient is at least $1 - \alpha$. We say that the set of confidence intervals (4.99) has simultaneous confidence coefficient $1 - \alpha$.

As noted in Section 3.4, the generality of the intervals (4.99) is a function of the initial choice of the matrix H. For example, with $H = I_p$, (4.99) becomes

$$a'\mu\text{: } a'\hat{\mu} \pm [pF(\alpha; p, N - p)s^2a'(W'W)^{-1}a]^{1/2}. \tag{4.100}$$

Confidence intervals on any linear function of μ, including the individual components of μ and the differences $\mu_i - \mu_j$, may be written using (4.100). Alternatively, $H\mu$ might correspond to the differences $\mu_i - \mu_j$. In that case, $v = p - 1$, and only linear functions of these differences may be considered. In particular, (4.99) with that choice of H will not yield intervals for the individual μ_i. As one might suspect, the generality of (4.100) must be paid for in wider intervals than might be obtained for specific functions using the individual intervals (3.111). This point will be illustrated in Chapters 5 and 6.

The Bonferroni intervals for a fixed set of k linear functions of μ—say, $a_i'\mu$, $i = 1, \ldots, k$—are given by

$$a_i'\mu\text{: } a_i'\hat{\mu} \pm [F(\alpha/k; 1, N - p)\widehat{\text{Var}}(a_i'\hat{\mu})]^{1/2}. \tag{4.101}$$

We now describe another method for constructing simultaneous confidence intervals.

4.4.1. Tukey's Studentized Range Intervals

A method for constructing simultaneous confidence intervals for linear functions of μ with the same experimentwise error rate philosophy as the Scheffé intervals was introduced by Tukey (see Miller, 1981). The method was initially described for contrasts on the balanced, unconstrained means model but has since been extended to include general linear functions for unbalanced and constrained models. The method is best understood by first describing a special case.

Consider the special case in which $\hat{\mu}_i$, $i = 1, \ldots, p$ are independently distributed as $N(\mu_i, k^2\sigma^2)$. For example, this will occur in the unconstrained model with $n_i = n$, $i = 1, \ldots, p$ and $k^2 = 1/n$. Suppose we are interested in linear contrasts on the μ_i—that is, linear functions, $a'\mu$, with $a'J = 0$. Let s^2 be an estimate of σ^2 with d degrees of freedom such that

$$ds^2/\sigma^2 \sim \chi^2(d) \tag{4.102}$$

and is independent of $\hat{\mu}_i$, $i = 1, \ldots, p$. (In the unconstrained model, $d = N - p$.) Then, a set of simultaneous confidence intervals on all such contrasts, with confidence coefficient $1 - \alpha$, is given by

$$a'\mu\colon a'\hat{\mu} \pm q(\alpha; p, d)ks \sum_i |a_i|/2. \tag{4.103}$$

Here $q(\alpha; p, d)$ is the upper 100α percentage point of the *studentized range* distribution with parameters p and d. This is the distribution of the range of *studentized* variables $z_i = u_i/(\sqrt{w/d})$, $i = 1, \ldots, p$, where u_i and w are independent, $u_i \sim N(0, 1)$, $i = 1, \ldots, p$, and $w \sim \chi^2(d)$. The range is defined as the difference between the maximum and the minimum of the z_i. Table C2 in Appendix C gives the values of $q(\alpha; p, d)$ for various values of α, p, and d.

To establish (4.103), we first note the following equivalent inequalities:

For variables z_i, $i = 1, \ldots, p$, and a constant, c,

$$\max_{i,j} |z_i - z_j| \leq c, \tag{4.104}$$

if and only if

$$|z_i - z_j| \leq c \tag{4.105}$$

for all $i \neq j = 1, \ldots, p$. Further, (4.104) is also equivalent to

$$|a'z| = \left|\sum a_i z_i\right| \leq c\sum |a_i|/2 \tag{4.106}$$

for all vectors a such that $a'J = 0$. The equivalence of (4.104) and (4.105) is obvious. Similarly, (4.106) trivially implies (4.104). It remains to show that (4.104) implies (4.106). To show this, let

$$\begin{aligned} a. &= \sum |a_i|/2 \\ a^+ &= \sum_{i \epsilon P} a_i \\ a^- &= -\sum_{i \epsilon N} a_i \end{aligned} \tag{4.107}$$

where $P = (i\colon a_i > 0)$ and $N = (i\colon a_i < 0)$. Then, it follows that

$$a. = a^+ = a^-. \tag{4.108}$$

Now

$$\begin{aligned} a'z &= \Big(a^- \sum_{i \epsilon P} a_i z_i + a^+ \sum_{j \epsilon N} a_j z_j\Big)/a. \\ &= \sum_{i \epsilon P} \sum_{j \epsilon N} a_i(-a_j)(z_i - z_j)/a.. \end{aligned} \tag{4.109}$$

Utilizing (4.105) and the triangle inequality, which says

$$\left|\sum b_i\right| \le \sum |b_i|, \tag{4.110}$$

we have

$$\begin{aligned} |a'z| &\le \sum_{i \epsilon P} \sum_{j \epsilon N} |a_i(-a_j)(z_i - z_j)|/a. \\ &\le a^+ a^- c/a. = a.c \end{aligned} \tag{4.111}$$

and hence (4.106) is established.

Since the range is defined as

$$Q = \max_{i,j} |z_i - z_j| \tag{4.112}$$

we have

$$\begin{aligned} 1 - \alpha &= \text{Prob}[Q \le q(\alpha; p, d)] \\ &= \text{Prob}\Big[|a'z| \le q(\alpha; p, d) \sum |a_i|/2\Big] \end{aligned} \tag{4.113}$$

for all a such that $a'J = 0$. Thus, (4.103) follows by defining

$$z_i = (\hat{\mu}_i - \mu_i)/(ks). \tag{4.114}$$

A special case of (4.103), for which the method was initially designed, is the pairwise comparison of means. Thus, confidence intervals on all differences $\mu_i - \mu_j$, $i \neq j = 1, \ldots, p$ are given by

$$\mu_i - \mu_j\colon \hat{\mu}_i - \hat{\mu}_j \pm q(\alpha; p, d)ks \tag{4.115}$$

since $\Sigma|a_i| = 2$.

The method has been extended to include general linear functions of μ without the constraint $a'J = 0$. In this case, the confidence intervals are given by

$$a'\mu\colon a'\hat{\mu} \pm q'(\alpha; p, d)ksa^* \tag{4.116}$$

where

$$a^* = \max(a^+, a^-) \tag{4.117}$$

and a^+ is the sum of the positive elements of a, and a^- is the sum of the absolute values of the negative elements of a. In (4.116), $q'(\alpha; p, d)$ is the upper 100α percentage point of the *augmented* studentized range distribution with parameters p and d. This is the distribution of the range of z_i, $i = 0, 1, \ldots, p$, where z_i, $i = 1, \ldots, p$ are as defined in the studentized range distribution, and $z_0 = 0$. (See Miller, 1981, for details.)

To this point, the Tukey method is a competitor for the Scheffé method for the special case of the balanced ($n_i = n$), unconstrained model. Spjøtvoll and Stoline (1973) developed the extension to include the unbalanced case. To describe this extension, assume that $\hat{\mu}_i \sim N(\mu_i, k_i^2\sigma^2)$, $i = 1, \ldots, p$. For general linear combinations, $a'\mu$, the intervals are given by

$$a'\mu\colon a'\hat{\mu} \pm q'(\alpha; p, d)sa^{**} \tag{4.118}$$

where a^{**} is defined as in (4.117) except that, in this case, a^+ is the sum of the positive k_ia_i, and a^- is the sum of the absolute values of k_ia_i for negative a_i.

The elementary proof of this result follows by defining new variables $\hat{\gamma}_i = \hat{\mu}_i/k_i$, $i = 1, \ldots, p$, which satisfy the previous requirement of equal variances with $k = 1$. Confidence intervals on $b'\gamma$ are thus given by (4.116) for any vector b. But $b'\gamma = a'\mu$ if we define $b_i = k_ia_i$. Application of (4.116) then yields (4.118). Note that we need the general result (4.116), even though $a'\mu$ may be a contrast, since $b'\gamma$ will not, in general, be a contrast. Special cases such as contrasts and pairwise comparisons may be developed directly from (4.118). The extension of (4.115) for unequal variances ($k_i^2 = 1/n_i$ in the unconstrained model) is given by

$$\mu_i - \mu_j\colon \hat{\mu}_i - \hat{\mu}_j \pm q'(\alpha; p, d)s[\max(k_i, k_j)]. \tag{4.119}$$

We are now prepared to extend the Tukey method to yield confidence intervals for the constrained model and, thus, make it as general as the Scheffé method. To do so, we now assume that the p-vector $\hat{\mu} \sim N(\mu, \sigma^2 K)$, where K is an arbitrary positive definite matrix. Note that, in the original method, we assumed $K = k^2 I$ and then extended that to the case $K = D_p(k_i^2)$. The extension to this case is analogous to the development of (4.118). Recall that we may write $K = CC'$ where C is a nonsingular matrix. Defining $\hat{\gamma} = C^{-1}\hat{\mu}$, we see that $\text{Var}(\hat{\gamma}) = \sigma^2 I$ and, hence, $\hat{\gamma}$ satisfies the independence and equal variance requirement leading to equation (4.116). Confidence intervals for $a'\mu$ are obtained from the intervals (4.116) for $b'\gamma$ with $b' = a'C$. Thus, we find the intervals for this general case are given by

$$a'\mu\colon\ a'\hat{\mu} \pm q'(\alpha; p, d)sa^{***} \tag{4.120}$$

where a^{***} is defined as in (4.117) except that, in this case, a^+ is the sum of the positive elements of the vector $a'C$, and a^- is the sum of the absolute value of the negative elements of $a'C$. Note that, since we are now in the constrained model, $d = N - r$ from Theorem 3.2.

While (4.120) is correct, it is somewhat disconcerting to realize that the intervals are not invariant to the choice of C and that this matrix is not unique. For example, one choice of C is based on the Cholesky decomposition given in Appendix A.I.9. Alternatively, one might consider the eigen analysis of K yielding the decomposition, as in Appendix A.I.7, given here as

$$K = T\Lambda T'. \tag{4.121}$$

Here, the columns of T are the eigenvectors of K, and $\Lambda = D_p(\lambda_i)$ where the λ_i are the corresponding eigenvalues. We then define

$$C = TD_p(\lambda_i^{1/2}). \tag{4.122}$$

Hochberg (1975), in his development of this approach, suggests that we should attempt to determine C to have a structure similar to that of K and, as near as possible, to a scalar matrix. This approach will be illustrated by example in Chapter 6, but no general techniques are available for achieving this structure.

Finally, we note that the use of the generalized Tukey method depends on the availability of tables of the augmented studentized range distribution. These tables are not generally available, but it has been noted that, for $p > 2$ and $\alpha \leq 0.05$, the upper percentage points of $q'(\alpha; p, d)$ do not differ appreciably from $q(\alpha; p, d)$.

To summarize, the Scheffé method is most general with $H = I_p$, but the intervals tend to be too conservative. Depending upon the needs of the analyst, other choices of H may be sufficient, and shorter intervals will result. The Tukey method for the balanced, independent case appears to be superior to the Scheffé method, but this may depend on the allowable choices of H in (4.99) for specific types of intervals. The results of Spjøtvoll and Stoline (1973) indicate that this superiority may extend to certain unbalanced cases. The generalization of the Tukey method to the case of correlated means may be of little value because of the lack of a specific decomposition of the K matrix. The Bonferroni method for a specified set of intervals may be superior to either of the more general methods.

For a complete discussion of these methods and the whole field of simultaneous inference, reference is made to the excellent text by Miller (1981).

4.4.2. Multiple Range Tests

In the preceding section, we focused on the concept of simultaneous confidence intervals on linear functions of μ as a means of making inferences. This discussion included the special case of pairwise comparisons of the means. It is quite natural to use these confidence intervals as tests of significance, rejecting the hypothesis if the interval does not contain the hypothesized value.

In this section, we shall discuss procedures that are primarily designed to test for differences in the means. In particular, we refer to what are commonly called *multiple range procedures* for mean comparisons. The basic idea follows from the studentized range method in Section 4.4.1. Recall that, if we had p means with estimates $\hat{\mu}_i$ and variances σ^2/n, then the confidence band on $\mu_i - \mu_j$ given in (4.115) has half-width $q(\alpha; p, N - p)s/\sqrt{n}$. As a test procedure, we would reject the hypothesis H: $\mu_i = \mu_j$ if $|\hat{\mu}_i - \hat{\mu}_j|$ exceeds this half-width.

The basic difference in the multiple range tests is that the standard for comparison changes as the means are being examined; that is, the critical value is given by $q(\alpha; k, N - p)$ where $k = p$ or $p - 1$, or . . . , or 2, according to the following procedure.

(a) Rank the sample means $\hat{\mu}_i$ by increasing order of magnitude. Assuming that they are labelled such that $\hat{\mu}_1 \leq \hat{\mu}_2 \leq, \ldots, \leq \hat{\mu}_p$, compare the difference, $\hat{\mu}_p - \hat{\mu}_1$, with $q(\alpha; p, N - p)s/\sqrt{n}$. If it is less, then none of the differences are significant. If it is greater, the difference $\mu_p - \mu_1$ is judged to be different from zero, and we go to step (b).

(b) Compare $\hat{\mu}_{p-1} - \hat{\mu}_1$ and $\hat{\mu}_p - \hat{\mu}_2$ with $q(\alpha; p - 1, N - p)s/\sqrt{n}$. If either difference is less than this value, there are no more significant differences. If either or both exceed the critical value, the corresponding differences are judged to be significant, and we proceed to step (c).

(c) If, for example, $\hat{\mu}_{p-1} - \hat{\mu}_1 > q(\alpha; p - 1, N - p)s/\sqrt{n}$, then compare $\hat{\mu}_{p-2} - \hat{\mu}_1$ and $\hat{\mu}_{p-1} - \hat{\mu}_2$ with $q(\alpha; p - 2, N - p)s/\sqrt{n}$ and proceed as above. The process is continued until all differences have been examined or determined to be nonsignificant. Those differences exceeding the critical value are judged to be significant.

This procedure is known as the *Newman-Keuls test*. A closely related procedure known as the *Duncan Multiple Range test* differs in that it recommends a variable significance level; that is, for comparing k means, we use the critical value $q(\alpha_k; k, N - p)s/\sqrt{n}$ where

$$\alpha_k = 1 - (1 - \alpha)^{k-1}. \tag{4.123}$$

This variable significance level has led to much discussion, since it offers little protection when k is large. (Note that, for $\alpha = 0.05$, $\alpha_5 = 0.19$ and $\alpha_{10} = 0.34$).

For this reason, the Duncan method will generally indicate more significant differences than the Newman-Keuls method.

It is important to emphasize the basic philosophy of multiple range tests; that is, the difference between any two means is significant provided the range of every subset of means that contain the given means is significant at the selected level. Thus, for example, if we terminate the procedure at the end of step (a), none of the differences would be judged significant. This conclusion may be reached even though, for example, $\hat{\mu}_p - \hat{\mu}_2$ might exceed the critical value $q(\alpha; p - 1, N - p)s/\sqrt{n}$. Such apparently contradictory results are confusing to the applied researcher.

We now consider an extension of the multiple range concept to the case of unequal cell frequencies. Kramer (1956) suggested that we proceed as above, with the exception that $s/\sqrt{n}$ be replaced by

$$s[(1/n_i + 1/n_j)/2]^{1/2} \tag{4.124}$$

where n_i and n_j are the cell frequencies associated with the means being considered. With different sample sizes, the possibility of contradictory results of the type noted above is greatly increased. This modification is based on the analogous result for the individual intervals for unequal sample sizes but has no theoretical basis.

An alternative procedure is to use the results of the extended Tukey procedure using the modified studentized range statistic. Thus, for example, in the Newman-Keuls procedure the differences would be compared to

$$q'(\alpha; k, N - p)s\{\max[\sqrt{(1/n_i)}, \sqrt{(1/n_j)}]\} \tag{4.125}$$

for $k = p, p - 1, \ldots, 2$ as before. The Duncan method would be extended in the same way using the modified significance level. Note that this suggestion is close to the test based on the simultaneous confidence intervals. The difference is that, in the latter, we fix $k = p$ as opposed to decreasing k and applying the basic philosophy described above. This modified multiple range procedure will tend to yield more significant differences than a test based on the simultaneous confidence intervals.

The problem of multiple comparisons has attracted a lot of attention, and it is unlikely that there will ever be a final resolution. The problem revolves around the questions of error rates and power; that is, we wish to be protected, in some sense, against declaring false significances and at the same time want to detect real differences. We shall not attempt to provide a discussion of these concepts. For further information, refer to Miller (1981) and papers mentioned in that book.

CHAPTER 4 EXERCISES

1. Using the data for Example 1.4 given in Table 1.4, compute the estimates of the mean rising times for the three flour types, and estimate σ^2 assuming the model (1.19).

2. **(a)** For Example 1.4, develop algebraic expressions for the numerator sum of squares for the hypothesis H_0: $\mu_1 = \mu_2 = \mu_3$ against the alternative that at least one of these equalities is violated. Use both the model reduction and the LaGrange Multiplier approaches to develop this expression.

(b) Use the data in Table 1.4 to test this hypothesis with $\alpha = 0.05$.

3. Consider the two-way classification model given by (4.17) for Example 4.4 for the special case $n_{ij} = n$ for all i and j.

(a) Use both the model reduction and the LaGrange Multiplier approaches to develop algebraic expressions for the estimates of μ and σ^2.

(b) Use these results to estimate the parameters for Example 1.7 using the data from Tables 1.4 and 1.5.

4. For Example 4.3, assuming $n_{ij} = n$, determine explicit expressions for the numerator sums of squares for the three hypotheses H_I, H_S, and H_R given by (4.83), (4.84), and (4.85) using

(a) model reduction and

(b) the general quadratic form in (4.67).

5. Consider Example 4.6 with $n_{ij} = n$, $(i, j) \neq (1, 1)$, and $n_{11} = 0$.

(a) Show that the estimates of μ obtained by using (i) the effective model (4.63), (ii) the reduced model (4.64), and (iii) the reduced model (4.18) are identical. Determine the estimate of σ^2.

(b) Suppose both $n_{11} = 0$ and $n_{23} = 0$. Write the effective model. Is it connected?

(c) Suppose $n_{11} = n_{22} = n_{23} = 0$. Is the model connected?

6. In many statistical methods texts, the problem of one missing cell in the two-factor model without interaction and $n_{ij} = 1$ for all i, j except $n_{11} = 0$ is treated as follows:

(a) The missing cell is estimated by

$$M = \frac{aT + bB - S}{(a - 1)(b - 1)}$$

where a is the number of rows and b is the number of columns, T is the sum of the observations in row 1, B is the sum of the observations in column 1, and S is the sum of all observations.

(b) This value M is used in place of y_{11}, and the analysis proceeds as if this value were the actual observation with the following exceptions: (i) The residual degrees of freedom are reduced by one, and (ii) for testing the row effect hypothesis, the numerator sum of squares is reduced by

$$Z = [B - (a - 1)M]^2/[a(a - 1)]$$

to get an "unbiased" numerator sum of squares for the test.

(1) For Example 4.6, show that M is the estimate of μ_{11} as obtained from the effective model.

(2) Now show that the numerator sum of squares obtained under (b) is correct.

7. Prove Theorem 4.7.

8. Use Example 1.4 to verify Theorem 4.8 for the contrasts

$$H_1' = (1\ 0\ -1) \qquad H_2' = (1\ -2\ 1).$$

9. Determine simultaneous 95 percent confidence intervals for all differences $\mu_i - \mu_j$ in Example 1.4 using the Scheffé, Tukey, and Bonferroni methods. Compare these with the individual intervals given by (3.111).

10. **(a)** Use the multiple range test in Section 4.4.2 to test for significant differences in the means in Example 1.4 with $\alpha = 0.05$.

(b) Contrast this result with the conclusions you would have reached using the confidence intervals from Exercise 9 to perform the test.

11. Establish the equivalence of:

(a) The estimates of μ given by (4.30) and (4.39).

(b) The estimates of σ^2 given by (4.32) and (4.39).

CHAPTER FIVE
THE ONE-WAY CLASSIFICATION MODEL

5.1. INTRODUCTION

In Section 4.1, we discussed several examples involving the test scores of college freshmen. In Example 4.2, the students were classified by sex. In Example 4.3, they were classified by two criteria, sex and race. These are two examples of what are known as *classification* or *factorial models*. In Example 4.2 the students are classified by a single factor, while in Example 4.3 there are two factors involved in the classification. Clearly, the number of factors may be greater than two. For example, the students could be further classified by their course of study (business, engineering, science, and so on), by their citizenship (U.S., Canadian, and so forth), and by many more factors depending on the nature of the study. In general, we shall be interested in experiments in which the data are classified according to K factors and shall refer to such models as *K-way classification* or *K-factor models*. We shall refer generally to the criteria within a factor as *levels* of that factor. Thus, males and females might be referred to as level one and level two of the sex factor. In other cases, the term *level* may refer to a quantitative measure, such as the amount of fertilizer, the baking time, and so on.

The general setting in which we will think of our data arising is as follows: We have available N experimental units that are essentially identical as far as we can determine. The "treatments" are randomly assigned to these units such that the ith treatment (or treatment combination) is administered to n_i units. The data consist of the observed responses. The complexity of the experiment is then involved only in the treatment structure, which may be a K-way classification, or K-factor, structure. This situation is known as a *completely randomized design* and is distinguished by the fact that the experimental units have no structure. In other cases, the experimental units may be grouped or "blocked" into units that would be expected to respond similarly.

In many cases, such as our test score example, it is difficult to think of the initial experimental units—that is, a group of students—to whom sex and race are assigned. We might well have thought of sex and race as blocking factors, such that students within a sex-by-race group would be expected to respond similarly. For the purpose of our analysis, this will be of little consequence. As long as we can assume that the data satisfy the assumptions of the cell means model, Definition 4.1, the analysis will remain the same. Later, in Chapter 8, we shall consider situations in which the nature of the experiment—for example, the blocking—will change the assumptions on the error structure. That problem will require additional theoretical development.

The development of the analysis for the general cell means model in Chapter 4 was done in the notation of the one-way classification model. The purpose of this chapter is to spell out some of the details of that analysis and illustrate them by example. In Chapters 6 and 7 we shall consider more complex treatment structures but still restrict ourselves to the completely randomized design or, equivalently, the general cell means model of Definition 4.1.

Our model here is just the unconstrained cell means model discussed in Chapter 4; that is, we may think of the data as representing observations on each of p populations with n_i observations on the ith population with mean μ_i. It is assumed that these are independent, random samples and that the data are normally distributed with common variance, σ^2. Thus, the model is written as

$$y_{ij} = \mu_i + e_{ij} \qquad \begin{array}{l} i = 1, \ldots, p \\ j = 1, \ldots, n_i, \end{array} \tag{5.1}$$

where y_{ij} denotes the jth response on the ith population with mean μ_i. The e_{ij} are independent random errors, distributed as $N(0, \sigma^2)$.

5.2. ESTIMATION OF PARAMETERS

From Section 4.2.1, we obtain the estimates and their distributions as follows:

$$\hat{\mu}_i = \bar{y}_{i\cdot} \sim N(\mu_i, \sigma^2/n_i) \tag{5.2}$$

$$s^2 = Q(\hat{\mu})/(N - p) \sim [\sigma^2/(N - p)]\chi^2(n - p) \tag{5.3}$$

where

$$Q(\hat{\mu}) = \sum_{i=1}^{p} \sum_{j=1}^{n_i} (y_{ij} - \bar{y}_{i\cdot})^2. \tag{5.4}$$

Further, $\hat{\mu}_i$, $i = 1, \ldots, p$ and $Q(\hat{\mu})$ are mutually independent.

5.3. TESTS OF HYPOTHESES

5.3.1. The Hypothesis of Equal Means

The test statistic for a linear hypothesis on μ for the unconstrained model was summarized in Theorem 4.5. It is instructive to spell out the details for the most common hypothesis—that is, the hypothesis of equal cell means

$$H_0\colon \mu_i = \mu_j \qquad i \neq j = 1, \ldots, p. \tag{5.5}$$

It will be convenient to express this hypothesis in matrix form. There are several ways to do this. First, recognizing that (5.5) is equivalent to

$$H_0\colon \mu_i = \mu_p \qquad i = 1, \ldots, p-1, \tag{5.6}$$

we may write the hypothesis as $H\mu = 0$, where

$$H = \Delta_p = (I_{p-1} | -J_{p-1}). \tag{5.7}$$

This matrix will appear frequently in later developments. In general, the notation Δ_a will denote an $(a-1) \times a$ matrix whose first $a-1$ columns are the identity matrix and whose last column consists of negative ones.

Alternatively, we may write (5.5) as

$$H_0\colon \mu_i = \sum_k n_k \mu_k / N \qquad i = 1, \ldots, p. \tag{5.8}$$

To write this in matrix form, we introduce the matrix

$$\Gamma_p = [I_p - J_p J_p' D_p(n_i)/N]. \tag{5.9}$$

The hypothesis, (5.8), may thus be written as $H_0\colon \Gamma_p \mu = 0$. The matrix Γ_p has p rows but is of rank $p-1$. Deleting the last row yields the desired, full row rank, hypothesis matrix. We shall see that this form will be useful for developing the test statistic and noncentrality parameter.

The numerator sum of squares for testing H_0—say, N_H—may be developed using either the model reduction approach or the general quadratic form from Theorem 4.5. Using the first approach, we use the hypothesis in the form of (5.6) and write the reduced model as

$$y_{ij} = \mu_p + e_{ij}. \tag{5.10}$$

In the general cell means model format, we have in this case $W = J_N$ and $\mu = \mu_p$, a scalar. Hence, we immediately obtain

$$\begin{aligned} \tilde{\mu}_p &= \bar{y}_{..} \\ Q(\tilde{\mu}) &= \sum_i \sum_j (y_{ij} - \bar{y}_{..})^2. \end{aligned} \tag{5.11}$$

The numerator sum of squares is then given as

$$\begin{aligned} N_H &= Q(\tilde{\mu}) - Q(\hat{\mu}) \\ &= \sum_i \sum_j (y_{ij} - \bar{y}_{..})^2 - \sum_i \sum_j (y_{ij} - \bar{y}_{i.})^2 \\ &= \sum_i \sum_j (\bar{y}_{i.} - \bar{y}_{..})^2. \end{aligned} \tag{5.12}$$

It is convenient to write (5.12) in matrix form. This follows from the general results of Theorem 4.5, but it may be inferred directly from (5.12) as

$$N_H = Y'[WD_p^{-1}(n_i)W' - J_N J_N'/N]Y \tag{5.13}$$

where $W = D_p(J_{n_i})$ from the original unconstrained model.

From Section 3.3.1 we recall that $N_H \sim \sigma^2\chi^2(p - 1, \lambda)$ where the noncentrality parameter, λ, may be obtained from the general expression (3.73) or from the relation (3.74) between $E(N_H)$ and λ. With the latter approach, we obtain, using (5.13) and Theorem 2.2,

$$\begin{aligned} E(N_H) &= (p - 1)\sigma^2 + \mu'W'(I - JJ'/N)W\mu \\ &= \sigma^2[(p - 1) + 2\lambda]. \end{aligned} \tag{5.14}$$

It then follows that

$$\begin{aligned} \lambda &= \mu'W'(I - JJ'/N)W\mu/(2\sigma^2) \\ &= \sum_i \sum_j (\mu_i - \mu^*)^2/(2\sigma^2) \end{aligned} \tag{5.15}$$

where

$$\mu^* = \sum_i n_i\mu_i/N. \tag{5.16}$$

The expressions for N_H and λ could have been obtained directly from the results of Section 3.3.1 using the general expressions (3.70) and (3.73). With that approach, it is more convenient to use the hypothesis matrix in (5.9) with the last row deleted. However, we have used the model reduction approach because of its simplicity. In general, we shall see that the model reduction approach is often more convenient in cases for which simple algebraic expressions are available for N_H and λ. In Chapter 6, we will note the computational advantages of this approach.

Finally, the test for hypothesis (5.5) is given by rejecting the hypothesis if

$$N_H/(p - 1)s^2 > F(\alpha; p - 1, N - p). \tag{5.17}$$

The power of the test is given by (2.27) as

$$\begin{aligned} \prod(\alpha; p - 1, N - p, \lambda) = \text{Prob}[F(p - 1, N - p, \lambda) \\ > F(\alpha; p - 1, N - p)]. \end{aligned} \tag{5.18}$$

The essential elements for this test are often displayed in tabular form in what is usually called an *Analysis of Variance* (*ANOVA*) *table*. This table is primarily a convenient way of summarizing the quantities for testing the hypothesis of equality of means; hence, we prefer to refer to it as an *Analysis of Means* (*AOM*) *table*.

Although this table is particularly simple in this case, it will become more detailed as we develop the analysis for complex models. We should, however, become familiar with the general structure of the table. Table 5.1 contains the results for our preliminary analysis of the one-way classification model. In this case, we need not make a particular distinction between the balanced ($n_i = n$) and unbalanced cases. The only difference is the variable range on the inner summation in quantities such as (5.04) and (5.12).

TABLE 5.1 Analysis of means table for the one-way classification model.

Description	*Hypothesis*	*df*	*SS*	*MS*	*EMS*
Treatment differences	$\mu_i = \mu_p$ $i = 1, \ldots, p-1$	$p - 1$	N_H	$N_H/(p-1)$	$\sigma^2 + \dfrac{\Sigma\Sigma(\mu_i - \mu^*)^2}{p-1}$
Residual	———	$N - p$	$Q(\hat{\mu})$	s^2	σ^2
Corrected total	———	$N - 1$	T	$T/(N-1)$	$\sigma^2 + \dfrac{\Sigma\Sigma(\mu_i - \mu^*)^2}{N-1}$
Mean	$\mu^* = 0$	1	M	M	$\sigma^2 + N\mu^{*2}$
Uncorrected total	———	N	U	U/N	$\sigma^2 + \dfrac{\Sigma\Sigma\mu_i^2}{N}$

In Table 5.1, the sums of squares, N_H and $Q(\hat{\mu})$, are given in (5.12) or (5.13) and (5.4). The remaining sums of squares are given by $T = Q(\tilde{\mu})$ as in (5.11) or

$$T = Y'(I - JJ'/N)Y,$$
$$M = Y'JJ'Y/N = N\bar{y}_{..}^2, \tag{5.19}$$

and

$$U = Y'Y. \tag{5.20}$$

Typically, only the first three rows of this table are shown. Row 4 gives the numerator information for testing the hypothesis H: $\mu^* = 0$ where μ^* is defined in (5.16). The test statistic is M/s^2, which is distributed as $F(1, N-p, \lambda^*)$ where $\lambda^* = \mu^{*2}/(2\sigma^2)$. This hypothesis is rarely tested. The column labeled "MS" (mean squares) is just the ratio of the sums of squares (SS) and their degrees of freedom (df). The last column gives the expected value of the mean squares (EMS). Recalling the relation (3.74) between the noncentrality parameter and the expected value of a noncentral chi-square variable, we see that this column provides a convenient reminder of the mean squares to be used in the test statistics for our hypotheses.

The fact that the sums of squares and the degrees of freedom in the first two rows add up to those in the third row allows us to apply Cochran's Theorem (Theorem 2.6) to establish the independence of N_H and $Q(\hat{\mu})$, which we have already done by directly applying Theorem 2.4. In later AOM tables, this addition property may not hold, but we shall see that this in no way invalidates the tests that will be developed. This situation will arise in more general unbalanced data models.

It is of interest to use this simple problem to relate the test statistic summarized in Table 5.1 to the likelihood ratio statistic that we first encountered in the development in Section 3.3. Note that, in this case, the (corrected) total sum of squares T is equal to $Q(\tilde{\mu})$, the residual sum of squares under the hypothesis. Thus, the likelihood ratio statistic (3.63) leads us to consider the ratio of the residual sum of squares to the (corrected) total sum of squares in Table 5.1. We may now develop the distribution of $\gamma^{2/N}$ by recalling the relation between the F and beta distributions. Thus, if $F \sim F(m, n)$, then

$$\beta = [1 + (m/n)F]^{-1} \sim \text{beta}(n/2, m/2).$$

[See, for example, Mood, Graybill, and Boes (1974).] In this situation,

$$\begin{aligned} \beta &= [1 + N_H/Q(\hat{\mu})]^{-1} \\ &= Q(\hat{\mu})/Q(\tilde{\mu}) \sim \text{beta}[(N - p)/2, (p - 1)/2]. \end{aligned} \tag{5.21}$$

Thus, a test which is equivalent to the F-test is given by rejecting the hypothesis (5.5) when β is less than the lower 100α percentage point of the beta distribution. The power of the test is computed by noting that, when the null hypothesis is not true, β follows a noncentral beta distribution [see Graybill (1961)]. Clearly, we could use either test statistic, but in practice the F statistic is almost universally preferred.

The distribution of $\gamma^{2/N}$ in the general case follows from (5.21) with appropriate changes in the degrees of freedom. It is emphasized that the denominator, $Q(\tilde{\mu})$, depends on the hypothesis tested.

5.3.2. Tests of Subhypotheses

The two hypotheses indicated in Table 5.1 represent a decomposition of the more general hypothesis H_0: $\mu_i = 0$, $i = 1, \ldots, p$. This decomposition is said to be orthogonal in the sense that the numerator sum of squares for this hypothesis, $\Sigma n_i \bar{y}_{i\cdot}^2$, is the sum of N_H and M. The hypotheses (5.5) and H^*: $\mu^* = 0$ also are orthogonal, since the row vectors of the hypothesis matrices Γ_p and $R_p(n_i)$ are orthogonal in the sense of equation (4.88). The application of Cochran's Theorem shows that N_H and M are independent, but, of course, the F-ratios are not. The discussion of Section 4.3.4 should be recalled if both hypotheses are to be tested.

While the hypothesis H^*: $\mu^* = 0$ is rarely tested, we may often consider decompositions of the hypothesis of equal means into subhypotheses; that is, we may consider subhypotheses designed to identify specific types of differences. More generally, we may wish to examine certain linear combinations of the means. This decomposition may be orthogonal in that the degrees of freedom and sums of squares add up to $p - 1$ and N_H, respectively. As noted earlier, the essential point is that the subhypotheses should be of interest to the researcher and should not be forced by the desire for orthogonality. In either case, care must be taken to ensure the significance level when many tests are conducted.

Often, the decomposition is into single degree of freedom hypotheses. In this case, we may prefer to use simultaneous confidence intervals, since they provide both the test and an indication of the value of the parameter function of interest.

5.4. SIMULTANEOUS CONFIDENCE INTERVALS

In the development of the Scheffé intervals, in Section 4.4.2 it was noted that the generality of the parameter functions for which confidence intervals can be written depended upon the initial choice of hypothesis matrix H. The most flexibility is obtained by choosing $H = I$, as in (4.100). With this choice, we obtain for the current unconstrained model the intervals

$$a'\mu: a'\hat{\mu} \pm [pF(\alpha; p, N - p)s^2 \textstyle\sum a_i^2/n_i]^{1/2}. \tag{5.22}$$

If we are interested only in linear functions of differences, $\mu_i - \mu_j$, we may use $H = \Delta_p$ as given in (5.7). Letting $\delta = \Delta_p \mu$, the intervals for linear functions of δ are given by (4.99) as

$$d'\delta: d'\hat{\delta} \pm [(p - 1)F(\alpha; p - 1, N - p)\widehat{\text{Var}}(d'\hat{\delta})]^{1/2} \tag{5.23}$$

where $d'\hat{\delta} = d'\Delta_p\hat{\mu} = a'\hat{\mu}$ and

$$\begin{aligned} \widehat{\text{Var}}(d'\hat{\delta}) &= s^2\left[\sum_{i=1}^{p-1} d_i^2/n_i + \left(\sum_{i=1}^{p-1} d_i\right)^2 \Big/ n_p\right] \\ &= s^2 \sum_{i=1}^{p} a_i^2/n_i. \end{aligned} \tag{5.24}$$

It is instructive to spell out some special cases. From (5.22), we may obtain interval estimates for μ_i and for $\mu_i - \mu_j$ as

$$\mu_i: \bar{y}_{i\cdot} \pm [pF(\alpha; p, N - p)s^2/n_i]^{1/2} \tag{5.25}$$

and

$$\mu_i - \mu_j: \bar{y}_{i\cdot} - \bar{y}_{j\cdot} \pm \{pF(\alpha; p, N - p)s^2[(1/n_i) + (1/n_j)]\}^{1/2}. \tag{5.26}$$

From (5.23), we cannot obtain interval estimates for μ_i, but for the differences $\mu_i - \mu_j$ we have

$$\mu_i - \mu_j: \bar{y}_{i\cdot} - \bar{y}_{j\cdot} \pm \{(p - 1)F(\alpha; p - 1, N - p)s^2[(1/n_i) + (1/n_j)]\}^{1/2}. \tag{5.27}$$

Our intuition suggests that the intervals (5.27) should be shorter than (5.26), and this is confirmed by the relation

$$pF(\alpha; p, N - p) > (p - 1)F(\alpha; p - 1, N - p). \qquad (5.28)$$

The analogous generalized Tukey intervals are obtained from (4.119) as

$$\mu_i - \mu_j\colon \bar{y}_{i\cdot} - \bar{y}_{j\cdot} \pm q'(\alpha; p, N - p)s[\max(1/\sqrt{n_i}, 1/\sqrt{n_j})]. \qquad (5.29)$$

The Bonferroni method requires that we specify the number of intervals of interest. Thus, if we are interested in all possible differences, $\mu_i - \mu_j$, we have $k = p(p - 1)/2$ in (4.101), and the resulting intervals are given by

$$\mu_i - \mu_j\colon \bar{y}_{i\cdot} - \bar{y}_{j\cdot} \pm [F(\alpha/k; 1, N - p)s^2(1/n_i + 1/n_j)]^{1/2}. \qquad (5.30)$$

An indication of the relative sizes of these intervals is given in the numerical example in the next section.

5.5. A NUMERICAL EXAMPLE

To illustrate the analysis of the one-way classification model and some of the general ideas introduced in Chapter 4, we present a typical numerical example.

One of the problems experienced by long-distance runners is the build-up of lactic acid in the muscles due to loss of fluid through perspiration. To combat this problem, runners are encouraged to drink fluids prior to, during, and after the run. In this study of 50 runners on a 10-mile course, five fluids were used. These consisted of water and two commercial "electrolytes," *A* and *B*, each prepared at two concentrations, (A1, A2) and (B1, B2). Each of the 50 runners, ten runners per fluid, was given a fixed amount of fluid prior to the run, at the 5-mile point, and at the finish. The response consisted of a measure of lactic acid build-up.

The data, in coded form, are shown in Table 5.2. The 50 runners were believed to be of comparable physical condition, but in spite of this only 41 runners completed

TABLE 5.2 Data for lactic acid example.

	Treatment					
Name	*Water*	*A1*	*A2*	*B1*	*B2*	
Label	1	2	3	4	5	
Sample mean ($\bar{y}_{i\cdot}$)	33.3	32.6	30.9	29.0	26.1	$\bar{y}_{\cdot\cdot} = 30.7$
Sample variance (s_i^2)	13.1	14.2	12.2	13.9	14.2	$s^2 = 13.37$
Sample size (n_i)	10	7	10	8	6	$N = 41$

the course. It was believed that failure to finish was due to problems not related to the fluid consumed or the lactic acid build-up.

The first step in our analysis is to test for differences in the mean response—that is, the hypothesis H_0: $\mu_i = \mu_j$, $i \neq j = 1, \ldots, 5$. The statistics for this test are given in the Analysis of Means Table 5.3. Note that $N_H = 243.4$, resulting in an F-ratio of $F = 4.55$ on 4 and 36 degrees of freedom. With $\alpha = 0.05$, the preselected size of the test, we find $F(0.05; 4, 36) = 2.64$; hence, the hypothesis is rejected with this level of significance. In fact, we find that the probability of a larger value of F under the null hypothesis is $p = 0.004$. We conclude that there is sufficient evidence to suspect different responses to the five treatments.

To examine the source of these differences, many researchers would consider a decomposition of the original four degrees of freedom hypothesis into four single degree of freedom orthogonal contrasts. In this case, the natural comparisons are described verbally as

1. water versus electrolytes,
2. A versus B,
3. A1 versus A2, and
4. B1 versus B2.

These verbal descriptions are, of course, not unique. The specific comparisons we have in mind, written in the form $a'\mu$, are, respectively,

$$\begin{aligned}
&\text{C1.} \quad \mu_1 - \sum_{i=2}^{5} n_i\mu_i/(N - n_1), \\
&\text{C2.} \quad \frac{n_2\mu_2 + n_3\mu_3}{n_2 + n_3} - \frac{n_4\mu_4 + n_5\mu_5}{n_4 + n_5}, \\
&\text{C3.} \quad \mu_2 - \mu_3, \\
&\text{C4.} \quad \mu_4 - \mu_5.
\end{aligned} \tag{5.31}$$

The sums of squares for these comparisons are given by

$$N_i = (a_i'\hat{\mu})^2 / a_i'D_p^{-1}(n_i)a_i \tag{5.32}$$

and these are shown in Table 5.3. The orthogonality is established by (4.88) and verified by noting that

$$N_H = \sum_i N_i. \tag{5.33}$$

With $F(0.05, 1, 36) = 4.12$, we note that, when testing the four single degree of freedom hypotheses, C1 $= 0$ and C2 $= 0$ would be rejected while C3 $= 0$ and C4 $= 0$ would not. The implication of this is that the electrolytes differ from water and that the two electrolytes differ from each other but that the two concentrations considered do not differ.

TABLE 5.3 Analysis of means for lactic acid data.

Description	*Hypothesis*	*df*	*SS*	*MS*	*F*
Fluid differences	$\mu_i = \mu_j$	4	243.4	60.85	4.55
Water versus electrolytes	C1	1	89.2	89.2	6.67
A versus B	C2	1	113.6	113.6	8.50
A1 versus A2	C3	1	11.9	11.9	0.89
B1 versus B2	C4	1	28.8	28.8	2.15
Residual	——	36	481.2	13.37	

Some comments on the first two comparisons, C1 and C2, are in order. Note that, if all 50 runners had completed the course, we would have considered the comparisons

$$\begin{aligned} &\text{C5.} \quad \mu_1 - \sum_{i=2}^{5} \mu_i/4 \\ &\text{C6.} \quad \frac{\mu_2 + \mu_3}{2} - \frac{\mu_4 + \mu_5}{2}. \end{aligned} \tag{5.34}$$

Since, in this example, the unequal cell frequencies were judged to be due to chance and not related to the treatment, to the relative frequency of use of the various fluids, or to any other assignable cause, it is natural to ask why the comparisons should depend on the cell sizes. Often, in such cases, the only argument is that C1 and C2 (with C3 and C4) provide orthogonal contrasts, while C5 and C6 do not. As noted earlier, this is not a valid argument. The essential point is to make the comparisons of interest to the researcher. In this example, the hypotheses C5 $= 0$ and C6 $= 0$ also would be rejected; hence, the same general conclusions would be made. The sums of squares are given by

$$N_5 = 99.8$$

and

$$N_6 = 132.0.$$

Recalling the problem of multiple F-ratios, we might consider the Bonferroni significance level $\alpha/4$ when testing the four contrasts. The critical value, corresponding to $\alpha = 0.05$, is $F(0.0125; 1, 36) = 6.92$. If this value is used in Table 5.3, only C2 is significant. On the other hand, both C5 and C6, with F-ratios of 7.46 and 9.87, respectively, would be significant.

Two additional comparisons which might be of interest in this study are

$$\begin{aligned} &\text{C7.} \quad \frac{\mu_2 + \mu_4}{2} - \frac{\mu_3 + \mu_5}{2} \\ &\text{C8.} \quad \mu_3 - \mu_2 - \mu_5 + \mu_4. \end{aligned} \tag{5.35}$$

In C7, we consider the difference between the high and low concentrations averaged over the two electrolytes. In C8, we compare the differences in concentrations for the two electrolytes. The sums of squares for testing these single degree of freedom hypotheses are

$$N_7 = 39.5$$

and

$$N_8 = 2.7.$$

The F-ratios of 2.96 and 0.20 suggest that the two concentrations used in the study resulted in small differences in the response and that the effect of concentration was the same for the two electrolytes.

As an alternative to the single degree of freedom tests in Table 5.3, we might present confidence intervals on these contrasts. To demonstrate the differences in the various types of confidence intervals, we shall consider the contrasts C1 and C2. Four types of intervals will be shown: the individual t-intervals based on (3.111), the Scheffé intervals based on (5.23), the Tukey intervals based on (4.118), and the Bonferroni intervals (4.101) with $k = 4$, all with $\alpha = 0.05$. Note that only the last three have simultaneous confidence coefficient 0.95 and that this is only approximate for the Bonferroni intervals.

In Table 5.4, we show the values of the parameter estimates, $a'\hat{\mu}$, and the half-widths of the four types of intervals. Note that, with the exception of the Tukey intervals, the half-widths are the same for the two contrasts. This is a consequence of the cell frequencies in this example and is not generally true. The Tukey intervals are most conservative, and the simultaneous intervals all cover zero except for the Bonferroni interval on C2. The indication of significance in the AOM Table 5.3, as supported by the t-intervals, is less emphatic when viewed in this simultaneous setting. If we had used the more general Scheffé intervals given by (5.22), the intervals would have been wider by a factor of 1.08 as implied by (5.28).

TABLE 5.4 Confidence intervals for C1 and C2.

		Half-widths			
Mean contrast	*Parameter estimates*	*t*	*Scheffé*	*Tukey*	*Bonferroni*
C1	3.44	2.71	4.34	5.38	3.51
C2	3.84	2.71	4.34	5.64	3.51

The contrasts C1 through C8 provide information on the nature of the responses, which may be sufficient for the analysis or may prompt further investigations. In many situations, we may wish to examine the differences in detail. To this end we show, in Table 5.5, confidence intervals on all possible differences, $\mu_i - \mu_j$. As in

Table 5.4, we illustrate four methods of constructing confidence intervals on the ten possible differences. In this case, the Bonferroni intervals are constructed for $k = 10$ as in (5.30), the Tukey intervals are given by (5.29), and the Scheffé intervals are given by (5.27). The t-intervals—which, we emphasize, do not have simultaneous confidence coefficient $(1 - \alpha)$—are constructed according to (3.111).

TABLE 5.5 Confidence intervals on differences in cell means.

		Half-widths ($\alpha = 0.05$)			
(i, j)	$\lvert\hat{\mu}_i - \hat{\mu}_j\rvert$	*t*	*Scheffé*	*Tukey*	*Bonferroni*
(1, 2)	0.7	3.66	5.86	5.61	5.40
(1, 3)	2.4	3.32	5.31	4.69	4.89
(1, 4)	4.3	3.52*	5.63	5.24	5.18
(1, 5)	7.2	3.82*	6.13*	6.06*	5.64*
(2, 3)	1.7	3.66	5.86	5.61	5.40
(2, 4)	3.6	3.84	6.15	5.61	5.66
(2, 5)	6.5	4.12*	6.60	6.06*	6.08*
(3, 4)	1.9	3.51	5.63	5.24	5.18
(3, 5)	4.8	3.82*	6.13	6.06	5.64
(4, 5)	2.9	4.01	6.41	6.06	5.90

*Intervals that do not cover zero.

In Table 5.5, we note that the Scheffé intervals are uniformly wider than the Tukey intervals, which is contrary to what we observed in Table 5.4. This illustrates the conjecture, made by Scheffé (1959), that the Tukey intervals would be narrower for differences in means but not so for general linear contrasts. The evidence supports the fact that electrolyte B at either concentration is superior to water. As anticipated, the t-intervals fail to cover zero more frequently than the others. The Bonferroni intervals, in this case, are comparable to the Tukey intervals. We recall that confidence intervals may be used to test hypotheses; for example, we would reject the hypothesis $\mu_i - \mu_j = 0$ if the confidence interval did not include zero.

Finally, we consider the multiple range tests on differences in means. We shall illustrate the methods with the Newman-Keuls test described in Section 4.4.2. Since the cell frequencies differ in our example, the generalized procedure, using equation (4.125), must be employed. The information required for this test is given in Table 5.6. Since the test is performed sequentially, not all of this information is required, but we show it for completeness. The values for $q'(0.05; k, 36)$ are approximated by $q(0.05; k, 36)$. The test proceeds as follows: Since $\hat{\mu}_5 - \hat{\mu}_1$ exceeds its critical value, we examine $\hat{\mu}_4 - \hat{\mu}_1$ and $\hat{\mu}_5 - \hat{\mu}_2$. Since the latter exceeds its critical value, we examine $\hat{\mu}_4 - \hat{\mu}_2$ and $\hat{\mu}_5 - \hat{\mu}_3$. Since neither of these are significant, we stop, and the differences $\mu_5 - \mu_1$ and $\mu_5 - \mu_2$ are declared significant. In this case, there would have been no contradiction had we continued with the remaining comparisons given in Table 5.6, but this feature is not guaranteed in general.

TABLE 5.6 Generalized Newman-Keuls test.

(i, j)	*k*	$\hat{\mu}_j - \hat{\mu}_i$	*Critical value*
(1, 5)	5	7.2	6.06*
(1, 4)	4	4.3	4.92
(2, 5)	4	6.5	5.68*
(2, 4)	3	3.6	4.78
(3, 5)	3	4.8	5.16
(1, 3)	3	2.4	4.00
(1, 2)	2	0.7	3.96
(2, 3)	2	1.7	3.96
(3, 4)	2	1.9	3.70
(4, 5)	2	2.9	4.28

*Significant differences.

We have used this simple example to illustrate many of the ideas discussed in Chapter 4. In later chapters, the examples will be less detailed, since the basic ideas for the cell means model do not change with the complexity of the model. This same example will serve to illustrate an interesting situation in Section 6.3.3.

CHAPTER 5 EXERCISES

1. For Example 1.4, with data given in Table 1.4, construct the Analysis of Means table appropriate for testing the hypothesis $\mu_1 = \mu_2 = \mu_3$ against the alternative that one of the equalities is not satisfied.
2. In Example 1.4, suppose that the three types of flour are, respectively, (i) white, (ii) rye, and (iii) whole wheat. It is conjectured that (a) rye and whole wheat flours respond similarly and (b) white flour responds differently from the other two. Formulate hypotheses for examining each of these conjectures, and test each at level $\alpha = 0.05$.
3. In Example 1.4, determine simultaneous 95 percent confidence intervals for the two contrasts C1: $\mu_2 - \mu_3$ and C2: $2\mu_1 - \mu_2 - \mu_3$ using the Scheffé and Tukey procedures. Use two different bases for the Scheffé intervals. Contrast these results with those obtained in Exercise 2.
4. The one-way classification model is often written in terms of different parameters, using the identity

$$\mu_i = \bar{\mu}. + (\mu_i - \bar{\mu}.)$$

and letting $\bar{\mu}. = \mu$ and $\mu_i - \bar{\mu}. = \alpha_i$. Thus,

$$E(y_{ij}) = \mu + \alpha_i. \tag{A}$$

Note that, for $i = 1, \ldots, p$, we appear to have $p + 1$ parameters, but they are related since $\alpha. = 0$.

(a) Ignoring the relation $\alpha. = 0$, write the model (A) in the matrix form $E(Y) = X\theta$ where $\theta' = (\mu, \alpha_1, \ldots, \alpha_p)$. Show that X does not have full column rank.

(b) Use the relation $\alpha. = 0$ to remove α_p from this model. Write the resulting model in matrix form, and show that the model matrix has rank p.

(c) The parameters α_i are called *main effects,* reflecting the fact that the α_i provide a measure of the effect of the ith treatment. Show that the hypothesis of zero main effects, H: $\alpha_i = 0$, $i = 1, \ldots, p$, is equivalent to the hypothesis of equal means. Note that model reduction is a natural approach for developing the sum of squares for this hypothesis.

5. Using the model obtained in Part (b) of Exercise 4, determine the estimates of μ and α_i, $i = 1, \ldots, p - 1$. Use these results to estimate μ_i, $i = 1, \ldots, p$.

6. Using the overparameterized matrix obtained in Part (a) of Exercise 4, suppose you remove the degeneracy by letting $\alpha_p = 0$.

(a) Describe the relation between the means, μ_i, and the parameters, μ, α_1, $\ldots, \alpha_{p-1}$.

(b) Show that the hypothesis H: $\alpha_i = 0$, $i = 1, \ldots, p - 1$ is equivalent to the hypothesis of equal means.

CHAPTER SIX
THE TWO-WAY CLASSIFICATION MODEL

6.1. INTRODUCTION

Continuing with our development of the analysis of K-factor classification models, we now consider the case in which the treatments can be described by two factors, as in the sex-by-race classification in Example 4.3. The extension from one to two factors introduces many new problems, such as constrained models, missing cells, and reparameterization for computational simplicity. The new concepts introduced in this chapter will generalize quite easily to higher-way classifications and will be illustrated along with other special problems in Chapter 7.

The two-way classification model without constraints is a simple, special case of the unconstrained cell means model in which the cell means for the populations are described using two subscripts. The model is given by

$$y_{ijk} = \mu_{ij} + e_{ijk} \qquad \begin{aligned} i &= 1, \ldots, a \\ j &= 1, \ldots, b \\ k &= 0, 1, \ldots, n_{ij}. \end{aligned} \tag{6.1}$$

Thus, y_{ijk} is the kth response on the (i, j)th population with mean μ_{ij}, and the errors are independent, $N(0, \sigma^2)$. The number of populations is $p = ab$. The range on k suggests that some populations may not be observed.

While constraints rarely arise in the one-way classification model, the two-way model frequently includes the constraints

$$\mu_{ij} - \mu_{kj} - \mu_{it} + \mu_{kt} = 0 \qquad \begin{aligned} i, k &= 1, \ldots, a \\ j, t &= 1, \ldots, b. \end{aligned} \tag{6.2}$$

This relation between the means is called the *constraint of no interaction*. The idea was introduced in Examples 1.7 and 4.4, where we noted that this constraint has a simple interpretation. Thus, in Example 4.4 this constraint implies that the difference in the mean responses for any two races is the same for both sexes or, equivalently, that the difference in the means for the two sexes is the same regardless of race. We shall see that, if the assumption is valid, then much stronger conclusions may be possible than otherwise.

The unconstrained model and the constrained model will be treated separately in this chapter. The latter will be called the *no-interaction model*. We shall also consider the missing-cell problem separately.

6.2. ESTIMATION OF PARAMETERS

6.2.1. The Unconstrained Model

Direct application of the results of Section 4.2.1, with $n_{ij} > 0$, yields

$$\hat{\mu}_{ij} = \bar{y}_{ij\cdot} \sim N(\mu_{ij}, \sigma^2/n_{ij}) \qquad \begin{aligned} i &= 1, \ldots, a \\ j &= 1, \ldots, b \end{aligned} \tag{6.3}$$

where the $\hat{\mu}_{ij}$ are independent, and independent of,

$$s^2 = Q(\hat{\mu})/(N - ab), \tag{6.4}$$

where $N = n_{\cdot\cdot}$ and

$$Q(\hat{\mu}) = \sum_i \sum_j \sum_k (y_{ijk} - \bar{y}_{ij\cdot})^2 \sim \sigma^2\chi^2(N - ab). \tag{6.5}$$

Clearly, if $n_{ij} = 0$, then μ_{ij} cannot be estimated, but otherwise the occurrence of missing cells causes no problems with parameter estimation in the unconstrained model.

6.2.2. The No-Interaction Model

The results of Section 4.2.2 may be applied using either the model reduction or the LaGrange Multiplier approach. For either method, it is convenient to have the constraints, (6.2), expressed in matrix form. As written in (6.2), however, there are several redundancies. A nonredundant set is given by

$$\mu_{ij} - \mu_{aj} - \mu_{ib} + \mu_{ab} = 0 \qquad \begin{aligned} i &= 1, \ldots, a - 1 \\ j &= 1, \ldots, b - 1. \end{aligned} \tag{6.6}$$

It is easily verified that (6.2) is equivalent to (6.6) and that there are no redundancies in (6.6). In matrix form we may write (6.6) as

$$(\Delta_a \otimes \Delta_b)\mu = 0 \tag{6.7}$$

where Δ_a and Δ_b are defined as in (5.7); that is,

$$\begin{aligned} \Delta_a &= (I_{a-1} | -J_{a-1}) \\ \Delta_b &= (I_{b-1} | -J_{b-1}). \end{aligned} \tag{6.8}$$

The symbol "$\otimes$" denotes the Kronecker product defined in Appendix A.I.8.

Alternatively, (6.2) is equivalent to the nonredundant set of constraints

$$\mu_{ij} - \bar{\mu}_{i\cdot} - \bar{\mu}_{\cdot j} + \bar{\mu}_{\cdot\cdot} = 0 \qquad \begin{array}{l} i = 1, \ldots, a-1 \\ j = 1, \ldots, b-1. \end{array} \tag{6.9}$$

Defining matrices M_a and M_b as

$$\begin{aligned} M_a &= (I_{a-1}|0) - J_{a-1}J_a'/a \\ M_b &= (I_{b-1}|0) - J_{b-1}J_b'/b, \end{aligned} \tag{6.10}$$

we may write (6.9) as

$$(M_a \otimes M_b)\mu = 0. \tag{6.11}$$

Both (6.7) and (6.11) will prove to be useful in our developments. As noted in Chapter 3, the model reduction method is computationally preferable since it reduces the dimensionality of the problem. The LaGrange Multiplier method was most useful for theoretical purposes but will provide solutions in some special cases. Both methods will be employed in our detailed treatment of this model. We shall see that this effort will be rewarded in Chapter 7 when we encounter more complex models and in the following chapters when we discuss models with general covariance structure.

6.2.2.1. Equal Cell Frequencies: $n_{ij} = n$

In this situation, known as the *balanced case,* the method of LaGrange Multipliers is particularly simple. Using the constraints in the form of (6.9), where we include redundant constraints for convenience by letting $i = 1, \ldots, a; j = 1, \ldots, b$, we see that the stationary equations (3.31) and (3.33) are

$$n\mu_{ij} - (\delta_{ij} - \bar{\delta}_{i\cdot} - \bar{\delta}_{\cdot j} + \bar{\delta}_{\cdot\cdot}) = y_{ij\cdot} \tag{6.12}$$

$$\mu_{ij} - \bar{\mu}_{i\cdot} - \bar{\mu}_{\cdot j} + \bar{\mu}_{\cdot\cdot} = 0 \qquad \begin{array}{l} i = 1, \ldots, a \\ j = 1, \ldots, b. \end{array} \tag{6.13}$$

To solve (6.12) and (6.13), determine μ_{ij} from (6.12) and substitute into (6.13) to obtain

$$(\delta_{ij} - \bar{\delta}_{i\cdot} - \bar{\delta}_{\cdot j} + \bar{\delta}_{\cdot\cdot})/n + (\bar{y}_{ij\cdot} - \bar{y}_{i\cdot\cdot} - \bar{y}_{\cdot j\cdot} + \bar{y}_{\cdot\cdot\cdot}) = 0. \tag{6.14}$$

Substituting (6.14) into (6.12) yields

$$\tilde{\mu}_{ij} = \bar{y}_{i\cdot\cdot} + \bar{y}_{\cdot j\cdot} - \bar{y}_{\cdot\cdot\cdot} \qquad \begin{array}{l} i = 1, \ldots, a \\ j = 1, \ldots, b. \end{array} \tag{6.15}$$

Writing the estimator in matrix form, as in (3.36), we have

$$\tilde{\mu} = A\hat{\mu} \tag{6.16}$$

where $\hat{\mu}$ is the vector of unconstrained estimates given by (6.3), and A is defined by (3.38). In this case, we have avoided the matrix inversion implied in the determination of A because of the simple structure of the stationary equations. The expression for A may be inferred directly from (6.15) and is given by

$$A = [(1/b)I_a \otimes J_b J_b'] + [(1/a)J_a J_a' \otimes I_b] - (1/ab)J_{ab}J_{ab}'. \tag{6.17}$$

From Theorem 4.2, we obtain

$$\tilde{\mu} \sim N[\mu, (\sigma^2/n)A]. \tag{6.18}$$

In particular, we see that

$$\mathrm{Var}(\tilde{\mu}_{ij}) = \sigma^2(a + b - 1)/abn. \tag{6.19}$$

Reference to (6.3), for $n_{ij} = n$, shows the gain in precision for estimating μ_{ij} caused by the constraints, since

$$\mathrm{Var}(\hat{\mu}_{ij}) = \sigma^2/n > \sigma^2(a + b - 1)/abn = \mathrm{Var}(\tilde{\mu}_{ij}). \tag{6.20}$$

The residual sum of squares is given by

$$\begin{aligned} Q(\tilde{\mu}) &= \sum_i \sum_j \sum_k (y_{ijk} - \tilde{\mu}_{ij})^2 \\ &= \sum_i \sum_j \sum_k (y_{ijk} - \bar{y}_{ij\cdot})^2 + \sum_i \sum_j \sum_k (\bar{y}_{ij\cdot} - \bar{y}_{i\cdot\cdot} - \bar{y}_{\cdot j\cdot} + \bar{y}_{\cdots})^2. \end{aligned} \tag{6.21}$$

From Theorem 4.2, we see that the unbiased estimator of σ^2 is given by

$$s^2 = Q(\tilde{\mu})/(nab - a - b + 1) \tag{6.22}$$

since $r = p - q = ab - (a - 1)(b - 1) = a + b - 1$. Further,

$$Q(\tilde{\mu}) \sim \sigma^2\chi^2(nab - a - b + 1). \tag{6.23}$$

In (6.21), the residual sum of squares is written as the sum of two components. The first is $Q(\hat{\mu})$, as given in (6.5), and the second arises as a consequence of the assumption of no interaction. To emphasize the importance of this observation, note that, in the special case, $n = 1$, $Q(\hat{\mu}) = 0$, and hence the estimate of σ^2 is entirely based on the no-interaction assumption. It should be emphasized that having one observation per cell does not justify the assumption of no interaction. Such an assumption should be based on the researcher's knowledge of the system being studied.

6.2.2.2. A Canonical Model and Canonical Forms

The LaGrange Multiplier solution in Section 6.2.2.1 for the balanced no-interaction model was particularly simple, but, unfortunately, this simplicity does not extend to the unbalanced case. In practice, we would prefer to work with the reduced model, even in the balanced case, for its computational advantages. However, as noted earlier, the reduced model suffers from notational complexities.

We shall see in this section that it is more convenient to transform to a new set of parameters that are linear functions of the cell means. Only two transformations will be considered, although many more are possible and may be preferred in special instances. These two transformations will both lead to the same general expression, which we shall call the *canonical model*. The special cases of this canonical model corresponding to the two transformations will be referred to as *canonical forms*. The first canonical form is suggested by the reduced model, and, while it is simple, it is not commonly used. The reasons for this will be apparent when we discuss the role of the canonical model in hypothesis testing. The second canonical form will be used frequently in the remainder of this book.

The reduced model. To write the reduced model as described in Section 4.2.2, we must identify a partition of μ such that the matrix G_1 in (4.36) is nonsingular. With the constraints written in the form of (6.6), an obvious partition of μ, for which G_1 is the identity, is suggested by writing (6.6) as

$$\mu_{ij} = \mu_{ib} + \mu_{aj} - \mu_{ab} \qquad \begin{matrix} i = 1, \ldots, a-1 \\ j = 1, \ldots, b-1. \end{matrix} \tag{6.24}$$

Thus, we define the vectors μ_1 and μ_2 in the reduced model as

$$\mu_1 = \begin{bmatrix} \mu_{ij} & \begin{matrix} i = 1, \ldots, a-1 \\ j = 1, \ldots, b-1 \end{matrix} \end{bmatrix} \tag{6.25}$$

and

$$\mu_2 = \begin{bmatrix} \mu_{ib} & i = 1, \ldots, a-1 \\ \mu_{aj} & j = 1, \ldots, b \end{bmatrix}. \tag{6.26}$$

Substituting from (6.24) into (6.1), we see that the mean vector for the reduced model is easily written in algebraic form as

$$E(Y) = W_R\mu_R = \begin{bmatrix} \mu_{ib} + \mu_{aj} - \mu_{ab} & \begin{matrix} i = 1, \ldots, a-1 \\ j = 1, \ldots, b-1 \end{matrix} \\ \mu_{ib} & \begin{matrix} i = 1, \ldots, a-1 \\ j = b \end{matrix} \\ \mu_{aj} & \begin{matrix} i = a \\ j = 1, \ldots, b \end{matrix} \end{bmatrix} \tag{6.27}$$

where $\mu_R = \mu_2$ as defined in (6.26). The matrix expression for W_R is straightforward but notationally clumsy, as noted in Example 4.4, so we shall not describe it. In the following, we shall see that it is not necessary.

It is easily verified that the reduced model yields the estimates

$$\tilde{\mu}_{ij} = \bar{y}_{i\cdot\cdot} + \bar{y}_{\cdot j\cdot} - \bar{y}_{\cdot\cdot\cdot} \tag{6.28}$$

for the components of μ_2. Substituting into (6.24), we see that the same expression, (6.28), applies for all i and j.

We now consider two reparameterizations of the reduced model which will lead to a more convenient notation for the reduced model and, in addition, will be especially helpful in the determination of sums of squares for testing certain hypotheses. The ideas presented here are closely related to those given by Bryce, Scott, and Carter (1980). Since we are interested only in special cases, the development here is much simpler and more intuitive.

First canonical form. From the expression for $E(Y)$ given in (6.27), we are led to consider the differences $\mu_{ib} - \mu_{ab}$ and $\mu_{aj} - \mu_{ab}$. Thus, we define the new parameters

$$\begin{aligned} \mu_r &= \mu_{ab} & & \\ \alpha_{r_i} &= \mu_{ib} - \mu_{ab} & \qquad i &= 1, \ldots, a-1, \\ \beta_{r_j} &= \mu_{aj} - \mu_{ab} & \qquad j &= 1, \ldots, b-1. \end{aligned} \tag{6.29}$$

The subscript, r, in (6.29) is used to identify the parameters in the first canonical form. This choice of subscript reflects the fact that this technique is frequently used to indicate qualitative variables in regression models. It then follows that the mean vector in (6.27) is given by

$$E(Y) = \begin{bmatrix} \mu_r + \alpha_{r_i} + \beta_{r_j} & i = 1, \ldots, a-1 \\ & j = 1, \ldots, b-1 \\ \mu_r + \alpha_{r_i} & i = 1, \ldots, a-1 \\ & j = b \\ \mu_r + \beta_{r_j} & i = a \\ & j = 1, \ldots, b-1 \\ \mu_r & i = a \\ & j = b \end{bmatrix} \tag{6.30}$$

To write this form of the model in matrix notation, let

$$\theta_r = \begin{bmatrix} \mu_r & \\ \alpha_{r_i} & i = 1, \ldots, a-1 \\ \beta_{r_j} & j = 1, \ldots, b-1 \end{bmatrix}. \tag{6.31}$$

Then the mean vector may be written as

$$E(Y) = X_r\theta_r. \tag{6.32}$$

Here the components of Y are in the natural order [see (1.22)] as opposed to the order implied by (6.30). The matrix X_r may be inferred from (6.30). Rather than do so directly, however, it is instructive to write it as a product of three matrices that are the basic building blocks of the canonical forms. Assuming the balanced case, $n_{ij} = n$, let

$$W = I_{ab} \otimes J_n \tag{6.33}$$

be the *counting matrix* for the cell means model. Let

$$X_0 = (J_{ab}|I_a \otimes J_b|J_a \otimes I_b) \tag{6.34}$$

denote the *design matrix,* and let P_r be the *parameter matrix* obtained by deleting column $(a + 1)$ and column $(a + b + 1)$ from I_{a+b+1}. Thus,

$$P_r = \text{Diag}\left(1, \begin{bmatrix} I_{a-1} \\ 0 \end{bmatrix}, \begin{bmatrix} I_{b-1} \\ 0 \end{bmatrix}\right). \tag{6.35}$$

Then, we have

$$X_r = WX_0P_r. \tag{6.36}$$

Note that P_r simply serves to delete columns $(a + 1)$ and $(a + b + 1)$ from X_0. The role of W is to reflect the number of responses on population (i, j). Thus, X_0P_r defines the basic structure of the model matrix for this choice of parameters.

The analysis of the model, with mean vector given by (6.32), follows directly from the results in Chapter 3 for the unconstrained model. It is instructive to spell out the details for estimating the parameters. The likelihood equations, analogous to (3.5) and (3.6), are

$$X_r'X_r\hat{\theta}_r = X_r'Y \tag{6.37}$$

and

$$\hat{\sigma}^2 = Q_r(\hat{\theta}_r)/N \tag{6.38}$$

where

$$Q_r(\theta_r) = (Y - X_r\theta_r)'(Y - X_r\theta_r).$$

The coefficient matrix in (6.37) is easily formed using the basic matrices in (6.36). Note that $W'W = nI$ and $X_0'X_0$ is easily computed using the rules for Kronecker products in Appendix A.I.8. Thus,

$$X_0'W'WX_0 = n\left[\begin{array}{c|c|c} ab & bJ_a' & aJ_b' \\ bJ_a & bI_a & J_aJ_b' \\ aJ_b & J_bJ_a' & aI_b \end{array}\right]. \tag{6.39}$$

Then, $X_r'X_r$ is given by pre- and post-multiplying (6.39) by P_r' and P_r, respectively, to obtain

$$X_r'X_r = n\left[\begin{array}{c|c|c} ab & bJ_{a-1}' & aJ_{b-1}' \\ bJ_{a-1} & bI_{a-1} & J_{a-1}J_{b-1}' \\ aJ_{b-1} & J_{b-1}J_{a-1}' & aI_{b-1} \end{array}\right] \tag{6.40}$$

with inverse matrix

$$(X_r'X_r)^{-1} = \tag{6.41}$$

$$(1/n)\left[\begin{array}{c|c|c} (a+b-1)/ab & -(1/b)J_{a-1}' & -(1/a)J_{b-1}' \\ -(1/b)J_{a-1} & (1/b)(I_{a-1} + J_{a-1}J_{a-1}') & 0 \\ -(1/a)J_{b-1} & 0 & (1/a)(I_{b-1} + J_{b-1}J_{b-1}') \end{array}\right].$$

Similarly, we obtain $X_r'Y$ by deleting $y_{a\cdot\cdot}$ and $y_{b\cdot\cdot}$ from

$$X_0'W'Y = \begin{bmatrix} y_{\cdots} & \\ y_{i\cdot\cdot} & i = 1, \ldots, a \\ y_{\cdot j\cdot} & j = 1, \ldots, b \end{bmatrix}. \tag{6.42}$$

The solution vector is then given by

$$\hat{\theta}_r = \begin{bmatrix} \bar{y}_{a\cdot\cdot} + \bar{y}_{\cdot b\cdot} - \bar{y}_{\cdots} & \\ \bar{y}_{i\cdot\cdot} - \bar{y}_{a\cdot\cdot} & i = 1, \ldots, a-1 \\ \bar{y}_{\cdot j\cdot} - \bar{y}_{\cdot b\cdot} & j = 1, \ldots, b-1 \end{bmatrix}. \tag{6.43}$$

The properties of $\hat{\theta}_r$ follow from Chapter 3 and are given by

$$\hat{\theta}_r = (X_r'X_r)^{-1}X_r'Y \sim N[\theta_r, \sigma^2(X_r'X_r)^{-1}]. \tag{6.44}$$

Using (6.29) and then (6.24), we obtain the estimates of μ_{ij} as given in (6.28).

The expression for the mean vector given in (6.32) is appealing for its simplicity in interpretation, as well as for convenience in setting up the likelihood equations. Unfortunately, the extension of this approach to more complex models that allow interactions may lead to confusion in hypothesis testing. For this reason we consider a second transformation.

Second canonical form. The next choice of parameters is less intuitive as a simplification of the reduced model, but it is motivated by an attempt to introduce a measure of the effects of the different levels of the two factors. Thus, we define

$$\begin{aligned} \mu_m &= \bar{\mu}_{..} & & \\ \alpha_{m_i} &= \bar{\mu}_{i.} - \bar{\mu}_{..} & \quad i &= 1, \ldots, a-1 \\ \beta_{m_j} &= \bar{\mu}_{.j} - \bar{\mu}_{..} & \quad j &= 1, \ldots, b-1 \end{aligned} \tag{6.45}$$

and let

$$\theta_m = \begin{bmatrix} \mu_m & \\ \alpha_{m_i} & i = 1, \ldots, a-1 \\ \beta_{m_j} & j = 1, \ldots, b-1 \end{bmatrix}. \tag{6.46}$$

The subscript, m, in (6.45) is used to identify the parameters in the second canonical form. This choice of subscript reflects the role of these parameters in the method of *marginal means,* to be described in Section 6.3.1.1. Utilizing the relation (6.9), we may now write the mean vector, (6.27), in terms of θ_m as

$$E(Y) = \begin{bmatrix} \mu_m + \alpha_{m_i} + \beta_{m_j} & i = 1, \ldots, a-1 \\ & j = 1, \ldots, b-1 \\ \mu_m + \alpha_{m_i} - \sum_{j=1}^{b-1} \beta_{m_j} & i = 1, \ldots, a-1 \\ & j = b \\ \mu_m - \sum_{i=1}^{a-1} \alpha_{m_i} + \beta_{m_j} & i = a \\ & j = 1, \ldots, b-1 \\ \mu_m - \sum_{i=1}^{a-1} \alpha_{m_i} - \sum_{j=1}^{b-1} \beta_{m_j} & i = a \\ & j = b \end{bmatrix}. \tag{6.47}$$

This vector may now be written, with the components of Y in the natural order, as

$$E(Y) = X_m \theta_m \tag{6.48}$$

where X_m may be inferred from (6.47). Following the approach for the previous canonical form, we find that we can write X_m as

$$X_m = WX_0P_m. \tag{6.49}$$

Here W and X_0 are as defined in (6.33) and (6.34), and the new parameter matrix is given by

$$P_m = \text{Diag}(1, \Delta_a', \Delta_b') \tag{6.50}$$

with Δ_a and Δ_b defined as in (6.8).

The form of the likelihood equations for θ_m follows from (6.39) and (6.42) by applying the parameter matrix, P_m. We obtain

$$X_m'X_m = n\begin{bmatrix} ab & 0 & 0 \\ 0 & b(I_{a-1} + J_{a-1}J_{a-1}') & 0 \\ 0 & 0 & a(I_{b-1} + J_{b-1}J_{b-1}') \end{bmatrix} \tag{6.51}$$

and

$$X_m'Y = \begin{bmatrix} y_{\cdots} & \\ y_{i\cdot\cdot} - y_{a\cdot\cdot} & i = 1, \ldots, a-1 \\ y_{\cdot j\cdot} - y_{\cdot b\cdot} & j = 1, \ldots, b-1 \end{bmatrix}. \tag{6.52}$$

The inverse matrix is given by

$$(X_m'X_m)^{-1} = 1/(nab)\begin{bmatrix} 1 & 0 & 0 \\ 0 & aI_{a-1} - J_{a-1}J_{a-1}' & 0 \\ 0 & 0 & bI_{b-1} - J_{b-1}J_{b-1}' \end{bmatrix}. \tag{6.53}$$

The solution vector is

$$\hat{\theta}_m = \begin{bmatrix} \bar{y}_{\cdots} & \\ \bar{y}_{i\cdot\cdot} - \bar{y}_{\cdots} & i = 1, \ldots, a-1 \\ \bar{y}_{\cdot j\cdot} - \bar{y}_{\cdots} & j = 1, \ldots, b-1 \end{bmatrix}. \tag{6.54}$$

Using (6.45), (6.9), and (6.24), we obtain the results previously given in (6.15).

We have developed the canonical form for two of the more common transformations, (6.29) and (6.45). For the two-way model without interaction, there is little to choose between the two transformations. The first is somewhat simpler in that it requires only deletion of columns, while the second has the advantage that it is independent of the partition of μ used to develop the reduced model. In the following chapters we shall use the second form almost exclusively.

We strongly emphasize that the canonical forms were developed primarily for notational and computational convenience. They are mathematically equivalent to the reduced model. For interpretation, however, we recommend that the cell means model be used. If a canonical form is used, it is essential that the relations between θ and μ be specified as in (6.29) or (6.45).

These two canonical forms are special cases of a general canonical model that is commonly encountered in the literature. This model enables us to establish the relation between the cell means model and the classical, overparameterized model found in many texts on linear models.

A general canonical model. The general canonical model is defined by writing the mean vector as

$$E(Y) = X_c\theta_c \tag{6.55}$$

where $X_c = WX_0$ with W and X_0 as defined in (6.33) and (6.34). The parameter vector θ_c of length $1 + a + b$ is given by

$$\theta_c = \begin{bmatrix} \mu & \\ \alpha_i & i = 1, \ldots, a \\ \beta_j & j = 1, \ldots, b \end{bmatrix}. \tag{6.56}$$

We immediately recognize that this model does not satisfy a fundamental requirement of Chapter 3, since the matrix X_c does not have full column rank. Indeed, the rank of X_c is the rank of X_0, which is $a + b - 1$, as is easily established from (6.34). The implication of this is that we have allowed too many parameters in θ_c. Reference to the two canonical forms reveals the sense in which the excess parameters should be related to the basic parameters of the canonical form. Thus, the first canonical form follows from the canonical model (6.55) by defining $\mu = \mu_{ab}$, $\alpha_i = \alpha_{r_i}$, $i = 1, \ldots, a - 1$; $\beta_j = \beta_{r_j}$, $j = 1, \ldots, b - 1$ and, in addition, defining $\alpha_a = \beta_b = 0$. This relation is specified by writing

$$\theta_c = P_r\theta_r \tag{6.57}$$

with P_r and θ_r defined in (6.35) and (6.31).

Similarly, the second canonical form follows from the canonical model by defining $\mu = \bar{\mu}..$; $\alpha_i = \alpha_{m_i}$, $i = 1, \ldots, a - 1$; $\beta_j = \beta_{m_j}$, $j = 1, \ldots, b - 1$ and, in addition, defining

$$\alpha_a = -\sum_{i=1}^{a-1} \alpha_i \qquad \beta_b = -\sum_{j=1}^{b-1} \beta_j. \tag{6.58}$$

Again, this relation is specified by writing

$$\theta_c = P_m\theta_m \tag{6.59}$$

with P_m and θ_m defined in (6.50) and (6.46).

Thus, we see that particular canonical forms arise from the canonical model by specifying a parameter matrix. It is instructive to examine the expression for the mean vector for an arbitrary parameter matrix P^* and associated parameter vector θ^*; that is, we write

$$\begin{aligned} E(Y) &= WX_0P^*\theta^* \\ &= X^*\theta^* \\ &= X_c\theta_c \\ &= W\mu. \end{aligned} \tag{6.60}$$

The second equality in (6.60) with $X^* = WX_0P^*$ is suitable for computing, since X^* has full column rank. The third equality is frequently used in the discussion of this model, where $X_c = WX_0$ and $\theta_c = P^*\theta^*$. Note that X_c does not have full column rank, or, equivalently, the parameter vector θ_c is not uniquely defined. In the last equality, with $\mu = X_0\theta_c$, we see the relation between the cell means model and the canonical model. We see from (6.34) that this last relation is given by

$$\mu_{ij} = \mu + \alpha_i + \beta_j. \tag{6.61}$$

The impact of this is that—given estimates of μ, α_i, $i = 1, \ldots, a - 1$ and β_j, $j = 1, \ldots, b - 1$ from the canonical analysis and, implicitly, of α_a and β_b from either $\alpha_a = \beta_b = 0$ or (6.58)—we immediately have the estimates of μ_{ij} without recourse to the relations, (6.24).

Finally, we note that some authors approach this problem from a different direction; that is, they begin with the canonical model (6.55) and then require that θ_c satisfy certain linear conditions. Imposing these conditions on the model leads to a full rank model, which is then subjected to the analysis in Chapter 3. While this approach is valid and will correspond to a canonical form with some parameter matrix, it is less intuitive. For example, imposing the conditions $\Sigma_i\alpha_i = \Sigma_j\beta_j = 0$ will lead to the second canonical form. Note that this is equivalent to defining the relations as in (6.45), but this fact is often not emphasized. It seems evident that explicit expressions for the reparameterizations are preferred. Still other authors prefer to attempt an analysis of the canonical model directly, using equations that are the analogs of the likelihood equations. Since $X_c'X_c$ is singular, they obtain a generalized inverse of $X_c'X_c$ (see Searle, 1971a) that leads to a solution of these normal equations. Again, this is equivalent to defining a specific relation between μ and θ_c. Here, the relation is even less clear and rarely spelled out. Further, it is not unique, since the generalized inverse is not unique.

Starting with the canonical model has led to unnecessary confusion in the literature. We shall find it convenient to use a canonical form—almost always the second one—for notational and computational purposes. The canonical model will never be used explicitly for computation.

6.2.2.3. Unequal Cell Frequencies: $n_{ij} \neq 0$

The case $n_{ij} = n$, treated in the last two sections, is clearly a special case of this section. The primary purpose of treating that case separately is that closed form expressions are available for the estimators. This is not the case with unbalanced data; hence, some attention must be given to the required computations. Using the model reduction approach, we shall obtain explicit solutions for a special case of the unbalanced problem. For the general case, we shall extend the canonical model and the canonical forms of the last section to include the unbalanced data situation.

The model reduction approach with a convenient partition of the parameter vector was described in Section 6.2.2.2. The following example illustrates this approach for a special class of problems.

EXAMPLE 6.1

To illustrate the reduced model and the complexities introduced by the unequal cell frequencies, we recall Example 4.4 for which $a = 2$ and $b = 3$ in (6.1) and (6.2). For convenience, we recall the coefficient matrix in the reduced model given by

$$W_R = \begin{bmatrix} J_{n_{11}} & J_{n_{11}} & 0 & -J_{n_{11}} \\ J_{n_{12}} & 0 & J_{n_{12}} & -J_{n_{12}} \\ J_{n_{13}} & 0 & 0 & 0 \\ 0 & J_{n_{21}} & 0 & 0 \\ 0 & 0 & J_{n_{22}} & 0 \\ 0 & 0 & 0 & J_{n_{23}} \end{bmatrix} \tag{6.62}$$

$$\mu_R = \begin{bmatrix} \mu_{13} \\ \mu_{21} \\ \mu_{22} \\ \mu_{23} \end{bmatrix}.$$

Using these expressions in the likelihood equations for the reduced model, we obtain, after some algebra, the solution

$$\tilde{\mu}_{ij} = \bar{y}_{\cdot j \cdot} + \bar{y}_{c_i} - \bar{y}_c \qquad (i, j) = (13, 21, 22, 23) \tag{6.63}$$

with

$$\begin{aligned} \bar{y}_{c_i} &= \sum_j h_{c_j}\bar{y}_{ij\cdot}/h_{c\cdot} \\ \bar{y}_c &= \sum_i n_{ij}\bar{y}_{c_i}/n_{\cdot j} \\ 1/h_{c_j} &= (1/2)\sum_i (1/n_{ij}) \\ h_{c\cdot} &= \sum_j h_{c_j}. \end{aligned} \tag{6.64}$$

Finally, $\tilde{\mu}_{11}$ and $\tilde{\mu}_{12}$ are obtained using $\tilde{\mu}_R$ and (6.24). The residual sum of squares is given by

$$Q(\tilde{\mu}) = Q(\hat{\mu}) + \sum_i \sum_j \sum_k (\bar{y}_{ij\cdot} - \bar{y}_{\cdot j\cdot} - \bar{y}_{c_i} + \bar{y}_c)^2. \qquad (6.65)$$

Here, $\hat{\mu}$ is the unconstrained estimator with components $\hat{\mu}_{ij} = \bar{y}_{ij\cdot}$. Note that $\tilde{\mu}$ and $Q(\tilde{\mu})$, in this example, agree with (6.9) and (6.21) in the special case $n_{ij} = n$.

We have introduced the general expression h_{c_j} for the harmonic mean of the elements in the jth column, as it will appear later in this chapter. The results in this example hold for $a = 2$ and any b. The extension of (6.63) for $a > 2$ is not evident. ■

The canonical model for unbalanced data is identical to that described in Section 6.2.2.2, recognizing that the counting matrix, W, is no longer given by (6.33) but by

$$W = D_{ab}(J_{n_{ij}}). \qquad (6.66)$$

Thus, the canonical model is

$$E(Y) = X_c\theta_c \qquad (6.67)$$

where

$$X_c = WX_0 \qquad \theta_c = P^*\theta^* \qquad (6.68)$$

with X_0 being defined by (6.34), P^* denoting a particular parameter matrix, and θ^* being the corresponding parameter vector. The parameter matrices, P_r and P_m, or any other convenient choice may be used. Again, we emphasize that the reparameterization should be carefully specified.

The normal equations for unbalanced data are more complex. The analog of (6.39) is given by the symmetric matrix

$$X_0'W'WX_0 = \left[\begin{array}{c|c|c} N & R_a(n_{i\cdot}) & R_b(n_{\cdot j}) \\ C_a(n_{i\cdot}) & D_a(n_{i\cdot}) & N_{ab} \\ C_b(n_{\cdot j}) & N'_{ab} & D_b(n_{\cdot j}) \end{array}\right] \qquad (6.69)$$

where the *incidence* matrix N_{ab} is given by

$$N_{ab} = \left(n_{ij} \qquad \begin{matrix} i = 1, \ldots, a \\ j = 1, \ldots, b \end{matrix}\right). \qquad (6.70)$$

The coefficient matrix is then formed by the appropriate parameter matrix. The right-hand side of the normal equations is determined by applying the parameter matrix to (6.42), where we recognize that now the components of this vector are given by, for example,

$$y_{\cdot\cdot\cdot} = \sum_{c=1}^{a} \sum_{j=1}^{b} \sum_{k=1}^{n_{ij}} y_{ijk}. \tag{6.71}$$

The distributional properties of the estimator follow as usual and are given by (6.44) for the appropriate choice of parameter matrix. The estimate of θ_c is given by (6.57) or (6.59), and, finally, the estimate of μ follows from the relation $\mu = X_0\theta_c$ or, equivalently, (6.61). Explicit expressions for the inverse of the coefficient matrix—and hence for the estimators—are not available, but numerical solutions of the normal equations are easily obtained.

EXAMPLE 6.2

To further illustrate the concepts of the reduced model and the canonical forms, consider the case $a = b = 3$. The reduced W matrix and associated mean vector are given by

$$W_R = \begin{bmatrix} J_{n_{11}} & 0 & J_{n_{11}} & 0 & -J_{n_{11}} \\ J_{n_{12}} & 0 & 0 & J_{n_{12}} & -J_{n_{12}} \\ J_{n_{13}} & 0 & 0 & 0 & 0 \\ 0 & J_{n_{21}} & J_{n_{21}} & 0 & -J_{n_{21}} \\ 0 & J_{n_{22}} & 0 & J_{n_{22}} & -J_{n_{22}} \\ 0 & J_{n_{23}} & 0 & 0 & 0 \\ 0 & 0 & J_{n_{31}} & 0 & 0 \\ 0 & 0 & 0 & J_{n_{32}} & 0 \\ 0 & 0 & 0 & 0 & J_{n_{33}} \end{bmatrix}$$

and

$$\mu_R = (\mu_{13}\,\mu_{23}\,\mu_{31}\,\mu_{32}\,\mu_{33})'.$$

The second canonical form is defined by the parameter matrix, (6.50), given here by

$$P_m = \begin{bmatrix} 1 & 0 & 0 & 0 & 0 \\ 0 & 1 & 0 & 0 & 0 \\ 0 & 0 & 1 & 0 & 0 \\ 0 & -1 & -1 & 0 & 0 \\ 0 & 0 & 0 & 1 & 0 \\ 0 & 0 & 0 & 0 & 1 \\ 0 & 0 & 0 & -1 & -1 \end{bmatrix}.$$

Thus,

$$\theta_m = (\mu, \alpha_1, \alpha_2, \beta_1, \beta_2)'$$

and

$$X_m = \begin{bmatrix} J_{n_{11}} & J_{n_{11}} & 0 & J_{n_{11}} & 0 \\ J_{n_{12}} & J_{n_{12}} & 0 & 0 & J_{n_{12}} \\ J_{n_{13}} & J_{n_{13}} & 0 & -J_{n_{13}} & -J_{n_{13}} \\ J_{n_{21}} & 0 & J_{n_{21}} & J_{n_{21}} & 0 \\ J_{n_{22}} & 0 & J_{n_{22}} & 0 & J_{n_{22}} \\ J_{n_{23}} & 0 & J_{n_{23}} & -J_{n_{23}} & -J_{n_{23}} \\ J_{n_{31}} & -J_{n_{31}} & -J_{n_{31}} & J_{n_{31}} & 0 \\ J_{n_{32}} & -J_{n_{32}} & -J_{n_{32}} & 0 & J_{n_{32}} \\ J_{n_{33}} & -J_{n_{33}} & -J_{n_{33}} & -J_{n_{33}} & -J_{n_{33}} \end{bmatrix}.$$

The structure of X_m should be noted and contrasted with X_r for this example. ∎

6.2.2.4. Missing Cells

We have noted, in the unconstrained model, that zero cell frequencies cause no problem with regard to estimation, since, if $n_{ij} = 0$, then we simply cannot estimate μ_{ij}. In the constrained model, and in particular the no-interaction model, it may be possible to estimate the means for unobserved cells through the linear relations describing the constraints. Thus, for example, if $n_{11} = 0$ and we have estimates for μ_{12}, μ_{21}, and μ_{22}, then μ_{11} is estimated by

$$\tilde{\mu}_{11} = \tilde{\mu}_{12} + \tilde{\mu}_{21} - \tilde{\mu}_{22}. \tag{6.72}$$

A simple example will illustrate that the constraints may not always yield estimates of the means for unobserved cells. Consider the 3 × 3 array shown in Table 6.1, with missing cells denoted by zero.

TABLE 6.1 An unconnected situation.

n_{11}	n_{12}	0
n_{21}	n_{22}	0
0	0	n_{33}

Upon inspection, we see that the no-interaction constraints are of no help in estimating the means for the missing cells. In this case, it is clear that we have two disjoint, or unconnected, experiments. The first consists of the 2 × 2 upper left corner in Table 6.1, and the second is just the single cell (3, 3).

Situations such as this motivated the concept of connected models. John (1971) defines a connected two-factor array as one in which it is possible to join any two observed cells by a row–column path of observed cells. Thus, cells (i, j) and (r, t) may be connected by the path

$$(i, j) \rightarrow (i, v) \rightarrow (u, v) \rightarrow (u, t) \rightarrow (r, t) \quad (6.73)$$

where all cells in the path are observed.

The motivation for this geometric definition is given by Graybill (1976), who relates the physical connectedness to estimation of parameters in the two-way classification, no-interaction model. In Chapter 4, Definition 4.3, we extended this concept to general cell means models and indicated in Section 4.2.3 two methods for establishing connectedness—that is, estimability of all components of μ. The first method requires the reduction of the constraint matrix, as in (4.47), to determine those constraints that are in effect on the observed cells, and the remaining constraints, (4.50), which reveal the relations between the observed and missing cells. The structure of these latter equations determines estimability. The second method is to determine the rank of the reduced model matrix W_R. The model is connected if and only if W_R has full column rank, as established in Theorem 4.4.

The results of Theorem 4.4 apply as well to the canonical models, since the matrices X_r or X_m, in (6.38) and (6.49), are obtained by postmultiplication of W_R by a nonsingular matrix. That is, we have the following theorem.

THEOREM 6.1
The model, in canonical form, is connected if and only if the canonical matrices X_r or X_m have full column rank, $a + b - 1$.

■

In this case, the estimate of μ is given, depending on the choice of the canonical form matrix, X^*, as

$$\hat{\mu} = X_0 P^* \hat{\theta}^* \quad (6.74)$$

where P^* and θ^* are the parameter matrix and the parameter vector corresponding to X^*. The estimate of σ^2 is given by the general results for the reduced model as

$$s^2 = Q(\hat{\theta})/(N - a - b + 1) \quad (6.75)$$

where

$$Q(\hat{\theta}) = Y'[I - X^*(X^{*\prime}X^*)^{-1}X^{*\prime}]Y. \quad (6.76)$$

The ranks of X_r and X_m are the same as the rank of W_R; hence, the comments of Section 4.2.3 on the nonfull rank case apply to the canonical model. Thus, we might attempt to invert $X_m'X_m$ by, say, the Sweep Method (Appendix A.I.10) and find that we invert only a portion of the matrix. Setting the remaining parameters to zero and applying (6.74) provides the estimates for the cell means that are uniquely estimable and assigns estimates to the remaining means that are unique only up to the relations (4.50).

EXAMPLE 6.3
To illustrate the problems with empty cells, consider the cell frequencies in the incidence matrix shown on Table 6.2.

TABLE 6.2 An example with missing cells.

n_{11}	n_{12}	n_{13}
n_{21}	n_{22}	0
0	0	n_{33}

Since $n_{23} = n_{31} = n_{32} = 0$, the appropriate W_R and X_m matrices are obtained, from Example 6.2 in Section 6.2.2.3, by deleting rows with subscripts (2, 3), (3, 1), and (3, 2). The resulting matrices are of full column rank, confirming the connectedness of the design, which in this case is easily verified by the row–column definition, (6.73).

In terms of the effective model, the G matrix in the form (4.47) is

$$G = \left[\begin{array}{ccc|cccccc} 1 & 0 & 0 & -1 & 1 & 0 & -1 & 0 & 0 \\ 0 & 1 & 0 & 1 & -1 & 0 & 0 & 0 & -1 \\ 0 & 0 & 1 & 1 & 0 & -1 & 0 & 0 & -1 \\ 0 & 0 & 0 & 0 & 1 & -1 & -1 & 1 & 0 \end{array}\right]. \tag{6.77}$$

Here, the columns are ordered as

$$(23, 31, 32, 13, 11, 12, 21, 22, 33). \tag{6.78}$$

The last row of the reduced G matrix gives the single effective constraint, $\mu_{11} - \mu_{12} - \mu_{21} + \mu_{22} = 0$. The first three rows define the unique relations between the missing and observed cells, hence verifying the connectedness.

■

The three equivalent approaches to estimating the cell means are now apparent. Using the reduced model yields estimates of μ_R, and the remaining μ_{ij} are estimated by (6.24). The canonical analysis yields estimates of μ, α_i, and β_j, and hence of μ_{ij} as defined by (6.61). The effective model yields estimates of the means for the observed cells, with the remaining estimates obtained from the defining relations in the G matrix.

EXAMPLE 6.4
A disconnected case is obtained by setting $n_{13} = 0$ in Table 6.2. From (6.77), we see that the effective constraint is the same, but now we do not have unique

expressions for the unobserved cells in terms of the observed cells. The best we can do is estimate linear functions of the unobserved means. For example,

$$\widehat{(\mu_{13} - \mu_{23})} = \hat{\mu}_{11} - \hat{\mu}_{21}. \tag{6.79}$$

Continuing with the analysis of the effective model, we can use the single effective constraint to obtain the reduced effective model defined by

$$W_{OR} = \begin{bmatrix} J_{n_{11}} & J_{n_{11}} & -J_{n_{11}} & 0 \\ J_{n_{12}} & 0 & 0 & 0 \\ 0 & J_{n_{21}} & 0 & 0 \\ 0 & 0 & J_{n_{22}} & 0 \\ 0 & 0 & 0 & J_{n_{33}} \end{bmatrix} \tag{6.80}$$

with

$$\mu_{OR} = (\mu_{12}, \mu_{21}, \mu_{22}, \mu_{33})'. \tag{6.81}$$

Alternatively, we could use either the reduced model or the canonical model, defined by (6.67) and (6.71), with $n_{13} = n_{23} = n_{31} = n_{32} = 0$. Both of these matrices are seen to be of rank 4. For example, the canonical matrix, X_m, is given by

$$X_m = \begin{bmatrix} J_{n_{11}} & J_{n_{11}} & 0 & J_{n_{11}} & 0 \\ J_{n_{12}} & J_{n_{12}} & 0 & 0 & J_{n_{12}} \\ J_{n_{21}} & 0 & J_{n_{21}} & J_{n_{21}} & 0 \\ J_{n_{22}} & 0 & J_{n_{22}} & 0 & J_{n_{22}} \\ J_{n_{33}} & -J_{n_{33}} & -J_{n_{33}} & -J_{n_{33}} & -J_{n_{33}} \end{bmatrix}. \tag{6.82}$$

Note that the sum of columns two and three is equal to the sum of columns four and five—hence the degeneracy.

As noted earlier, we can delete one of the columns involved in the degeneracy and proceed with the canonical analysis. For example, we may delete the last column or, equivalently, set $\beta_2 = 0$. The estimates for the observed cells given by

$$\begin{aligned} \hat{\mu}_{12} &= \hat{\mu} + \hat{\alpha}_1 \\ \hat{\mu}_{21} &= \hat{\mu} + \hat{\alpha}_2 + \hat{\beta}_1 \\ \hat{\mu}_{22} &= \hat{\mu} + \hat{\alpha}_2 \\ \hat{\mu}_{33} &= \hat{\mu} - \hat{\alpha}_1 - \hat{\alpha}_2 - \hat{\beta}_1 \end{aligned} \tag{6.83}$$

are seen to agree with those obtained from the effective model. We can compute estimates for all of the cell means using the relation

$$\hat{\mu}_{ij} = \hat{\mu} + \hat{\alpha}_i + \hat{\beta}_j \qquad \begin{aligned} i &= 1, 2, 3 \\ j &= 1, 2, 3 \end{aligned} \tag{6.84}$$

with $\hat{\beta}_2 = 0$, $\hat{\alpha}_3 = -\hat{\alpha}_1 - \hat{\alpha}_2$, and $\hat{\beta}_3 = -\hat{\beta}_1 - \hat{\beta}_2$. This yields the correct estimate of μ_{11} as obtainable from the effective constraint. The estimates of the remaining cell means for which $n_{ij} = 0$ are not unique but satisfy the relations given in the G matrix (6.77). The danger with this automated procedure is that the analyst cannot distinguish between unobserved cells for which the means are uniquely estimable and those for which the estimates are arbitrary, subject to the linear relations in G. Thus, for the unconnected case, we recommend analysis by the effective model. (Note that, if $n_{11} = 0$ in this example, there would have been no effective constraint, but μ_{11} would have been uniquely estimable.)

■

EXAMPLE 6.5

In Examples 6.3 and 6.4, the single effective constraint was evident from the incidence matrix in Table 6.2. A more interesting example is given by the incidence matrix in Table 6.3.

TABLE 6.3 Incidence matrix for Example 6.5.

n_{11}	n_{12}	0
n_{21}	0	n_{23}
0	n_{32}	n_{33}

Here, we find no obvious interaction constraints; however, the reduction of the constraint matrix as in (4.47) yields

$$G = \begin{bmatrix} 1 & 0 & 0 & 0 & -1 & 0 & 0 & 1 & -1 \\ 0 & 1 & 0 & 0 & 0 & 0 & -1 & -1 & 1 \\ 0 & 0 & 1 & 0 & 0 & -1 & 1 & 0 & -1 \\ 0 & 0 & 0 & 1 & -1 & -1 & 1 & 1 & -1 \end{bmatrix} \tag{6.85}$$

where the cells are ordered as

$$(13, 22, 31, 11, 12, 21, 23, 32, 33). \tag{6.86}$$

The last row of G gives the effective constraint,

$$\mu_{11} - \mu_{12} - \mu_{21} + \mu_{23} + \mu_{32} - \mu_{33} = 0. \tag{6.87}$$

Examination of the first three rows of (6.85) shows that the model is connected.

■

6.2.2.5. The Randomized Block Design

In the introduction to Chapter 5, we suggested that implicit in the K-factor models is the assumption of a completely randomized design; that is, we can, at least conceptually, think of assigning the various treatments to a homogeneous set of experimental units. We noted that assigning sex and/or race as treatments seemed a bit abstract, but, on the other hand, it was easy to visualize the sex-by-race combinations as populations to be sampled, and hence the model was appropriate.

The possibility of nonhomogeneous experimental units leads us to a new model called the *randomized block design*. To illustrate, suppose we are interested in comparing the effect of three different fertilizers on a particular crop. For the experiment, we have available a field with twelve plots. If the plots were essentially identical, we might assign the plots at random, four per fertilizer, and analyze the results as a one-way classification. Suppose, however, that it is known that there is a general fertility trend across the field. The three northernmost plots are most fertile, and the fertility decreases down to the three southernmost plots. It is assumed that the field is laid out in a 3×4 rectangular array.

Since we do not have similar experimental units with regard to the response—say, yield—it is natural to attempt comparisons of the fertilizers on plots of equal fertility. The field is divided into "blocks," with the three most fertile plots in block 1 and the three least fertile plots in block 4. The three treatments are assigned at random within each block. The experimental layout might appear as in Table 6.4.

TABLE 6.4 Randomized block design.

		Blocks			
		1	2	3	4
	1	F2	F3	F2	F1
Plots	2	F1	F2	F3	F3
	3	F3	F1	F1	F2

Note that this design provides a comparison of each fertilizer at each fertility level.

To develop a model for this experiment, we note that the blocking variable, fertility level, is not conceptually different from the factors, sex and race, in the examples from Chapter 4; that is, we could think of this problem as a two-way classification in which we have available twelve homogeneous plots. The twelve "treatment combinations"—that is, all combinations of fertilizers and fertility levels—are then assigned at random to the plots. The two-way classification model, (6.1), would then be appropriate.

The first problem we encounter with this assumption is that we have only one observation per cell; hence, we obtain no estimate of σ^2. The second observation is that we are probably not interested in the blocking factor. It is introduced only to provide comparisons of fertilizers that would not be affected by basic fertility differences in the plots. The block differences are only a function of this particular field.

The problem of estimating σ^2 may be resolved by assuming that there is no treatment by block interaction—that is, by assuming that the difference in the response of any two fertilizers is the same in any block. We are thus in the no-interaction model with $n_{ij} = n = 1$, and, as noted in Section 6.2.2.1, this does allow an estimate of σ^2.

This illustration is typical of situations in which we have only one observation per cell, but it is logical to assume the constrained model. We emphasize that $n = 1$ does not justify the assumption of no interaction; this must be based on the researcher's prior knowledge of such experiments.

From this discussion, we see that the randomized block design can be analyzed using the results of this chapter even though the physical set-up of the experiment is distinct from the completely randomized design.

An alternative approach to this problem, suggested by the fact that these particular blocks are not of interest, is to argue that we have a simple one-way classification on the three treatments but that the observations within a block are correlated while the observations between blocks are not correlated. This assumption violates the basic assumptions of the simple cell means model in Definition 4.1 but suggests a natural extension to the case where $\text{Var}(Y) = V \neq \sigma^2 I$. We shall return to a discussion of this generalization in Chapter 8 and, in particular, in Example 8.6.

6.2.3. A Canonical Form for the Unconstrained, Two-Factor Model

We saw in Section 6.2.1 that parameter estimation in the unconstrained model is particularly simple and, apart from the use of double subscripts, equivalent to the one-way classification analysis. The constrained model is more complex, and the desire to simplify the presentation and avoid notational problems leads us to the use of canonical forms. Recall that the canonical forms consist of writing

$$\begin{aligned} E(Y) &= X^* \theta^* \\ &= W X_0 P^* \theta^* \\ &= X_c \theta_c \end{aligned} \tag{6.88}$$

where

$$\begin{aligned} X^* &= W X_0 P^* \\ X_c &= W X_0 \\ \theta_c &= P^* \theta^*. \end{aligned} \tag{6.89}$$

The different canonical forms arise by choosing the parameter matrix, P^*. The cell means are related to the parameters by

$$\begin{aligned} \mu &= X_0 P^* \theta^* \\ &= X_0 \theta_c. \end{aligned} \tag{6.90}$$

We shall now extend this concept to the unconstrained model. This model will be useful in our discussion of hypothesis testing in Section 6.3. The reparameterization suggested by the second canonical form is most frequently used and is the only one we shall consider. The reasons for this will be apparent later.

Following (6.45), we define the *marginal means* parameters as follows. (For simplicity, we drop the subscript m on the parameters.)

$$\begin{aligned} \mu &= \bar{\mu}_{..} & & \\ \alpha_i &= \bar{\mu}_{i.} - \bar{\mu}_{..} & i &= 1, \ldots, a-1 \\ \beta_j &= \bar{\mu}_{.j} - \bar{\mu}_{..} & j &= 1, \ldots, b-1 \\ (\alpha\beta)_{ij} &= \mu_{ij} - \bar{\mu}_{i.} - \bar{\mu}_{.j} + \bar{\mu}_{..} & i &= 1, \ldots, a-1 \\ & & j &= 1, \ldots, b-1. \end{aligned} \tag{6.91}$$

Let θ_m be the vector of parameters in (6.91), and write the vector of means as

$$E(Y) = X_m\theta_m. \tag{6.92}$$

The matrix X_m is given by

$$X_m = WX_0P_m \tag{6.93}$$

where W is the counting matrix and X_0 and P_m are the extensions of the design matrix (6.34) and the parameter matrix (6.50). These extensions are given by defining

$$X_0 = (J_{ab}|I_a \otimes J_b|J_a \otimes I_b|I_a \otimes I_b) \tag{6.94}$$

and

$$P_m = \text{Diag}(1, \Delta_a', \Delta_b', \Delta_a' \otimes \Delta_b'). \tag{6.95}$$

Finally, defining

$$\begin{aligned} \alpha_a &= -\sum_{i=1}^{a-1} \alpha_i & & \\ \beta_b &= -\sum_{j=1}^{b-1} \beta_j & & \\ (\alpha\beta)_{ib} &= -\sum_{j=1}^{b-1} (\alpha\beta)_{ij} & i &= 1, \ldots, a-1 \\ (\alpha\beta)_{aj} &= -\sum_{i=1}^{a-1} (\alpha\beta)_{ij} & j &= 1, \ldots, b \end{aligned} \tag{6.96}$$

we may extend θ_m to the vector θ_c by the relation

$$\theta_c = P_m\theta_m \tag{6.97}$$

and write the general canonical model as

$$E(Y) = X_c\theta_c. \tag{6.98}$$

The matrices and parameters in (6.98) are related as in (6.89) and (6.90). Note that the canonical model in Section 6.2.2.2 is a special case of this model for which $(\alpha\beta)_{ij} = 0$ for all i, j.

At this point, it may seem strange to introduce this more complex model as an alternative to the simple cell means model. There are essentially two reasons: the first is for computing efficiency in the development of test statistics, and the second is to introduce the general design matrix X_0, which will be useful when discussing problems with more general error structures. We shall continue to use the cell means model as our basic means of developing concepts and interpreting results.

While either the cell means or the canonical form can be used with all $n_{ij} \neq 0$, the canonical form leads to complications if some of the cells are empty. In the cell means model, we noted that μ_{ij} cannot be estimated in the unconstrained case if $n_{ij} = 0$. The remaining parameters, denoted by μ_0, in the effective model are estimated, as usual, by the sample cell means.

In the canonical model, X_m is no longer of full column rank with empty cells. A common practice is to set $(\alpha\beta)_{ij} = 0$ if $n_{ij} = 0$ and $i \neq a, j \neq b$. This approach will result in the correct estimates, $\hat{\mu}_{ij} = \hat{\mu} + \hat{\alpha}_i + \hat{\beta}_j + \widehat{(\alpha\beta)}_{ij}$ for observed cells. It also will yield the "estimate" $\hat{\mu}_{ij} = \hat{\mu} + \hat{\alpha}_i + \hat{\beta}_j$ for unobserved cells. This estimate is not justified by the data but is a consequence of the fact that the assumption $(\alpha\beta)_{ij} = 0$ is equivalent, in view of (6.91), to adding the constraint $\mu_{ij} = \bar{\mu}_{i\cdot} + \bar{\mu}_{\cdot j} - \bar{\mu}_{\cdot\cdot}$ to the model. The analyst using this procedure must ask whether this constraint is reasonable. We shall see that this practice will have more serious consequences when the canonical model is used for hypothesis testing.

If $n_{ij} = 0$ for $i = a$ or $j = b$, the procedure is of no help. One suggestion, to give X_m full column rank, is to impose

$$\begin{aligned} \sum_{j=1}^{b-1} (\alpha\beta)_{ij} &= 0 \qquad \text{if } n_{ib} = 0 \\ \sum_{i=1}^{a-1} (\alpha\beta)_{ij} &= 0 \qquad \text{if } n_{aj} = 0. \end{aligned} \tag{6.99}$$

Solving for one of the $(\alpha\beta)_{ij}$ with $n_{ij} \neq 0$ and eliminating it will yield the desired full column rank matrix. This procedure is rarely used in practice.

6.3. TESTS OF HYPOTHESES

The more general structure on the means in the two-factor model leads to more interesting hypotheses for our consideration. However, given a statement of the hypothesis, the results of Chapter 4 are applied to develop the test statistic, and no new concepts arise. The complexities introduced in this section arise as we attempt to develop either algebraic expressions for the sums of squares or numerical methods for computing these quantities so as to avoid using the general quadratic forms. Our primary approach will be through model reduction, but we shall use the LaGrange Multiplier method to develop some special results.

There has been some confusion in the literature, and especially in computer packages, as to how one should analyze the two-factor model with unequal cell frequencies. In this chapter, we shall develop an analysis based on the theory in Chapter 4. We shall also describe other methods that have been used and relate them to the general theory.

6.3.1. The Unconstrained Model

In the unconstrained, two-factor model, (6.1), we may consider testing any linear hypothesis on the μ_{ij}. In this section, we provide a detailed discussion of some of the most commonly tested hypotheses. Motivated by our discussion of the no-interaction model, a natural hypothesis to consider in the unconstrained case is whether those constraints might be valid. That is, we consider testing the hypothesis

$$H_{AB}\colon \mu_{ij} - \mu_{kj} = \mu_{it} - \mu_{kt} \qquad \begin{array}{l} i, k = 1, \ldots, a \\ j, t = 1, \ldots, b. \end{array} \tag{6.100}$$

Alternative expressions for this hypothesis are given by (6.6) and (6.9), and matrix expressions are given by (6.7) and (6.11).

To indicate the interpretation of this constraint, we show in Figure 6.1 a typical plot of the cell means μ_{ij} for the case $a = b = 3$. If the constraint, (6.100), is satisfied, the three lines will have parallel segments. A wide variety of situations can exist, ranging from the parallel case through cases where the line segments actually intersect.

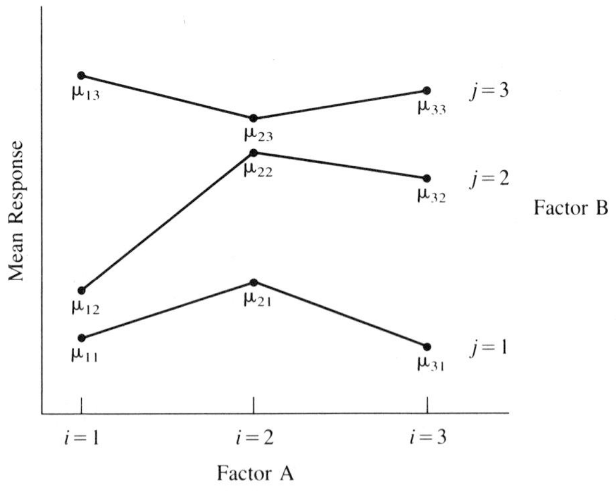

FIGURE 6.1
Plot of cell means for the two-factor model

Usually, the primary question of interest is whether there are differences in the responses to the different levels of the factors. There are many ways that this question can be stated. A natural one is to compare the mean levels of one factor averaged over all levels of the second factor. Thus, for the first factor, the hypothesis is

$$H_A\colon \bar{\mu}_{i\cdot} = \bar{\mu}_{k\cdot} \qquad i, k = 1, \ldots, a \tag{6.101}$$

and, for the second factor,

$$H_B\colon \bar{\mu}_{\cdot j} = \bar{\mu}_{\cdot t} \qquad j, t = 1, \ldots, b. \tag{6.102}$$

It is emphasized that we can test these hypotheses, sometimes called *main effect tests,* quite apart from our conclusions on the interaction hypothesis. The value of our conclusions on the main effect tests will depend on (a) the particular application and (b) the nature of the interaction. Thus, in our sex-by-race example, we may be interested in the relative performance of males and females and recognize that in the use of these results there will be a mixture of races. In this case, comparison of the performance of the two sexes averaged over the races is meaningful. In other applications, it may be appropriate to consider a weighted average based on the relative sizes of the races. If random sampling were used, the relative sizes of the populations could be indicated by the cell frequencies.

The nature of the interaction is also important. Thus, in Figure 6.1, we might observe a significant interaction because of the lack of parallelism, but level three of factor B—which gives the highest average response, in the sense of (6.102)—is actually highest for all levels of factor A. This should be contrasted with the case in which level three might be highest on the average, but, since the line segments intersect, the level of factor B yielding highest response would depend on the level of factor A. In any case, it must be recognized that the testing of general hypotheses—such as H_A, H_B, and H_{AB}—is only a first step in the analysis. Decompositions into subhypotheses and the examination of confidence regions and confidence intervals will provide a more detailed understanding of the problem.

6.3.1.1. Development of Test Statistics: $n_{ij} \neq 0$

We shall develop the test statistics for the main effect and interaction hypotheses for the general case of unequal but nonzero cell frequencies and note the special features of the balanced case, $n_{ij} = n$. Although missing cells caused no difficulty in the estimation problem for the unconstrained model, we shall see that this is a source of confusion in hypothesis testing, and we shall discuss this case separately in Section 6.3.1.7.

Test for interaction. The essential results for developing this test statistic were given in Section 6.2.2, where we examined the model restricted by the assumption of no interaction. Thus, in accordance with the general theory developed in Section 4.3.1, the test statistic is given by

$$N_{AB}/[(a-1)(b-1)s^2] \sim F[(a-1)(b-1), N-ab, \lambda_{AB}] \tag{6.103}$$

where s^2 is given by (6.4). The numerator is given by

$$N_{AB} = Q_{AB}(\tilde{\mu}) - Q(\hat{\mu}) \tag{6.104}$$

where $Q(\hat{\mu})$ is given by (6.5), and $Q_{AB}(\tilde{\mu})$ is the residual sum of squares, given by (6.21) for $n_{ij} = n$ and, for general n_{ij}, by

$$\begin{aligned} Q_{AB}(\tilde{\mu}) &= (Y - W\tilde{\mu})'(Y - W\tilde{\mu}) \\ &= (Y - WX_0\hat{\theta}_c)'(Y - WX_0\hat{\theta}_c). \end{aligned} \tag{6.105}$$

Here, $\tilde{\mu} = X_0\hat{\theta}_c$ from the canonical analysis, and $\tilde{\mu}$ is the constrained estimate of μ that may be computed directly as in Chapter 4 or by using the canonical, no interaction model from Section 6.2.2.2.

Although easily calculated, $Q_{AB}(\tilde{\mu})$ and N_{AB} do not have simple algebraic expressions for unequal cell frequencies. Similarly, the noncentrality parameter, λ_{AB}, given by (3.73) does not have a simple algebraic expression.

It is instructive to consider the special case, $n_{ij} = n$. From (6.5), (6.21), and (6.104) we see that

$$N_{AB} = \sum_i \sum_j \sum_k (\bar{y}_{ij\cdot} - \bar{y}_{i\cdot\cdot} - \bar{y}_{\cdot j\cdot} + \bar{y}_{\cdot\cdot\cdot})^2 \tag{6.106}$$

which, from the general theory, is distributed as $\chi^2[(a - 1)(b - 1), \lambda_{AB}]$ with

$$\lambda_{AB} = n/(2\sigma^2)\sum_i \sum_j (\mu_{ij} - \bar{\mu}_{i\cdot} - \bar{\mu}_{\cdot j} + \bar{\mu}_{\cdot\cdot})^2. \tag{6.107}$$

The degrees of freedom, $(a - 1)(b - 1)$, associated with N_{AB} are given by the rank of the hypothesis matrix associated with H_{AB} given by (6.7).

The results of Example 6.1 in Section 6.2.2.3 also yield a simple expression for N_{AB} in the special case, $a = 2$, but, beyond that, closed form algebraic expressions are not available at this time.

Tests for main effects. The results for testing the row effect hypothesis, H_A, will be developed in detail. The analogous results for the column effect hypothesis, H_B, follow by interchanging subscripts. In this case, explicit expressions for the constrained estimates, the test statistics, and the noncentrality parameters will be given.

The hypothesis, H_A, given in (6.101) can also be written as

$$H_A\colon \bar{\mu}_{i\cdot} - \bar{\mu}_{a\cdot} = 0 \qquad i = 1, \ldots, a - 1. \tag{6.108}$$

The hypothesis matrix is given by

$$H = \Delta_a \otimes (1/b)J_b' \tag{6.109}$$

with Δ_a defined by (6.8). Direct application of the results in Section 4.2.2 gives the estimate of μ, subject to (6.108), as

$$\tilde{\mu} = A\hat{\mu} \tag{6.110}$$

where

$$A = I - (W'W)^{-1}H'[H(W'W)^{-1}H']^{-1}H \tag{6.111}$$

and $\hat{\mu}$ is the unconstrained estimator.

Following the notation developed in Example 6.1, Section 6.2.2.3, we denote the harmonic mean of the cell frequencies in a given row of the incidence matrix as h_{r_i}, $i = 1, \ldots, a$, where

$$(1/h_{r_i}) = (1/b)\sum_j (1/n_{ij}). \tag{6.112}$$

Then, we may show that

$$bH(W'W)^{-1}H' = D_{a-1}^{-1}(h_{r_i}) + (1/h_{r_a})J_{a-1}J'_{a-1} \tag{6.113}$$

with inverse

$$(1/b)[H(W'W)^{-1}H']^{-1} = D_{a-1}(h_{r_i}) - (1/h_{r\cdot})C_{a-1}(h_{r_i})R_{a-1}(h_{r_i}). \tag{6.114}$$

Here

$$h_{r\cdot} = \sum_i h_{r_i}. \tag{6.115}$$

Substituting (6.114) into (6.110) and (6.111) and simplifying yields

$$\tilde{\mu}_{ij} = \bar{y}_{ij\cdot} - [h_{r_i}/(bn_{ij})]\left(\sum_k \bar{y}_{ik\cdot} - \bar{y}_r\right) \tag{6.116}$$

where

$$\bar{y}_r = \sum_i (h_{r_i}/h_{r\cdot})\sum_j \bar{y}_{ij\cdot}. \tag{6.117}$$

The residual sum of squares, under H_A, is then given by

$$\begin{aligned} Q_A(\tilde{\mu}) &= \sum_i \sum_j \sum_k (y_{ijk} - \tilde{\mu}_{ij})^2 \\ &= Q(\hat{\mu}) + \sum_i (h_{r_i}/b)\left(\sum_j \bar{y}_{ij\cdot} - \bar{y}_r\right)^2. \end{aligned} \tag{6.118}$$

Thus, for testing H_A, Theorem 4.5 yields the test statistic

$$N_A/[(a - 1)s^2] \sim F(a - 1, N - ab, \lambda_A) \tag{6.119}$$

where

$$N_A = \sum_i (h_{r_i}/b)\left(\sum_j \bar{y}_{ij\cdot} - \bar{y}_r\right)^2. \tag{6.120}$$

The degrees of freedom, $(a - 1)$, are obtained as the rank of H in (6.109). The noncentrality parameter is given by (3.73) as

$$\lambda_A = b\sum_i h_{r_i}(\bar{\mu}_{i\cdot} - \bar{\mu}_r)^2/(2\sigma^2) \tag{6.121}$$

where

$$\bar{\mu}_r = \sum_i h_{r_i}\bar{\mu}_{i\cdot}/h_{r\cdot}. \tag{6.122}$$

The analogous results for the column hypothesis, H_B, are

$$\bar{\bar{\mu}}_{ij} = \bar{y}_{ij\cdot} - [h_{c_j}/(an_{ij})]\left(\sum_k \bar{y}_{kj\cdot} - \bar{y}_c\right) \tag{6.123}$$

$$N_B = \sum_j (h_{c_j}/a)\left(\sum_i \bar{y}_{ij\cdot} - \bar{y}_c\right)^2 \tag{6.124}$$

$$\lambda_B = a\sum_j h_{c_j}(\bar{\mu}_{\cdot j} - \bar{\mu}_c)^2/(2\sigma^2) \tag{6.125}$$

where

$$(1/h_{c_j}) = (1/a)\sum_i (1/n_{ij}) \tag{6.126}$$

$$h_{c\cdot} = \sum_j h_{c_j} \tag{6.127}$$

$$\bar{y}_c = \sum_j (h_{c_j}/h_{c\cdot})\sum_i \bar{y}_{ij\cdot} \tag{6.128}$$

and

$$\bar{\mu}_c = \sum_j h_{c_j}\bar{\mu}_{\cdot j}/h_{c\cdot}. \tag{6.129}$$

The test statistic for H_B is

$$N_B/[(b-1)s^2] \sim F(b-1, N-ab, \lambda_B). \tag{6.130}$$

The results of this analysis are summarized in the Analysis of Means Table 6.5. For reference, we shall refer to this as the *marginal means analysis*. This terminology, which is fairly common, reflects the nature of the hypotheses, H_A and H_B.

TABLE 6.5 Analysis of means table for the two-way classification model (marginal means hypotheses).

Description	*Hypothesis*	*df*	*SS*	*MS*	*EMS*
Row differences	H_A	$a-1$	N_A	$\dfrac{N_A}{(a-1)}$	$\sigma^2 + \dfrac{2\sigma^2\lambda_A}{(a-1)}$
Column differences	H_B	$b-1$	N_B	$\dfrac{N_B}{(b-1)}$	$\sigma^2 + \dfrac{2\sigma^2\lambda_B}{(b-1)}$
Interaction	H_{AB}	$(a-1)(b-1)$	N_{AB}	$\dfrac{N_{AB}}{(a-1)(b-1)}$	$\sigma^2 + \dfrac{2\sigma^2\lambda_{AB}}{(a-1)(b-1)}$
Residual	———	$N-ab$	$Q(\hat{\mu})$	s^2	σ^2

For the special case of equal cell frequencies, the results follow by setting $n_{ij} = n$ in Table 6.5. For the record, the essential results are as follows.

Under H_A

$$\tilde{\mu}_{ij} = \bar{y}_{ij\cdot} - \bar{y}_{i\cdot\cdot} + \bar{y}_{\cdot\cdot\cdot} \tag{6.131}$$

$$N_A = bn \sum_i (\bar{y}_{i\cdot\cdot} - \bar{y}_{\cdot\cdot\cdot})^2 \tag{6.132}$$

$$\lambda_A = bn \sum_i (\bar{\mu}_{i\cdot} - \bar{\mu}_{\cdot\cdot})^2/(2\sigma^2). \tag{6.133}$$

Under H_B

$$\tilde{\mu}_{ij} = \bar{y}_{ij\cdot} - \bar{y}_{\cdot j\cdot} + \bar{y}_{\cdot\cdot\cdot} \tag{6.134}$$

$$N_B = an \sum_j (\bar{y}_{\cdot j\cdot} - \bar{y}_{\cdot\cdot\cdot})^2 \tag{6.135}$$

$$\lambda_B = an \sum_j (\bar{\mu}_{\cdot j} - \bar{\mu}_{\cdot\cdot})^2/(2\sigma^2). \tag{6.136}$$

Under H_{AB}

$$\tilde{\mu}_{ij} = \bar{y}_{i\cdot\cdot} + \bar{y}_{\cdot j\cdot} - \bar{y}_{\cdot\cdot\cdot} \tag{6.137}$$

$$N_{AB} = n \sum_i \sum_j (\bar{y}_{ij\cdot} - \bar{y}_{i\cdot\cdot} - \bar{y}_{\cdot j\cdot} + \bar{y}_{\cdot\cdot\cdot})^2 \tag{6.138}$$

$$\lambda_{AB} = n \sum_i \sum_j (\mu_{ij} - \bar{\mu}_{i\cdot} - \bar{\mu}_{\cdot j} + \bar{\mu}_{\cdot\cdot})^2/(2\sigma^2). \tag{6.139}$$

A useful alternative expression for these quadratic forms, in the case $n_{ij} = n$, may be obtained by utilizing the matrices described in the canonical forms in Section 6.2.3. From the expression for X_0 in (6.94) define

$$\begin{aligned} X_A &= W(I_a \otimes J_b) \\ X_B &= W(J_a \otimes I_b) \\ X_{AB} &= W. \end{aligned} \tag{6.140}$$

Then we may write

$$\begin{aligned} N_A &= Y'(X_A X_A' - J_N J_N'/a)Y/bn \\ N_B &= Y'(X_B X_B' - J_N J_N'/b)Y/an \\ N_{AB} &= Y'(X_{AB} X_{AB}' - X_A X_A'/b - X_B X_B'/a + J_N J_N'/ab)Y/n \\ Q(\hat{\mu}) &= Y'(I_N - WW'/n)Y. \end{aligned} \tag{6.141}$$

Note that the sums of squares in (6.141) add up to the total sum of squares,

$$\begin{aligned} T &= \sum_i \sum_j \sum_k (y_{ijk} - \bar{y}_{\cdot\cdot\cdot})^2 \\ &= Y'(I - J_{ab} J_{ab}'/abn)Y. \end{aligned} \tag{6.142}$$

Hence, Theorem 2.6 shows that N_A, N_B, and N_{AB} are independent chi-square variates. In the general case, $n_{ij} \neq n$, they are not independent, but the distribution of the test statistics is not affected. In either case, the problem of multiple tests and associated significance levels, discussed in Section 3.3.3, should be considered.

The expected mean squares in Table 6.5 are convenient reminders of the appropriate ratios to use for testing these three hypotheses.

6.3.1.2. Computational Considerations

The computations required for developing the numerator sums of squares in Table 6.5 may be done efficiently in terms of the canonical model described in Section 6.2.3. Note that setting $(\alpha\beta)_{ij} = 0$ in that model yields the canonical model for the no-interaction situation. The difference in the residual sum of squares for these two models gives N_{AB}. In that same spirit, note from (6.91) that the hypothesis H_A is equivalent to the condition $\alpha_i = 0$, $i = 1, \ldots, a - 1$ in the canonical model, (6.92). Thus, fitting that model with $\alpha_i = 0$, $i = 1, \ldots, a - 1$ yields the residual sum of squares $Q_A(\tilde{\mu})$ in (6.118), and the test statistic N_A follows as usual. Similarly, setting $\beta_j = 0$, $j = 1, \ldots, b - 1$ in (6.92) will yield $Q_B(\tilde{\mu})$ and, hence, N_B. These computations are conveniently performed using the Sweep Operator described in Appendix A.I.10.

To describe the procedure, recall the canonical form for the unconstrained model, given in Section 6.2.3, with parameter vector θ_m defined by (6.91) and coefficient matrix X_m defined in (6.93). We write X_m with the natural partitioning as

$$X_m = (J_N | X_{m_A} | X_{m_B} | X_{m_{AB}}) \tag{6.143}$$

where

$$\begin{aligned} X_{m_A} &= W(I_a \otimes J_b)\Delta'_a \\ X_{m_B} &= W(J_a \otimes I_b)\Delta'_b \\ X_{m_{AB}} &= W(I_a \otimes I_b)(\Delta'_a \otimes \Delta'_b). \end{aligned} \tag{6.144}$$

The essential quantities are described in terms of the matrix

$$S = \begin{bmatrix} X'_m X_m & X'_m Y \\ Y' X_m & Y' Y \end{bmatrix}. \tag{6.145}$$

After "sweeping" on all but the last column, we have

$$S_F = \begin{bmatrix} (X'_m X_m)^{-1} & \hat{\theta}_m \\ -\hat{\theta}'_m & Q(\hat{\theta}_m) \end{bmatrix} \tag{6.146}$$

where $\hat{\theta}_m$ is the vector of estimates of θ_m in the unconstrained model, and $Q(\hat{\theta}_m)$ is the residual sum of squares. Sweeping again on any column excludes the corresponding variable from the model. Thus, if we sweep on the $(a - 1)(b - 1)$ columns associated with $(\alpha\beta)_{ij}$, we obtain the matrix, denoted by S_{AB}, corresponding to the fit of the no-interaction model. The residual sum of squares, Q_{AB}, is found in the lower right-hand corner. The estimates of μ, α_i, and β_j are found in the first $a + b - 1$ components of the last column of S_{AB}. The numerator sum of squares for testing H_{AB} is determined as usual by (6.104).

Returning to (6.146) and sweeping on the $(a - 1)$ columns associated with α_i, we obtain the matrix S_A. The lower right-hand corner of this matrix is Q_A, and the constrained estimates of θ_m are found in the last column, excluding rows 2 through a. Similarly, starting with (6.146) and sweeping on the $(b - 1)$ columns associated with β_j will yield the residual sum of squares, Q_B, and the corresponding vector of estimates. The numerator sums of squares for testing H_A and H_B are then obtained as usual.

We now see the value of the canonical form for the unconstrained model. It is primarily a device for simplifying the computations associated with hypothesis testing. We emphasize that the canonical form described in Section 6.2.3 is designed for testing the main effect and interaction hypotheses as we have defined them. The same computational procedure applied to a different canonical form will develop the test statistics for different hypotheses.

To illustrate this point, suppose we extend the first canonical form of Section 6.2.2.2 to the unconstrained model by defining

$$(\alpha\beta)_{r_{ij}} = \mu_{ij} - \mu_{ib} - \mu_{aj} + \mu_{ab} \qquad \begin{array}{l} i = 1, \ldots, a-1 \\ j = 1, \ldots, b-1. \end{array} \tag{6.147}$$

The corresponding vector, θ_r, consists of μ_{ab}, α_{r_i}, $i = 1, \ldots, a-1$; β_{r_j}, $j = 1, \ldots, b-1$; and $(\alpha\beta)_{r_{ij}}$ as defined in (6.147). The coefficient matrix is given by

$$X_r = WX_0P_r \tag{6.148}$$

where X_0 is defined in (6.94), and P_r is defined to be

$$P_r = \text{Diag}\left(1, \begin{bmatrix} I_{a-1} \\ 0 \end{bmatrix}, \begin{bmatrix} I_{b-1} \\ 0 \end{bmatrix}, \begin{bmatrix} I_{a-1} \\ 0 \end{bmatrix} \otimes \begin{bmatrix} I_{b-1} \\ 0 \end{bmatrix}\right). \tag{6.149}$$

If we apply the above procedure to the matrix S defined in terms of X_r, rather than X_m, we find that the sum of squares for interaction is correct, but the sums of squares for the row and column effects are appropriate for testing the hypotheses

$$\begin{array}{ll} \mu_{ib} = \mu_{ab} & i = 1, \ldots, a-1 \\ \mu_{aj} = \mu_{ab} & j = 1, \ldots, b-1 \end{array} \tag{6.150}$$

respectively. While these are perfectly valid hypotheses, it is not likely that they would be considered as general measures of row and column effects.

The point we wish to stress is that computational devices such as the one described in this section are not to be used blindly. The hypothesis of interest should be formulated, and then an appropriate computing method should be applied to develop the test statistic.

6.3.1.3. The R-Notation

The computational procedure described in Section 6.3.1.2 suggests a useful notation for describing the numerator sums of squares for testing certain hypotheses. In particular, this procedure is appropriate for hypotheses that correspond to, setting to zero, subsets of the coefficients in a particular canonical model. The notation will be described for the canonical form of the two-factor model, but it will be extended to more complex models in Chapter 7.

Let

$$R[\mu, \alpha, \beta, (\alpha\beta)] = Y'Y - Q(\hat{\theta}) \tag{6.151}$$

denote the *regression sum of squares* for fitting the unconstrained model. Let

$$R(\mu, \alpha, \beta) = Y'Y - Q_{AB}(\hat{\theta}) \tag{6.152}$$

$$R[\mu, \beta, (\alpha\beta)] = Y'Y - Q_A(\hat{\theta}) \tag{6.153}$$

$$R[\mu, \alpha, (\alpha\beta)] = Y'Y - Q_B(\hat{\theta}) \tag{6.154}$$

denote the regression sums of squares for fitting the constrained models with, respectively, $(\alpha\beta)_{ij} = 0$, $i = 1, \ldots, (a - 1)$, $j = 1, \ldots, (b - 1)$; $\alpha_i = 0$, $i = 1, \ldots, (a - 1)$; and $\beta_j = 0$, $j = 1, \ldots, (b - 1)$. The difference between (6.151) and any of (6.152) through (6.154) will be denoted by

$$R(\theta_1 \mid \theta_2) = R(\theta_1, \theta_2) - R(\theta_2). \tag{6.155}$$

Here, θ_1 and θ_2 represent subsets of the complete set of parameters, θ. The numerator sums of squares for testing row, column, and interaction effects are then denoted by

$$N_A = R[\alpha \mid \mu, \beta, (\alpha\beta)] \tag{6.156}$$

$$N_B = R[\beta \mid \mu, \alpha, (\alpha\beta)] \tag{6.157}$$

$$N_{AB} = R[(\alpha\beta) \mid \mu, \alpha, \beta]. \tag{6.158}$$

This notation must be used with care since the hypotheses associated with these sums of squares depend on the particular canonical form used. Thus, for example, if we use the second canonical form, described in Section 6.2.3, then $R(\alpha \mid \mu, \beta, \alpha\beta)$ is appropriate for testing H_A: $\mu_{i\cdot} = \mu_{a\cdot}$, $i = 1, \ldots, (a - 1)$. On the other hand, if the first canonical form, described in the previous section, is used, $R(\alpha \mid \mu, \beta, \alpha\beta)$ is appropriate for testing the hypothesis H_A: $\mu_{ib} = \mu_{ab}$, $i = 1, \ldots, a - 1$. The user of standard computer programs, which are based on canonical models, should be aware of the choice of reparameterization since the sums of squares are typically generated by the method of Section 6.3.1.2. [For additional comments on the R-notation, see Speed and Hocking (1976).]

6.3.1.4. Other Row and Column Hypotheses

The researcher should feel free to test any hypotheses of interest and should not be restricted by the available computer programs. If standard programs are used, however, it is imperative that the output include a specification of the hypotheses associated with the sums of squares displayed in the Analysis of Means table. We have seen two possibilities in Section 6.3.1.2, depending on the choice of canonical form. Another procedure that is commonly used is based on a sequential approach.

In this *sequential method,* we first test for interaction. If this test is not significant at some specified level—say, $\alpha = 0.25$—the model is assumed to satisfy the no-interaction constraints. (Bozivich, Bancroft, and Hartley, 1956, made this suggestion for the case of the random model. It seems reasonable to apply it to the fixed effects model.) The column effect sum of squares is then defined by

$$R(\beta \mid \mu, \alpha) = R(\mu, \alpha, \beta) - R(\mu, \alpha). \tag{6.159}$$

If this test is not significant, the model is further simplified to a one-way classification on the row factor. The row-effect sum of squares is then defined by

$$R(\alpha \mid \mu) = R(\mu, \alpha) - R(\mu). \tag{6.160}$$

These sums of squares are then displayed in an AOM table such as that shown in Table 6.6.

TABLE 6.6 Analysis of means for the sequential sums of squares method.

Hypothesis	*df*	*Sums of Squares*
H_A^{**}	$a - 1$	$R(\alpha \mid \mu)$
H_B^*	$b - 1$	$R(\beta \mid \mu, \alpha)$
H_{AB}	$(a - 1)(b - 1)$	$R[(\alpha\beta) \mid (\mu, \alpha, \beta)]$
Residual	$N - ab$	$Q(\hat{\mu})$

The sums of squares in Table 6.6 are easily generated by the sweep method, as in Section 6.3.1.2. Thus, after having swept out the $(\alpha\beta)$ columns to obtain S_{AB} and, hence, Q_{AB} and N_{AB}, we would not return to S_F; rather, we would continue to sweep out the β columns. This would yield $R(\beta \mid \mu, \alpha)$, as in (6.159). Finally, the α columns would be swept out to obtain $R(\alpha \mid \mu)$, as in (6.160). It is of interest to ask what hypotheses correspond to these sums of squares. The interaction sum of squares properly tests H_{AB} as defined in (6.6). The hypotheses tested by the other sums of squares depend on whether the change in models is justified. Thus, $R(\beta \mid \mu, \alpha)$ will test for column effects in the no-interaction model. This hypothesis will be developed in Section 6.3.2. If the further reduction is justified, $R(\alpha \mid \mu)$ will test the one-way hypothesis on row effects discussed in Chapter 5.

The practice of changing the model based on the data is certainly questionable and has been the subject of much debate. If the model is changed, then it is natural to consider modifying the denominator sum of squares as well; that is, if it is judged that the no-interaction assumption is tenable, then the interaction and residual sum of squares in Table 6.6 might be pooled as in the no-interaction model, Section 6.2.2, to form a new estimate of σ^2.

Frequently, Table 6.6 is simply used as an alternative to Table 6.5, with the first two rows being measures of row and column effects without being more specific as to the hypotheses being tested in the unconstrained model. Hocking, Hackney, and Speed (1978) have shown that these hypotheses are, respectively,

$$H_A^{**}: \sum_j n_{ij}\mu_{ij}/n_{i\cdot} = \sum_j n_{aj}\mu_{aj}/n_{a\cdot} \qquad i = 1, \ldots, (a-1) \qquad (6.161)$$

and

$$H_B^{*}: \sum_i n_{ij}\mu_{ij} = \sum_i \sum_k n_{ij}n_{ik}\mu_{ik}/n_{i\cdot} \qquad j = 1, \ldots, (b-1). \qquad (6.162)$$

We note that, if $n_{ij} = n$, then H_a^{**} is equivalent to H_A, and H_B^* is equivalent to H_B. Thus, these two hypotheses represent generalizations of the hypotheses tested with equal cell frequencies. The researcher should ask whether hypotheses involving the cell frequencies are appropriate and, if so, whether either of these expressions seems reasonable.

Since the labeling of the two factors is arbitrary, it is natural to ask why the row and column hypotheses should differ. For this reason, some analysts recommend a partially sequential procedure in which the row effect sum of squares is computed by

$$R(\alpha|\mu, \beta) = R(\mu, \alpha, \beta) - R(\mu, \beta). \qquad (6.163)$$

The associated hypothesis is given as an analog of H_B^* with subscripts reversed; that is,

$$H_A^{*}: \sum_j n_{ij}\mu_{ij} = \sum_j \sum_k n_{ij}n_{kj}\mu_{kj}/n_{\cdot j} \qquad i = 1, \ldots, a-1. \qquad (6.164)$$

For reference, we summarize this partially sequential approach in an AOM table as shown in Table 6.7. This analysis is often referred to as the *experimental design method* or the *method of fitting constants*. However, we shall try to avoid using such names for methods that are not well defined; we prefer to specify the hypotheses being tested. The researcher can then decide which, if any, of these hypotheses should be tested. (For a complete discussion of these and other methods and their relation to existing computer packages, see Speed and Hocking, 1976, and Francis, 1973.)

TABLE 6.7 Analysis of means for the partially sequential method.

Hypothesis	*df*	*Sums of Squares*
H_A^*	$a-1$	$R(\alpha\|\mu, \beta)$
H_B^*	$b-1$	$R(\beta\|\mu, \alpha)$
H_{AB}	$(a-1)(b-1)$	$R(\alpha\beta\|\mu, \alpha, \beta)$
Residual	$N-ab$	$Q(\hat{\mu})$

As noted, Tables 6.6 and 6.7 are identical to Table 6.5 for the case of equal cell frequencies. The philosophy in Table 6.5 is that the hypotheses of interest—namely, H_A and H_B—should not change because of the unequal cell frequencies. Tables 6.6 and 6.7 offer two different sets of hypotheses that might be considered when the cell frequencies are reflective of the population sizes. Elston and Bush (1964) suggest additional hypotheses that might be considered as measures of main effects.

One of the currently popular computer packages, SAS (Statistical Analysis System, Barr et al., 1976), offers all of these analyses under the General Linear Models procedure (PROC GLM). Thus, SAS Type I corresponds to the sequential method in Table 6.6, SAS Type II is the partially sequential method in Table 6.7, and SAS Type III agrees with the marginal means analysis described in Table 6.5. While the computations are not performed as described here, the results are the same. Further, PROC GLM provides the user with a numerical statement of the hypotheses tested by the three methods. (Of course, the algebraic statements given here are more easily interpreted for this simple two-way situation.) For the missing cell problem, SAS offers a fourth analysis, Type IV, which we shall not discuss (see Freund, 1980).

The philosophy of the BMD package (Dixon, 1977) is to use the marginal means analysis and the effective hypotheses with missing cells.

As more complex models are considered, the number of potential analyses increases rapidly. Again, we emphasize that the analysis should be selected by the researcher and not left to the vagaries of a particular computer program.

6.3.1.5. The Power of Competing Tests

The question of which test statistic to use when testing for main effects was raised in Sections 6.3.1.1 and 6.3.1.4 where we observed that there are several different formulations of the hypotheses referred to as the *main effect hypotheses*. The appropriate likelihood ratio statistics were given for each of the hypotheses. The choice of which, if any, of these hypotheses to use must be based on the needs of the researcher, but, once the choice is made, the test statistic is determined.

In general, it makes no sense to ask which procedure is "best." Once the hypothesis is specified, the associated test statistic is best for that hypothesis in the sense of likelihood ratio tests. Therefore, arguments as to which procedure is best must be based on the choice of hypothesis. It is instructive to attempt comparisons of special cases in which the parameters may satisfy, or nearly satisfy, certain relations. We shall describe one such situation that may shed some light on the problem.

Suppose the unconstrained model is assumed but, in fact, the no-interaction constraint is satisfied; that is, the researcher has no prior knowledge to justify any constraints on the model, but the relation (6.2) actually holds. In this case, it can be shown that both H_A and H_A^* reduce to the simpler hypothesis

$$H_{A_0}\colon \mu_{ij} = \mu_{aj} \qquad i = 1, \ldots, a-1 \text{ for any } j = 1, \ldots, b. \tag{6.165}$$

(Note that H_a^{**} is not equivalent to (6.165) under the no-interaction constraint.)

Since both hypotheses are equivalent in this case, the decision on the choice of numerator sum of squares can now be based on the power of the competing tests. We shall show that the test based on $R(\alpha|\mu, \beta)$ is uniformly more powerful than the one based on N_A with the analogous result for the column hypotheses. This fact is intuitively plausible since $R(\alpha|\mu, \beta)$ will be shown to be the correct numerator for the likelihood ratio test in the unconstrained model in Section 6.3.2.

To establish this result and, in fact, a more general one, we note from Hocking, Hackney, and Speed (1978) that both H_A and H_A^* are special cases of the general row effect hypothesis

$$H_A(c)\colon \sum_j c_{ij}\mu_{ij} = \sum_j \sum_k c_{ij}c_{kj}\mu_{kj}/c_{.j} \qquad i = 1, \ldots, a-1. \tag{6.166}$$

Note that H_A is the special case of $H_A(c)$ with $c_{ij} = 1$ for all i, j, and H_A^* is the special case in which $c_{ij} = n_{ij}$ for all i, j. It can be shown that $H_A(c)$ is equivalent to H_{A_0}, under the no-interaction constraint, for $c_{ij} = 1$, for $c_{ij} = n_{ij}$, and for any other choice of c_{ij} such that the hypothesis matrix has rank $a - 1$.

To compare the power of the tests using $R(\alpha|\mu, \beta)$ or N_A, we shall develop the noncentrality parameter for the general hypothesis $H_A(c)$ and then compare the noncentralities under the no-interaction constraint. The comparison of power by comparing the magnitudes of noncentralities is based on the results of Ghosh (1973) as summarized in Section 2.3.3.

The noncentrality parameter associated with the hypothesis $H_A(c)$ is denoted by $\lambda(c)$ and is defined by

$$\lambda(c) = [H(c)\mu]'[H(c)(W'W)^{-1}H(c)']^{-1}H(c)\mu/(2\sigma^2) \tag{6.167}$$

where $H(c)$ denotes the hypothesis matrix for $H_A(c)$. The essential result of this section is summarized in the next theorem. (See also Burdick and Herr, 1980).

THEOREM 6.2
For $c_{ij} = 1$ and $c_{ij} = n_{ij}$, the hypothesis $H_A(c)$ is equivalent to the hypothesis H_{A_0} when the means satisfy (6.2). Further, when (6.2) holds, we have

$$\lambda(1) \leq \lambda(n).$$

Proof: The details of this proof are somewhat tedious, so we shall merely sketch the essential ideas. To do so, let

$$\begin{aligned} M(c) &= H(c)(W'W)^{-1}W' \\ P(c) &= M(c)'[M(c)M(c)']^{-1}M(c). \end{aligned} \tag{6.168}$$

Then (6.167) may be written as

$$\begin{aligned} \lambda(c) &= \mu'W'P(c)W\mu/(2\sigma^2) \\ &= \|P(c)W\mu\|^2/(2\sigma^2). \end{aligned} \tag{6.169}$$

The second form for $\lambda(c)$ in (6.169) follows since $P(c)$ is idempotent and it conveniently expresses $\lambda(c)$ as a function of the squared length of the vector $P(c)W\mu$. (We use the notation $\|a\|^2 = a'a$ to denote the squared length of the vector, a.) The essential step in the proof is to show that, when (6.2) holds, then

$$P(c)P(n)W\mu = P(c)W\mu \tag{6.170}$$

where $P(n)$ denotes $P(c)$ evaluated at $c_{ij} = n_{ij}$. Now the triangle equality allows us to write

$$\|P(c)P(n)W\mu\|^2 + \|[I - P(c)]P(n)W\mu\|^2 = \|P(n)W\mu\|^2 \tag{6.171}$$

from which we have the inequality

$$\|P(c)P(n)W\mu\|^2 \leq \|P(n)W\mu\|^2. \tag{6.172}$$

Thus, when (6.2) holds, we use (6.170) to obtain

$$\|P(c)W\mu\|^2 \leq \|P(n)W\mu\|^2. \tag{6.173}$$

Letting $\lambda_0(c)$ denote the noncentrality parameter when (6.2) holds, (6.173) establishes that

$$\lambda_0(c) \leq \lambda_0(n) \tag{6.174}$$

for any c_{ij} such that $H(c)$ has rank $a - 1$ and, in particular, for $c_{ij} = 1$ for all i, j.

■

This theorem leaves us with a dilemma. If (6.2) is actually true, then we should use $R(\alpha|\mu, \beta)$, rather than N_A, in the numerator of our test statistic for testing the hypothesis H_A in (6.101). On the other hand, if (6.2) is false, then N_A may be more appropriate on two counts. First, the probability of rejecting H_A, when it is true, may be substantially greater than the desired value—say, 0.05—if $R(\alpha|\mu, \beta)$ is used. Second, the probability of rejecting H_A, when it is false, may be less if $R(\alpha|\mu, \beta)$ is used than if N_A is used. Clearly, these results depend on the degree of imbalance in the cell frequencies. The extreme differences in the size and power of the tests occur with extreme differences in the n_{ij}.

A discussion of the special case $a = b = 2$ is given by Hocking and Speed (1980) and by Littell and Lynch (1983). Gosslee and Lucas (1965) give some limited power computations.

As one might expect, $R(\alpha|\mu, \beta)$ yields a more powerful test than N_A with small bias if (6.2) is "nearly" true, but this is difficult to quantify. In any case, specific recommendations on which numerator to use will depend on a knowledge of the interaction contrasts. Since this will rarely be available, it seems safer to use N_A

(and N_B for column effects) since it does yield an unbiased test. The loss of power may be a less serious error.

6.3.1.6. Main Effect Tests in the Unconstrained Model

In the introduction to Section 6.3.1, we discussed the interpretation of hypotheses such as H_A and H_B in the unconstrained model. In Section 6.3.1.4, we encountered hypotheses such as H_A^* and H_a^{**} that evolved through a sequential approach to the problem. In view of the potential for disagreement with regard to the Analysis of Means table that provides our preliminary analysis, we shall make a few summary comments.

At the heart of the problem is the way in which we react to the test for interaction. The options are as follows:

I. H_{AB} is *significant.*
  a. Do not perform main effect tests.
  b. Consider main effect hypotheses.

II. H_{AB} is *not significant.*
  a. Start over assuming the no-interaction model.
  b. Continue with the original model.

Some statisticians adhere to the philosophy that we should not test for main effects in the presence of interaction. Thus, if H_{AB} is significant, they would choose Option (I.a.) and draw no general conclusions about main effects. Of course, they might examine specific contrasts or differences. There is nothing to prevent us from testing main effects, and situations such as Figure 6.1 indicate that these tests can be useful. With severe interactions, main effect tests may not indicate differences even though there may be important information in the experiment. Thus, whether we choose Option (I.a.) or (I.b.), we should continue with a detailed analysis such as that described in Section 4.4. The details for this analysis are developed in the next section.

Option (II.a.) appears to be at the center of the controversy. It is this situation that leads to the hypothesis H_A^*. It is not really this hypothesis that is entertained, but, rather, the hypothesis H_{A_0} given in (6.165). $R(\alpha \mid \mu, \beta)$ is the correct numerator sum of squares for this hypothesis in the no-interaction model. If we go all the way with the model change, we should pool the residual and interaction sums of squares to estimate σ^2 as in Section 6.2.2.3. As a compromise, some analysts recommend using $R(\alpha \mid \mu, \beta)$ for the numerator but keep the original residual sum of squares, $Q(\hat{\mu})$, for the denominator. The results are interpreted as if the hypotheses were H_{A_0}.

Changing the model based on the data is at best a questionable practice. It seems that the best procedure is to stay with the original model but possibly entertain the stronger conclusion implied by H_{A_0}. The choice of N_A or $R(\alpha \mid \mu, \beta)$ for the numerator might be based on the level of significance of the test for interaction. The discussion in Section 6.3.1.5 suggests that this might be reasonable, but general guidelines for the decision are not too satisfying. (See Bozivich, Bancroft, and Hartley, 1956.)

Finally, we note that, if the unconstrained model is used when the constraint is valid, the estimates of the μ_{ij} are unbiased but do not have minimum variance. Conversely, if the constrained model is used when the constraint is not valid, the estimates are biased but have smaller variance.

6.3.1.7. Confidence Intervals

In Section 4.4, we described general expressions for simultaneous confidence intervals using the Scheffé, Tukey, and Bonferroni methods. In this section, we will spell out these intervals for the two-factor model.

As noted, the Scheffé intervals depend on the hypothesis matrix used to generate them. For illustration, we will develop four sets of intervals of decreasing generality. The four hypothesis matrices to be used are

$$\begin{aligned} H_1 &= I_{ab} & H_2 &= \Delta_a \otimes I_b \\ H_3 &= \Delta_a \otimes (1/b)J_b' & H_4 &= \Delta_a \otimes \Delta_b. \end{aligned} \tag{6.175}$$

Note that H_1 corresponds to the hypothesis $\mu = 0$; H_2 corresponds to the hypothesis $\mu_{ij} = \mu_{aj}$, $i = 1, \ldots, (a - 1), j = 1, \ldots, b$; H_3 corresponds to H_A, the main effect hypothesis, (6.101); and H_4 corresponds to H_{AB}, the interaction hypothesis, (6.100).

The set of intervals determined by H_1 is

$$c'\mu\colon c'\hat{\mu} \pm [abF(\alpha; ab, N - ab)s^2 \sum_i \sum_j c_{ij}^2/n_{ij}]^{1/2} \tag{6.176}$$

where $c' = (c_{ij}, i = 1, \ldots, a, j = 1, \ldots, b)$. Note that these linear functions include (1) individual cell means, μ_{ij}, (2) differences of the type $\mu_{ij} - \mu_{kj}$ or $\mu_{ij} - \mu_{it}$, which we might call *simple effects,* (3) measures of main effects, $\bar{\mu}_{i\cdot} - \bar{\mu}_{a\cdot}$ or $\bar{\mu}_{\cdot j} - \bar{\mu}_{\cdot b}$, and (4) measures of interaction, $\mu_{ij} - \mu_{kj} - \mu_{it} + \mu_{kt}$.

A more restrictive set of intervals may be constructed using H_2 in (6.175), which has rank $b(a - 1)$. The intervals are given by

$$c'H\mu\colon c'H\hat{\mu} \pm \{b(a - 1)F[\alpha; b(a - 1), N - ab]s^2c'H(W'W)^{-1}H'c\}^{1/2}. \tag{6.177}$$

In this expression let $c' = (c_{ij}, i = 1, \ldots, (a - 1), j = 1, \ldots, b)$, and note that

$$c'H\mu = \sum_{i=1}^{a-1} \sum_{j=1}^{b} c_{ij}(\mu_{ij} - \mu_{aj}) \tag{6.178}$$

and

$$c'H(W'W)^{-1}H'c = \sum_{i=1}^{a-1} \sum_{j=1}^{b} c_{ij}^2/n_{ij} + \sum_j c_{\cdot j}^2/n_{aj}. \tag{6.179}$$

The linear functions, (6.178), allow us to examine (1) the simple effects, $\mu_{ij} - \mu_{kj}$, (2) the row main effects, $\bar{\mu}_{i\cdot} - \bar{\mu}_{k\cdot}$, and (3) the measures of interaction. For the simple effects, set $c_{ij} = 1$, set $c_{kj} = -1$, and set all other components of c to zero. Then (6.179) becomes $(1/n_{ij} + 1/n_{kj})$. For the marginal row mean comparisons, $\bar{\mu}_{i\cdot} - \bar{\mu}_{k\cdot}$, set $c_{ij} = 1/b, j = 1, \ldots, b$, set $c_{kj} = -1/b, j = 1, \ldots, b$, and set the remaining components to zero. In this case, (6.179) reduces to

$$(1/b)^2 \sum_j [(1/n_{ij}) + (1/n_{kj})] = (1/b)[(1/h_{r_i}) + (1/h_{r_k})] \tag{6.180}$$

where h_{r_i} and h_{r_k} are the harmonic means defined in (6.112). For the interaction contrast, $\mu_{ij} - \mu_{kj} - \mu_{it} + \mu_{kt}$, let $c_{ij} = c_{kt} = 1$ and $c_{kj} = c_{it} = -1$ and note that (6.179) is given by $(1/n_{ij} + 1/n_{kj} + 1/n_{it} + 1/n_{kt})$.

Using H_3, we obtain the intervals

$$c'H\mu\colon c'H\hat{\mu} \pm [(a - 1)F(\alpha; a - 1, N - ab)s^2c'H(W'W)^{-1}H'c]^{1/2}. \tag{6.181}$$

Here, $c' = (c_1, \ldots, c_{a-1})$

$$c'H\mu = \sum_{i=1}^{a-1} c_i(\bar{\mu}_{i\cdot} - \bar{\mu}_{a\cdot}) \tag{6.182}$$

$$c'H\hat{\mu} = \sum_{i=1}^{a-1} c_i[(1/b)\sum_j \bar{y}_{ij\cdot} - (1/b)\sum_j \bar{y}_{aj\cdot}] \tag{6.183}$$

and

$$c'H(W'W)^{-1}H'c = (1/b)\left(\sum_{i=1}^{a-1} c_i^2/h_{r_i} + c_\cdot^2/h_{r_a}\right). \tag{6.184}$$

These intervals are designed specifically for differences, $\bar{\mu}_{i\cdot} - \bar{\mu}_{k\cdot}$. To obtain them, we set $c_i = 1$ and $c_k = -1$, and (6.184) reduces to (6.180).

Finally, using H_4, we obtain the intervals

$$c'H\mu\colon c'H\hat{\mu} \pm \{(a - 1)(b - 1)F[\alpha; (a - 1)(b - 1), N - ab]s^2c'H(W'W)^{-1}H'c\}^{1/2}. \tag{6.185}$$

Here, $c' = [c_{ij}, i = 1, \ldots, (a - 1), j = 1, \ldots, (b - 1)]$

$$c'H\hat{\mu} = \sum_{i=1}^{a-1}\sum_{j=1}^{b-1} c_{ij}(\bar{y}_{ij\cdot} - \bar{y}_{ib\cdot} - \bar{y}_{aj\cdot} + \bar{y}_{ab\cdot}) \tag{6.186}$$

and, with $c_{ij} = 1$ and $c_{kt} = 0$ for $k, t \neq i, j$,

$$c'H(W'W)^{-1}H'c = (1/n_{ij} + 1/n_{ib} + 1/n_{aj} + 1/n_{ab}). \tag{6.187}$$

To summarize, we note that $c'H\hat{\mu}$ and $\text{Var}(c'H\hat{\mu})$ are identical for those comparisons that can be made by two or more of the sets of intervals. The difference in the lengths of the intervals rests on the multiplier and numerator degrees of freedom in the F statistic in (6.176), (6.177), and (6.181). The question of which intervals to use depends on the needs of the analyst. The intervals become narrower as their range of application becomes more restrictive. We also note that the obvious analogs of H_2 and H_3 can be used for column comparisons.

The Tukey studentized range intervals are given by (4.118). Thus, with $c' = (c_{ij}, i = 1, \ldots, a, j = 1, \ldots, b)$ we have

$$c'\mu\text{: } c'\hat{\mu} \pm q'(\alpha; ab, N - ab)s \max(c^+, c^-) \tag{6.188}$$

where $c^+ = \Sigma_{ij}c_{ij}/\sqrt{n_{ij}}$, for $c_{ij} > 0$, and $c^- = -\Sigma_{ij}c_{ij}/\sqrt{n_{ij}}$ for $c_{ij} < 0$. Thus, for row differences within a column, $\mu_{ij} - \mu_{kj}$, we have $c^+ = 1/\sqrt{n_{ij}}$ and $c^- = 1/\sqrt{n_{kj}}$. For marginal mean differences, $\bar{\mu}_{i\cdot} - \bar{\mu}_{k\cdot}$, we have

$$c^+ = (1/b)\sum_j 1/\sqrt{n_{ij}} \qquad c^- = (1/b)\sum_j 1/\sqrt{n_{kj}}. \tag{6.189}$$

These intervals and other concepts will be illustrated by numerical example in the next section. The Bonferroni intervals follow directly from (4.101).

6.3.1.8. A Numerical Example

The analysis of the two-factor, unconstrained model will be illustrated by an experiment on weight gains of laboratory animals under nine feeding treatments in a completely randomized design. The nine treatments are defined in terms of two factors, each at three levels, as follows:

Factor A
  Source of Protein: Beef, Pork, Grain
Factor B
  Amount of Protein: Low, Medium, High

The treatment combinations are assigned at random to the experimental units. Initially, there were six units per treatment combination, but, for various reasons unrelated to the treatments, the data available for analysis had unequal cell frequencies. These frequencies are shown in Table 6.8.

TABLE 6.8 Cell frequencies for the protein study.

		Amount		
		Low	Medium	High
Source	Beef	4	5	6
	Pork	3	4	5
	Grain	4	6	6

The cell mean estimates, $\bar{y}_{ij\cdot}$, are given in Table 6.9.

TABLE 6.9 Cell mean estimates for the protein study.

		Amount		
		Low	Medium	High
Source	Beef	76.0	86.8	101.8
	Pork	83.3	89.5	98.2
	Grain	83.8	83.5	86.2

In our preliminary analysis, we are interested in testing for the source-by-amount interaction as well as examining for source or amount differences. With regard to the latter, the unequal cell frequencies open the question of the appropriate main effect hypotheses as discussed in Section 6.3.1.4. In Table 6.10, we show the AOM table for the marginal means analysis as developed in Section 6.3.1.1, Table 6.5. In Tables 6.11 and 6.12, we show the first two rows of the alternative analyses given in Tables 6.6 and 6.7. The interaction and residual rows are the same as in Table 6.10.

TABLE 6.10 Marginal means analysis for protein study.

Description	*Hypothesis*	*df*	*SS*	*MS*	*F*	$P > F$
Source	H_A	2	240	120	2.2	0.13
Amount	H_B	2	1462	731	13.2	0.00
Interaction	H_{AB}	4	677	169	3.1	0.03
Residual	—	34	1888	56		

TABLE 6.11 Sequential analysis of protein data.

Description	*Hypothesis*	*df*	*SS*	*MS*	*F*	$P > F$
Source	H_A^{**}	2	390	195	3.5	0.04
Amount	H_B^*	2	1470	735	13.2	0.00

TABLE 6.12 Partially sequential analysis of protein data.

Description	*Hypothesis*	*df*	*SS*	*MS*	*F*	$P > F$
Source	H_A^*	2	356	178	3.2	0.05
Amount	H_B^*	2	1470	735	13.2	0.00

The last column in Tables 6.10 through 6.12 gives the probability of obtaining a larger value for the F-ratio under the null hypothesis. These numbers are commonly known as p-values. To indicate their interpretation, note that the interaction hypothesis has a p-value of 0.03. This means that the interaction hypothesis would have been declared significant if we had conducted the test with $\alpha = 0.03$. The effect of amount differences is highly significant, whether measured by H_B as in Table 6.10 or by H_B^* as in Tables 6.11 and 6.12. The effect of the different protein sources as measured by the three different hypotheses is not so clear. The weighted hypotheses, H_A^* and H_A^{**}, would have indicated significance with $\alpha = 0.05$ or $\alpha = 0.04$, while the marginal mean hypothesis is less emphatic with $p = 0.13$. A study of the hypotheses reveals that the difference is due to higher weightings on some of the more extreme cell means. In this case, it is difficult to justify a hypothesis with weights based on cell frequencies. This example, which is not atypical, emphasizes our point that it is not sufficient to describe a hypothesis by a descriptive phrase such as "differences in source of protein." A precise statement of how these differences are measured should accompany the analysis.

We now turn to a more detailed analysis of the protein data. In general, this may be done in terms of subhypotheses on one or more degrees of freedom or in terms of confidence intervals on linear functions of the parameters. We choose the latter approach.

Another useful aid to the analysis is a plot of the sample cell means for one factor as a function of the other factor. In Figure 6.2, we plot the sample cell means for the three sources of protein as a function of the amount of protein, and in Figure 6.3 we

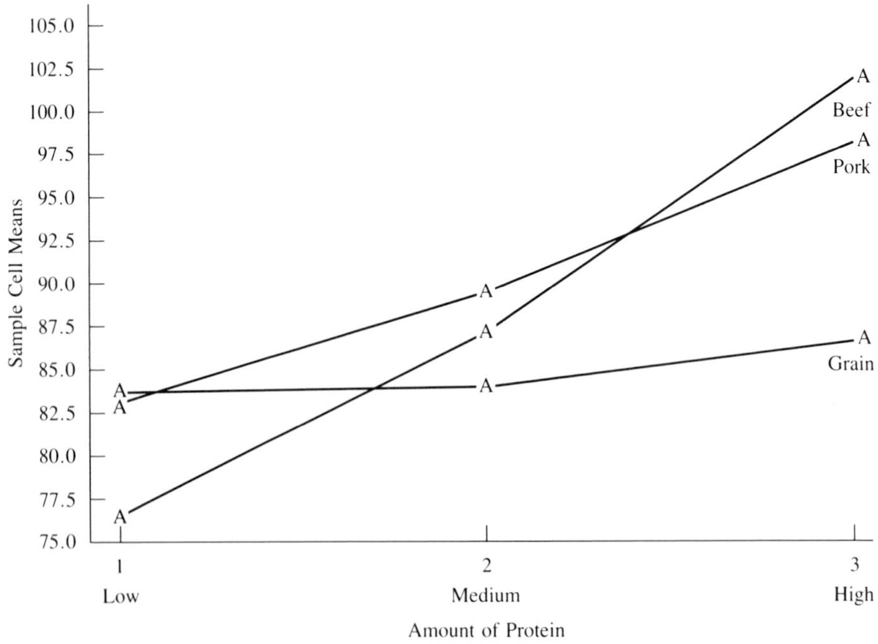

FIGURE 6.2.
Plot of sample cell means by amount for protein study

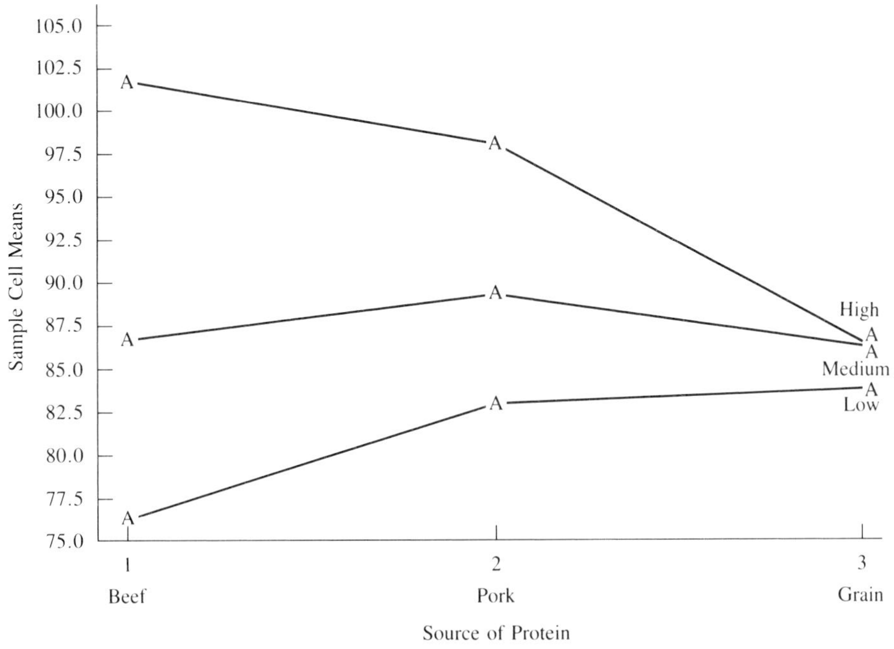

FIGURE 6.3.
Plot of sample cell means by source for protein study

plot the means as a function of the source of protein. Note that the points have been joined for ease in visualizing the differences. This does not imply that we can interpolate for intermediate values. Several things are suggested by these plots. First, weight gain appears to increase with amount of protein but not at the same rate for each source. Also, beef and pork perform similarly, while grain is less effective and seems nearly independent of amount. The significant interaction noted in the AOM table may be due primarily to the behavior of grain as a source of protein. We shall exhibit simultaneous confidence intervals on linear functions of μ that are designed to support these observations. The linear functions of interest, written in the form $c'\mu$, are as follows:

1. Interaction between beef and grain at low and medium amounts.
   C1. $\mu_{31} - \mu_{11} - \mu_{32} + \mu_{12}$
2. Beef versus pork averaged over all amounts.
   C2. $\bar{\mu}_{2\cdot} - \bar{\mu}_{1\cdot}$
3. Meat sources versus grain averaged over all amounts.
   C3. $\bar{\mu}_{1\cdot} + \bar{\mu}_{2\cdot} - 2\bar{\mu}_{3\cdot}$
4. Quadratic effect of response to meat sources.
   C4. $\dfrac{(\mu_{13} + \mu_{23})}{2} - (\mu_{12} + \mu_{22}) + \dfrac{(\mu_{11} + \mu_{21})}{2}$
5. High versus low amounts for meat sources.
   C5. $\dfrac{(\mu_{13} + \mu_{23})}{2} - \dfrac{(\mu_{11} + \mu_{21})}{2}$

In Table 6.13, we show several sets of simultaneous confidence intervals for these functions. To write simultaneous Scheffé intervals for these five functions, we must use the general intervals defined by (6.176). These intervals are labeled Scheffé (general) in Table 6.13. For comparison, we show the Tukey intervals based on (6.188), labeled Tukey (general) in Table 6.13.

TABLE 6.13 Confidence intervals for protein study.

		Half-width ($\alpha = 0.05$)			
Comparison	$c'\hat{\mu}$	*Scheffé (general)*	*Scheffé (separate)*	*Tukey (general)*	*Tukey (separate)*
C1	10.9	30.7	4.1*	33.0	—
C2	0.6	13.0	7.5	17.7	10.8
C3	4.1	21.2	12.3	32.8	20.1
C4	3.6	27.3	23.4	33.7	—
C5	20.4	16.1*	13.8*	18.8*	—

For these two sets of intervals, only the interval for C5 does not contain zero, indicating a significant difference between the high and low amounts of meat sources. Both sets of intervals are quite wide, reflecting the conservatism of these procedures. The Tukey intervals are somewhat wider, supporting the conjecture of Scheffé (1953). As noted by O'Neill and Wetherill (1971), both procedures use the experimentwise error rate, so we anticipate similar results.

The second set of Scheffé intervals is based on the concept of controlling the error rate within each of the sets of intervals based on H_2, H_3, and H_4 in (6.175). Thus, we may use H_4 for C1, H_3 for C2 and C3, and H_2 for C4 and C5. The intervals are seen to be substantially narrower. The comparison, C1, is judged significant in this context. A similar approach could be applied to the Tukey method. For example, we can apply the general result, (4.118), to the estimates of the marginal means, $\bar{\mu}_{i\cdot}$, for C2 and C3. The half-widths of the resulting intervals for these two contrasts are shown in the last column of Table 6.13. Note that the widths of the intervals are decreased by the ratio $q'(0.05; 3, 34)/q'(0.05, 9, 34)$. As anticipated, these intervals are narrower than those obtained by the general Tukey method but wider than those obtained from the separate Scheffé method.

The comparisons C1 through C5 are typical of the linear functions examined in two-factor experiments. When written in terms of cell means, we can determine the precise nature of the comparison. Thus, for example, C3—comparing meat sources to the grain source—can be misleading, since it represents the comparison of the averages over the three amounts. It might be better to examine this question at each amount of protein. Thus, with the highest amount of protein, we find the interval based on H_2 given by

$$\frac{(\mu_{13} + \mu_{23})}{2} - \mu_{33}\text{: } 13.8 \pm 14.3.$$

While this 95 percent confidence interval also covers zero, there is fairly strong evidence that the meat sources are more effective than grain at the highest amount.

Since this example is only for illustration, we shall not pursue it further, but it should be clear that many other comparisons are of interest, some developed *a priori* and some suggested by the data. The need for the protection offered by the simultaneous intervals is evident.

6.3.1.9. Missing Cells in the Unconstrained Model

We have noted that zero cell frequencies cause no conceptual problems with regard to estimation in the unconstrained model, since μ_{ij} is estimated by $\bar{y}_{ij}.$ if $n_{ij} > 0$, and μ_{ij} cannot be estimated if $n_{ij} = 0$. A similar situation holds with regard to testing hypotheses; that is, we can test hypotheses only on linear functions of the means of observed cells. The same requirement is placed on linear functions for which we can write confidence intervals. The analysis should thus consist of an examination of meaningful linear functions of the means of observed cells.

Assuming that a preliminary analysis of row, column, and interaction effects is desired, we ask how these effects should be defined. Implicit in the Analysis of Means table produced by any computing procedure is a definition of these effects. One objective of this section is to define these effects. If these definitions are not acceptable, then we must consider different ones. As we attempt to identify linear functions to measure these effects, we realize that we are entering the second phase of the analysis; that is, we are actually bypassing the preliminary investigations for main effects and interactions and are considering specific linear functions for subhypotheses or confidence intervals. In this sense, the Analysis of Means table may be of little value when we have missing cells.

In Section 4.3.3, we presented a general philosophy on hypothesis testing in which we eliminated the missing cells from the model and from the hypothesis matrix to obtain the effective model and the effective hypothesis, as in Definition 4.4. It is of interest to note that all of the common procedures use the effective hypothesis for the interaction test. Usually this hypothesis consists of those constraints of the form $\mu_{ij} - \mu_{kj} - \mu_{it} + \mu_{kt} = 0$ that can be written on observed cells, but there are more interesting situations, such as that in Example 4.5. The effective hypotheses corresponding to H_A and H_B, defined in (6.101) and (6.102), are particularly simple. For example, to test for row effects, we would use the hypothesis H: $\bar{\mu}_{i}. = \bar{\mu}_{k}.$ for all rows i and k in which there are no missing cells. No comparisons are made for rows with missing cells.

Computationally, the sums of squares for the effective hypotheses corresponding to H_A, H_B, and H_{AB} are easy to generate using the computing procedure described in Section 6.3.1.2; that is, we apply the Sweep Operator to the matrix S defined in (6.145). However, since in this case $X_m'X_m$ is singular, the Sweep procedure will terminate after completing $ab - m$ sweeps where m is the number of missing cells. The correct residual sum of squares is found as usual in the lower right corner of the swept S matrix (6.146). To obtain the sum of squares for the effective interaction hypothesis, compute $R[(\alpha\beta) \mid \mu, \alpha, \beta]$ as usual by sweeping on the $(\alpha\beta)$ columns of (6.146). To generate the sum of squares for the effective hypothesis corresponding to H_A, sweep on the α columns of (6.146) and any $(\alpha\beta)$ columns not previously entered. Because of the elimination of the α columns, some of the $(\alpha\beta)$ columns are now free to enter. The modified $R[\alpha \mid \mu, \beta, (\alpha\beta)]$ is then computed as usual. For the effective hypothesis corresponding to H_B, compute the modified $R[\beta \mid \mu, \alpha, (\alpha\beta)]$ in the same way.

A computing procedure quite similar to the one just described is often used (see Section 6.2.3). In this case, the rows and columns corresponding to $(\alpha\beta)_{ij}$, for which $n_{ij} = 0$, are first deleted from S. Note that, as a result of the definition of the canonical model, setting $(\alpha\beta)_{ij} = 0$ is equivalent to assuming that $\mu_{ij} - \bar{\mu}_{i\cdot} - \bar{\mu}_{\cdot j} + \bar{\mu}_{\cdot\cdot} = 0$. (For the canonical form as we have defined it in Section 6.2.3, we do not have columns for $(\alpha\beta)_{aj}$, $j = 1, \ldots, b$, or $(\alpha\beta)_{ib}$, $i = 1, \ldots, a$, so this method will not work if any of these n_{ij} are zero.) With this reduced S matrix, we compute $R[(\alpha\beta) \mid \mu, \alpha, \beta]$, $R[\alpha \mid \mu, \beta, (\alpha\beta)]$, and $R[\beta \mid \mu, \alpha, (\alpha\beta)]$ as usual. $R[(\alpha\beta) \mid \mu, \alpha, \beta]$ will be the same as in the effective interaction hypothesis, but the others will differ. This follows since any $(\alpha\beta)$ columns that might have entered in the previous procedure have been deleted from S. It was shown by Hocking, Hackney, and Speed (1978) that the row effect hypothesis corresponding to this procedure is given by $H_A(c)$, defined in (6.166) with $c_{ij} = 1$ if $n_{ij} \neq 0$ and $c_{ij} = 0$ if $n_{ij} = 0$. Defining $H_B(c)$ by interchanging the roles of subscripts yields the column hypothesis. This procedure corresponds to the SAS Type III analysis for the two-factor model but differs for more than two factors.

A third procedure is to compute $R(\alpha \mid \mu, \beta)$ and $R(\beta \mid \mu, \alpha)$ as described in Section 6.3.1.4. In this case the hypotheses are given by H_A^* and H_B^* in (6.164) and (6.162).

To illustrate the results of these three procedures, we provide three simple examples for the case $a = b = 3$. The incidence matrices are given by Table 6.14.

TABLE 6.14 Incidence matrices for Example 6.6.

	(a)			(b)			(c)	
0	n_{12}	n_{13}	n_{11}	n_{12}	0	n_{11}	n_{12}	n_{13}
n_{21}	n_{22}	n_{23}	n_{21}	0	n_{23}	n_{21}	n_{22}	0
n_{31}	n_{32}	n_{33}	0	n_{32}	n_{33}	0	0	n_{33}

EXAMPLE 6.6

In Table 6.14(a) we have one missing cell, and the effective interaction hypothesis with three degrees of freedom is given by

$$\begin{aligned} H_{AB_E}\colon\ & \mu_{12} - \mu_{13} - \mu_{22} + \mu_{23} = 0 \\ & \mu_{21} - \mu_{22} - \mu_{31} + \mu_{32} = 0 \\ & \mu_{21} - \mu_{23} - \mu_{31} + \mu_{33} = 0. \end{aligned} \tag{6.190}$$

This may be obtained by inspection or by the row reduction method described in Section 4.3.3. The effective row hypothesis corresponding to H_A is

$$H_{A_E}\colon \bar{\mu}_{2\cdot} = \bar{\mu}_{3\cdot}. \tag{6.191}$$

This single degree of freedom hypothesis tests "as much as possible" of the desired hypothesis H_A. Since this hypothesis ignores the first level of factor A, the conclusion from this main effect test could differ substantially from those associated with other procedures. In an attempt to include information about the first level of factor A, we might adjoin another contrast to H_{A_E}, such as

$$H_{A_1}: \frac{(\mu_{12} + \mu_{13})}{2} = \frac{(\mu_{22} + \mu_{23} + \mu_{32} + \mu_{33})}{4}. \tag{6.192}$$

This compares level one of factor A with the average of levels two and three of factor A. The comparison is made for the average of the second and third levels of factor B.

Our earlier comment that the Analysis of Means table may be inappropriate with missing cells should now be clear. As we attempt to decide on the appropriate hypothesis to consider, we are really formulating contrasts to be considered for subhypotheses or confidence intervals, thus bypassing the need for the preliminary hypotheses.

The two other computational procedures test two degree of freedom hypotheses. The second procedure, which yields $H_A(c)$ for $c_{ij} = 1$, $n_{ij} \neq 0$, and $c_{ij} = 0$ otherwise, in this case tests H_{A_E} along with H_{A_1}, described in (6.191) and (6.192). The third procedure tests H_A^* as defined in (6.164).

In Table 6.14(b), we show the same incidence matrix as in Table 6.3 for Example 6.5; hence, the single degree of freedom effective hypothesis for interaction is the single interaction constraint developed there. Thus,

$$H_{AB_E}: \mu_{11} - \mu_{21} - \mu_{12} = \mu_{33} - \mu_{23} - \mu_{32}. \tag{6.193}$$

Since there is one missing cell in each row and column, there is no effective hypothesis corresponding to H_A or H_B. For this example, it seems reasonable to conclude that there are no general main effect hypotheses. Looking at row comparisons, we see that the only reasonable functions to examine are $\mu_{11} - \mu_{21}$, $\mu_{12} - \mu_{32}$, and $\mu_{23} - \mu_{33}$. Simultaneous confidence intervals on these functions provide the information on row differences. The standard hypothesis-testing procedures are difficult to justify as measures of row (column) effects. For example, the second computational procedure yields the row hypothesis

$$\begin{aligned} H_A(1)\colon \mu_{11} - \mu_{21} &= \mu_{32} - \mu_{12} \\ \mu_{11} - \mu_{21} &= \mu_{23} - \mu_{33}. \end{aligned} \tag{6.194}$$

Table 6.14(c) gives the same incidence matrix as Table 6.2; hence, the single degree of freedom interaction hypothesis is given by

$$H_{AB_E}: \mu_{11} - \mu_{12} - \mu_{21} + \mu_{22} = 0. \tag{6.195}$$

Again, there is no effective hypothesis for either row or column effects. The second computing procedure yields, for row effects, the hypothesis

$$H_A(1)\colon \mu_{11} + \mu_{12} = \mu_{21} + \mu_{22}$$
$$\mu_{13} = \mu_{33}. \tag{6.196}$$

While these two contrasts are reasonable comparisons to make, analysts should be aware that they are measuring row effects in this particular way. It would seem preferable to specify these contrasts and specify single degree of freedom tests or simultaneous confidence intervals.

If $n_{13} = 0$ in Table 6.14(c), we recall from Example 6.4 that the model is not connected. In this case, the effective hypothesis for interaction is unchanged, and there is no effective hypothesis for rows or columns. The row effect hypothesis for the second computing procedure is given by the first constraint under $H_A(1)$ in (6.196).

■

The development of confidence intervals with missing cells follows from the results in Section 6.3.1.7; that is, we can develop the Scheffé intervals based on any hypothesis matrix. In this case, $H = I_{ab-m}$ is a natural candidate since it is completely general and, further, any other choice of H is clearly problem dependent. Similarly, the Tukey intervals are defined in the obvious way as

$$a'\mu\colon a'\hat{\mu} \pm q'(\alpha; ab - m, N - ab + m)s[\max(a^+, a^-)] \tag{6.197}$$

where

$$a^+ = \sum_{ij} a_{ij}/\sqrt{n_{ij}} \quad \text{for } a_{ij} > 0,\ n_{ij} > 0$$
$$a^- = \sum_{ij} |a_{ij}|/\sqrt{n_{ij}} \quad \text{for } a_{ij} < 0,\ n_{ij} > 0. \tag{6.198}$$

6.3.2. The No-Interaction Model

The two-way classification model, without interaction, is defined as in (6.1) with the no-interaction constraints defined in several equivalent forms as (6.2), (6.6), or (6.9). Equivalently, after reducing the model by the constraints, we arrived at the canonical forms developed in Section 6.2.2.2. In this section, we shall use the second canonical form.

The preliminary hypotheses to be considered for this model are concerned only with row and column effects. Following the reasoning in Section 6.3.1, natural measures of these effects are given by H_A and H_B as defined in (6.101) and (6.102). However, in the no-interaction model, these hypotheses have a much stronger interpretation. To see this, note that by applying the constraints in the form (6.6) or (6.9) we may write,

$$\bar{\mu}_{i\cdot} - \bar{\mu}_{\cdot\cdot} = \mu_{ij} - \bar{\mu}_{\cdot j}. \tag{6.199}$$

Thus, H_A given by the left side of (6.199) is equivalent under the no-interaction constraint to

$$H_{A_0}\colon \mu_{ij} - \bar{\mu}_{\cdot j} = 0 \qquad i = 1, \ldots, a - 1 \text{ for any } j. \tag{6.200}$$

[Note that this hypothesis is equivalent to (6.165).] Similarly, H_B is equivalent to

$$H_{B_0}\colon \mu_{ij} - \bar{\mu}_{i\cdot} = 0 \qquad j = 1, \ldots, b - 1 \text{ for any } i. \tag{6.201}$$

The implication of these hypotheses should be emphasized. While acceptance of H_A implies that the responses to the different levels of factor A, when averaged over factor B, are the same, acceptance of H_{A_0} implies that the responses to the different levels of factor A are the same for all levels of factor B. Conversely, rejection of H_A might lead us to the conclusion that, for instance, level one of factor A was superior on the average to the other levels of factor A but not necessarily superior for each level of factor B. The analogous conclusion based on the no-interaction model would be that level one of factor A is superior at all levels of factor B. The strength of this conclusion makes the no-interaction model very attractive but, of course, does not justify its assumption.

In this section we shall develop the details of the analysis of this model.

6.3.2.1. Development of Test Statistics: $n_{ij} \neq 0$

We have seen in Sections 6.2.2.2 and 6.2.2.3 that, except for the case of equal cell frequencies or the special case $a = 2$ in Example 6.1, there is not a simple expression for the vector of estimates or for the residual sum of squares for the no-interaction model. Of course, these are easily computed using the reduced or canonical models. As in Chapter 4, we let $\tilde{\mu}$ and $Q(\tilde{\mu})$ denote the vector of estimates and the residual sum of squares.

To develop the test statistic for H_{A_0}, note that under this hypothesis the reduced model is just the one-way classification model

$$y_{ijk} = \bar{\mu}_{\cdot j} + e_{ijk} \tag{6.202}$$

with estimates

$$\tilde{\tilde{\mu}}_{\cdot j} = \bar{y}_{\cdot j\cdot} \sim N(\bar{\mu}_{\cdot j}, \sigma^2/n_{\cdot j}) \tag{6.203}$$

and residual sum of squares

$$Q_A(\tilde{\tilde{\mu}}) = \sum\sum\sum(y_{ijk} - \bar{y}_{\cdot j\cdot})^2. \tag{6.204}$$

The test statistic for H_{A_0} is then

$$N_A/[(a - 1)s^2] \sim F(a - 1, N - a - b + 1, \lambda_A) \tag{6.205}$$

where

$$N_A = Q_A(\tilde{\tilde{\mu}}) - Q(\tilde{\mu}) \tag{6.206}$$

and

$$s^2 = Q(\tilde{\mu})/(N - a - b + 1). \tag{6.207}$$

The noncentrality parameter is given in general by (4.70) but does not have a simple expression in this situation.

These computations are efficiently performed in terms of the second canonical model. Noting from (6.199) that H_{A_0} is equivalent to the hypothesis $\alpha_i = 0$, $i = 1, \ldots, a-1$ in that model, we compute $Q(\tilde{\mu}) = Y'Y - R(\mu, \alpha, \beta)$ by sweeping on all columns. By then sweeping again on the α columns, we obtain $Q_A(\tilde{\tilde{\mu}}) = Y'Y - R(\mu, \beta)$. We then have

$$N_A = R(\alpha | \mu, \beta) = R(\mu, \alpha, \beta) - R(\mu, \beta). \tag{6.208}$$

In the special case $n_{ij} = n$, $\tilde{\mu}$ and $Q(\tilde{\mu})$ are given by (6.15) and (6.21), and we have

$$N_A = bn \sum (\bar{y}_{i\cdot\cdot} - \bar{y}_{\cdot\cdot\cdot})^2 \tag{6.209}$$

with noncentrality parameter

$$\lambda_A = bn \sum (\bar{\mu}_{i\cdot} - \bar{\mu}_{\cdot\cdot})^2/(2\sigma^2). \tag{6.210}$$

The analogous results for testing H_{B_0} follow by the appropriate interchange of subscripts.

The results for the general case of unequal cell frequencies are summarized in Table 6.15.

TABLE 6.15 Analysis of means table for the two-factor, no-interaction model.

Description	*Hypothesis*	*df*	*SS*	*EMS*
Row differences	H_{A_0}	$a - 1$	$R(\alpha \| \mu, \beta)$	$\sigma^2 + 2\sigma^2\lambda_A/(a-1)$
Column differences	H_{B_0}	$b - 1$	$R(\beta \| \mu, \alpha)$	$\sigma^2 + 2\sigma^2\lambda_B/(b-1)$
Residual	—	$N - a - b + 1$	$Q(\tilde{\mu})$	σ^2

In the special case, $n_{ij} = n$, we may use the definitions of X_A, X_B, and X_{AB} in (6.140) to write the sum of squares in Table 6.15 as follows

$$\begin{aligned} N_A &= R(\alpha | \mu, \beta) = bn \sum (\bar{y}_{i\cdot\cdot} - \bar{y}_{\cdot\cdot\cdot})^2 \\ &= (1/bn)Y'(X_A X_A' - (1/a)J_N J_N')Y \end{aligned} \tag{6.211}$$

$$\begin{aligned} N_B &= R(\beta | \mu, \alpha) = an \sum (\bar{y}_{\cdot j\cdot} - \bar{y}_{\cdot\cdot\cdot})^2 \\ &= (1/an)Y'(X_B X_B' - (1/b)J_N J_N')Y \end{aligned} \tag{6.212}$$

and

$$\begin{aligned} Q(\tilde{\mu}) &= \sum \sum \sum (y_{ijk} - \bar{y}_{i\cdot\cdot} - \bar{y}_{\cdot j\cdot} + \bar{y}_{\cdots})^2 \\ &= Y'[I - (1/bn)X_A X_A' - (1/an)X_B X_B' + (1/abn)J_N J_N']Y. \end{aligned} \tag{6.213}$$

The matrix expressions for these quadratic forms are particularly convenient for working out the expected values of the mean squares. Note that the numerator sums of squares for row and column effects in the constrained model are identical to those in the unconstrained model. This is a consequence of the orthogonality of the submatrices in the canonical model with balanced data.

There is little disagreement among practitioners as to the AOM table to be used for the no-interaction model with unequal cell frequencies. Table 6.15, which is based on quite reasonable row and column effect hypotheses, agrees with the partially sequential method in Section 6.3.1.4 and is almost exclusively recommended. The only potential contender is based on the sequential method. In that case, the row effect sum of squares would be computed as $R(\alpha \mid \mu)$, and the hypothesis tested is H_A^{**}, given in (6.161), subject to the constraint (6.9). The numerator sum of squares for this hypothesis is

$$N_A^{**} = \sum n_{i.} (\bar{y}_{i\cdot\cdot} - \bar{y}_{\cdots})^2. \tag{6.214}$$

Note that, for $n_{ij} = n$, (6.214) is identical to (6.211); hence, all three methods agree in this special case.

A common special case of this model occurs when $n_{ij} = 1$ for all cells. In the unconstrained model, this case does not yield an estimate of σ^2, but with the constraint an estimate with $(a-1)(b-1)$ degrees of freedom is available, as noted in (6.21). No new concepts are introduced by this special case. The analysis in Table 6.15 is applicable. Again we emphasize that insufficient data, that is $n_{ij} = 1$, do not justify the assumption of no interaction. In the next section, we describe a test that may be used to justify the no-interaction model.

6.3.2.2. A Test for Interaction with $n_{ij} = 1$

In many cases, because of physical, economic, or other restrictions, we do not have the luxury of more than one observation on each treatment combination. At the same time, we do not have sufficient justification to assume the no-interaction model. Since we do not have the information to conduct the test for interaction as described in Section 6.3.1.1, we ask whether there is anything we can do to provide some support for the constrained model. The answer is a qualified yes. A test was proposed by Tukey (1949) and discussed in general in Graybill (1976). The test is known as Tukey's *one degree of freedom test for additivity*.

To understand the basis for this test, we temporarily think of the cell means as values of a function of two variables—say, $f(u, v)$. Such a function is said to be additive if we can write $f(u, v)$ as the sum of two functions of a single variable—that is,

$$f(u, v) = f_1(u) + f_2(v). \tag{6.215}$$

A property enjoyed by additive functions is that a change in the value of one variable has the same effect for all values of the other variable. Thus, using (6.215) for values u_1, u_2 and v_1, v_2, we have

$$f(u_1, v_1) - f(u_2, v_1) = f(u_1, v_2) - f(u_2, v_2). \tag{6.216}$$

Comparing (6.216) with (6.2), we note that the concept of additivity is analogous to saying that the factors do not interact. The analogy $f(u_i, v_j) = u_{ij}$ establishes the relation. In a typical two-factor experiment, there need not be a scale associated with the levels of the two factors. However, the concepts of nonadditivity and interaction can be related with or without a scale on the factors.

To examine for nonadditivity, we may write $f(u, v)$, assuming differentiability, in a Taylor series as

$$\begin{aligned} f(u, v) = f(u_0, v_0) &+ [\partial f/\partial u](u - u_0) + [\partial f/\partial v](v - v_0) \\ &+ (1/2)[\partial^2 f/\partial u^2](u - u_0)^2 + [\partial^2 f/\partial u \partial v](u - u_0)(v - v_0) \\ &+ (1/2)[\partial f^2/\partial v^2](v - v_0)^2 + \dots . \end{aligned} \tag{6.217}$$

[In (6.217), all derivatives are evaluated at (u_0, v_0).] Note that the nonadditivity is due to the cross-product terms in u and v.

Returning to the analogy with cell means, we write the approximation

$$\mu_{ij} \doteq \bar{\mu}_{..} + (\bar{\mu}_{i.} - \bar{\mu}_{..}) + (\bar{\mu}_{.j} - \bar{\mu}_{..}) + \partial(\bar{\mu}_{i.} - \bar{\mu}_{..})(\bar{\mu}_{.j} - \bar{\mu}_{..}). \tag{6.218}$$

The terms involving only u in (6.217) correspond to $(\bar{\mu}_{i.} - \bar{\mu}_{..})$, and the terms involving only v correspond to $(\bar{\mu}_{.j} - \bar{\mu}_{..})$. The remaining terms are approximated by the single cross-product term in (6.218). Thus, as an approximation to the unconstrained cell means model, we use

$$E(y_{ijk}) \doteq \mu + \alpha_i + \beta_j + \delta z_{ij} \tag{6.219}$$

where α_i and β_j are as defined in the second canonical form with $\alpha. = \beta. = 0$ and

$$z_{ij} = (\bar{\mu}_{i.} - \bar{\mu}_{..})(\bar{\mu}_{.j} - \bar{\mu}_{..}). \tag{6.220}$$

[Mandel (1971) proposed a general two-factor model that includes the model, (6.219), as a special case. Snee (1982) discusses the application of that general model.]

With this approximation, the test for additivity is

$$H\colon \delta = 0. \tag{6.221}$$

Unfortunately, (6.219) is not a linear model since both δ and z_{ij} are unknown parameters. To develop an approximate test statistic, we temporarily assume that z_{ij} is known and develop the test statistic under that assumption. We then approximate z_{ij} by using estimates of μ_{ij} in (6.220).

Under this assumption, (6.219) is a linear regression model; hence, standard procedures apply for testing (6.221). (This model is often discussed under the heading of the Analysis of Covariance, a topic that will be described in Chapter 7.) The test statistic is based, as usual, on the comparison of the residual sum of squares for the model, (6.219), and the additive model, $\delta = 0$. The residual sum of squares under the hypothesis is given by (6.21) with $n = 1$; that is,

$$Q_H = \sum \sum (y_{ij} - \bar{y}_{i\cdot} - \bar{y}_{\cdot j} + \bar{y}_{\cdot\cdot})^2. \tag{6.222}$$

It can be shown, using standard regression procedures, or by the results of Section 7.4, that the numerator sum of squares for testing H: $\delta = 0$ is given by

$$\begin{aligned} N_H &= Q_H - Q(\hat{\mu}, \hat{\alpha}, \hat{\beta}, \hat{\delta}) \\ &= \left(\sum \sum z_{ij}^* y_{ij}^*\right)^2 \Big/ \left(\sum \sum z_{ij}^{*2}\right) \end{aligned} \tag{6.223}$$

where

$$\begin{aligned} z_{ij}^* &= z_{ij} - \bar{z}_{i\cdot} - \bar{z}_{\cdot j} + \bar{z}_{\cdot\cdot} \\ y_{ij}^* &= y_{ij} - \bar{y}_{i\cdot} - \bar{y}_{\cdot j} + \bar{y}_{\cdot\cdot}. \end{aligned} \tag{6.224}$$

From (6.222) and (6.223), we can compute $Q(\hat{\mu}, \hat{\alpha}, \hat{\beta}, \hat{\delta})$ and, hence, write the test statistic as

$$F = N_H/[Q(\hat{\mu}, \hat{\alpha}, \hat{\beta}, \hat{\delta})/(ab - a - b)]. \tag{6.225}$$

Note that, under H: $\delta = 0$, this statistic is distributed as $F(1, ab - a - b)$ conditional on z_{ij}, but, since the distribution does not depend on z_{ij}, this is also the unconditional distribution. The noncentral distribution does depend on the z_{ij}. We can thus use (6.225) to provide an exact test in the model (6.219) of size α for the hypothesis H: $\delta = 0$.

It remains to approximate z_{ij} as defined by (6.220). A simple procedure is to replace μ_{ij}, in (6.220), by its estimate from the no-interaction model, as given by (6.15); that is,

$$\tilde{\mu}_{ij} = \bar{y}_{i\cdot} + \bar{y}_{\cdot j} - \bar{y}_{\cdot\cdot}. \tag{6.226}$$

We then have

$$\hat{z}_{ij} = (\bar{y}_{i\cdot} - \bar{y}_{\cdot\cdot})(\bar{y}_{\cdot j} - \bar{y}_{\cdot\cdot}). \tag{6.227}$$

It follows that $\bar{z}_{i\cdot} = \bar{z}_{\cdot j} = \bar{z}_{\cdot\cdot} = 0$; hence, the numerator sum of squares, (6.223), is given after some simplification as

$$N_H = \left(\sum \sum \hat{z}_{ij} y_{ij}\right)^2 \Big/ \sum \sum \hat{z}_{ij}^2. \tag{6.228}$$

Using (6.228) in (6.225), the hypothesis of no interaction is rejected if F exceeds $F(\alpha;\, 1,\, ab - a - b)$.

Since (6.218) provides only an approximation to the general cell mean structure, it is possible that this test, designed for (6.222), may lead to erroneous conclusions. However, it does seem to perform fairly well in practice and is preferable to ignoring the possibility of interaction. Snee (1982) provides a good example and discussion of other potential explanations for the apparent nonadditivity.

If the hypothesis is rejected, a general analysis of the two-factor model is impossible, although in some cases a transformation on the data may remove the nonadditivity. (See Snedecor and Cochran, 1967, or Atkinson, 1982.)

6.3.2.3. Simultaneous Confidence Intervals

Simultaneous confidence intervals for the constrained model follow from the general theory in Chapter 4, but it is instructive to spell out some of the details.

The Scheffé intervals are given by (4.99) for any hypothesis matrix H. For example, we might consider $H = I$ for arbitrary linear functions of μ and H as determined by H_{A_0} or H_{B_0} in (6.200) or (6.201) for linear functions of row or column effects.

In practice, the computations for the constrained model will usually be done in terms of one of the canonical forms. Using the second canonical form, the Scheffé intervals on linear functions of θ_m are given by

$$c'\theta_m:\ c'\hat{\theta}_m \pm [(a + b - 1)F(\alpha;\, a + b - 1,\, N - a - b + 1)s^2c'(X_m'X_m)^{-1}c]^{1/2} \tag{6.229}$$

where θ_m is given by (6.46) and X_m by (6.49). To relate to linear functions of the cell means, note that, in view of (6.45) and (6.199), the row effect parameters are

$$\alpha_i = \mu_{ij} - \bar{\mu}_{\cdot j} \qquad i = 1, \ldots, a - 1 \text{ for any } j. \tag{6.230}$$

Similarly, the column effect parameters are

$$\beta_j = \mu_{ij} - \bar{\mu}_{i\cdot} \qquad j = 1, \ldots, b - 1 \text{ for any } i. \tag{6.231}$$

Thus, linear functions of row or column effects are simply expressed in terms of α_i or β_j. For confidence intervals on individual cell means we recall the general relation between μ_{ij} and θ_m given in (6.61).

If we are interested only in row effects, note that H_{A_0}, (6.200), is equivalent to the hypothesis H_{A_0}: $\alpha_i = 0,\ i = 1, \ldots, (a - 1)$. Thus, we can use the general Scheffé procedure, (4.99), with

$$H = (0|I_{a-1}|0) \tag{6.232}$$

yielding simultaneous confidence intervals on linear functions of the α_i as

$$c'H\theta_m = \sum_{i=1}^{a-1} c_i\alpha_i = \sum_{i=1}^{a-1} c_i(\mu_{ij} - \bar{\mu}_{\cdot j}):\ \sum_{i=1}^{a-1} c_i\hat{\alpha}_i \pm [(a - 1)F(\alpha;\, a - 1,\, N - a - b + 1)s^2c'H(X_m'X_m)^{-1}H'c]^{1/2}. \tag{6.233}$$

To illustrate, consider the special case, $n_{ij} = n$, in which $(X_m'X_m)^{-1}$ is given by (6.53), and hence

$$H(X_m'X_m)^{-1}H' = (1/abn)(aI_{a-1} - J_{a-1}J_{a-1}'). \qquad (6.234)$$

Recalling $\hat{\theta}_m$ as given by (6.54), we have a confidence interval on α_i given by

$$\begin{aligned} \alpha_i &= \mu_{ij} - \bar{\mu}_{\cdot j}\colon \bar{y}_{i\cdot\cdot} - \bar{y}_{\cdots} \\ &\pm [(a-1)F(\alpha; a-1, N-a-b+1)s^2(a-1)/abn]^{1/2}. \end{aligned} \qquad (6.235)$$

For simple differences we obtain

$$\begin{aligned} \alpha_i - \alpha_k &= \mu_{ij} - \mu_{kj}\colon \bar{y}_{i\cdot\cdot} - \bar{y}_{k\cdot\cdot} \\ &\pm [(a-1)F(\alpha; a - 1, N-a-b+1)2s^2/bn]^{1/2}. \end{aligned} \qquad (6.236)$$

As usual, the difference between this interval and that obtained by the general expression (6.229) is that $(a-1)F[\alpha; (a-1), N-a-b+1]$ is replaced by $(a+b-1)F(\alpha; a+b-1, N-a-b+1)$. The intervals, such as (6.233), are narrower, but the simultaneous confidence coefficients apply only to linear functions of row effects. The analogous expressions can be developed for column effects using $H_{B_0}\colon \beta_j = 0, j = 1, \ldots, (b-1)$. As discussed in the example in Section 6.3.1.8, we might use such intervals for row and column effects as an alternative to the more conservative general intervals given by (6.229).

Finally, we illustrate the use of (6.229) for the case $n_{ij} = n$ to obtain a confidence interval for μ_{ij} as

$$\begin{aligned} &\mu_{ij}\colon \bar{y}_{i\cdot\cdot} + \bar{y}_{\cdot j\cdot} - \bar{y}_{\cdots} \\ &\pm [(a+b-1)F(\alpha; a+b-1, N-a-b+1)s^2(a+b-1)/nab]^{1/2}. \end{aligned} \qquad (6.237)$$

The Tukey studentized range intervals for general linear functions of θ_m require the use of (4.120), since the components of $\hat{\theta}_m$ are correlated even in the case of balanced data. Before illustrating the use of this general result, we observe that in the balanced case we may easily write the Tukey intervals for row differences (and by analogy for column differences). To do so, note that, for $i = 1, \ldots, a$, the marginal means $\bar{y}_{i\cdot\cdot}$ are independently distributed as

$$\bar{y}_{i\cdot\cdot} \sim N(\bar{\mu}_{i\cdot}, \sigma^2/bn). \qquad (6.238)$$

Further, since $\bar{\mu}_{i\cdot} - \bar{\mu}_{k\cdot} = \mu_{ij} - \mu_{kj} = \alpha_i - \alpha_k$, we may use the simple Tukey procedure (4.115) to write the confidence intervals

$$\mu_{ij} - \mu_{kj}\colon \bar{y}_{i\cdot\cdot} - \bar{y}_{k\cdot\cdot} \pm q(\alpha; a, N-a-b+1)(s^2/bn)^{1/2}. \qquad (6.239)$$

These intervals should be contrasted with (6.236).

If we are interested in more general linear functions of $\bar{\mu}_{i\cdot}$, we may further exploit (6.238) and use (4.116) to obtain

$$\sum c_i\bar{\mu}_{i\cdot}: \sum c_i\bar{y}_{i\cdot\cdot} \pm q'(\alpha; a, N - a - b + 1)(s^2/bn)^{1/2}c^* \qquad (6.240)$$

where c^* is defined as in (4.117). This enables us to write intervals such as (6.235).

Finally, to illustrate the general Tukey procedure for correlated variables, consider the balanced case in which

$$\hat{\theta}_m \sim N(\theta_m, \sigma^2 K) \qquad (6.241)$$

with $K = (X_m'X_m)^{-1}$ as in (6.53). The intervals follow by using any factorization, $K = CC'$, but, since C is not unique, some thought must be given to the choice of factorization. One requirement is that the results should exhibit certain consistencies. For example, intervals on $\alpha_i - \alpha_k$ should have the same width for any choice of i and k. It can be seen that neither the Cholesky nor the eigen analysis factorizations have this property. The suggestion by Hochberg (1975) that we seek C in the form of K does offer reasonable intervals.

It is easily verified that C may be written as

$$\begin{aligned} C = (1/\sqrt{n})\text{Diag}\{1/\sqrt{ab}, (1/\sqrt{b})[I_{a-1} - (1/(\sqrt{a(a \pm 1)})J_{a-1}, J_{a-1}'], \\ (1/\sqrt{a}[I_{b-1} - (1/\sqrt{b(b \pm 1)})J_{b-1}\, J_{b-1}']\}. \end{aligned} \qquad (6.242)$$

Thus, confidence intervals on linear functions of θ_m are given by

$$c'\theta_m: c'\hat{\theta}_m \pm q'(\alpha; a + b - 1, N - a - b + 1)s\, c^* \qquad (6.243)$$

where $c^* = \max(c^+, c^-)$, c^+ is the sum of the positive elements of $c'C$, and c^- is the sum of the absolute values of the negative elements of $c'C$. Thus, for comparison with (6.236) and (6.239) we have

$$\begin{aligned} \alpha_i - \alpha_k = \mu_{ij} - \mu_{kj}: \bar{y}_{i\cdot\cdot} - \bar{y}_{k\cdot\cdot} \\ \pm q'(\alpha; a + b - 1, N - a - b + 1)(2s^2/bn)^{1/2}. \end{aligned} \qquad (6.244)$$

These intervals are slightly wider than those given by (6.239). The advantage of these more general intervals is that, for example, with $c'\theta_m = \mu + \alpha_i + \beta_j$, we can write confidence intervals on individual cell means. Also, the simultaneous confidence coefficient is valid for row and column effects, whereas in (6.239) it is valid for row effects only.

While this extension of the Tukey procedure is acceptable for the balanced case where a reasonable factorization of K was easy, it seems doubtful that this procedure would be useful in the unbalanced case. For the unbalanced case, the Scheffé intervals, (6.229), or special cases such as (6.233) seem preferable.

6.3.2.4. Missing Cells in the No-Interaction Model

In the unconstrained model, hypotheses that could be tested with all $n_{ij} > 0$ may not be testable with missing cells. In Section 4.3.3, we saw that, for constrained models that are connected, we can test any hypothesis. In the connected, two-way, no-interaction model, the row and column effect hypotheses, H_{A_0} and H_{B_0}, are unaffected by missing cells. The essential reason for this is that in the connected case these hypotheses may be written in terms of observed cells. To illustrate, consider H_{A_0} that is equivalent to the hypothesis

$$H_{A_0}\colon \mu_{ij} - \mu_{kj} = 0 \qquad i, k = 1, \ldots, a \text{ for any } j. \tag{6.245}$$

The original definition of connectedness as given in Section 6.2.2.3 in terms of row–column paths such as (6.73) was motivated by the need to determine $a - 1$ differences in (6.245) that are equivalent to H_{A_0}. Thus, for example, in Table 6.2 we cannot write (6.245) for a given j in terms of observed cells, but we can write

$$\begin{aligned} \mu_{11} - \mu_{21} &= 0 \\ \mu_{13} - \mu_{33} &= 0. \end{aligned} \tag{6.246}$$

In view of the no-interaction constraint, (6.246) is equivalent to H_{A_0}. This hypothesis could be formally developed as the effective hypothesis corresponding to (6.245). In this case, since the model is connected, they are equivalent.

An unconnected case is given in Example 6.4 and illustrated by Table 6.2 with $n_{13} = 0$. In this case, the effective hypothesis is just the first comparison in (6.246). In general, with designs that are not connected, the best we can do is apply the analysis to connected subsets of the original design.

In terms of the canonical reparameterizations, the computations proceed just as in the case of no missing cells. Thus, using the second canonical form, H_{A_0} is equivalent to $\alpha_i = 0$, $i = 1, \ldots, (a - 1)$ in the connected case. We have seen that the model has full rank and that the sum of squares is given by $R(\alpha \mid \mu, \beta)$.

If the model is not connected, we have seen that X_m does not have full column rank and that the computations must be modified. We will illustrate with the two approaches suggested in Section 6.3.1.2. Again, we use Example 6.4 to illustrate the ideas.

EXAMPLE 6.7

As in Example 6.4, we first fit the model by setting $\beta_2 = 0$; that is, we use

$$X_m^* = \begin{bmatrix} J_{n_{11}} & J_{n_{11}} & 0 & J_{n_{11}} \\ J_{n_{12}} & J_{n_{12}} & 0 & 0 \\ J_{n_{21}} & 0 & J_{n_{21}} & J_{n_{21}} \\ J_{n_{22}} & 0 & J_{n_{22}} & 0 \\ J_{n_{33}} & -J_{n_{33}} & -J_{n_{33}} & -J_{n_{33}} \end{bmatrix} \tag{6.247}$$

to obtain the residual sum of squares, $Q(\tilde{\theta}_m)$, as described in Section 6.3.1.2. Next, we return to the original X_m matrix, including the β_2 column, and to test row effects we set $\alpha_1 = \alpha_2 = 0$, yielding

$$X_m^{**} = \begin{bmatrix} J_{n_{11}} & J_{n_{11}} & 0 \\ J_{n_{12}} & 0 & J_{n_{12}} \\ J_{n_{21}} & J_{n_{21}} & 0 \\ J_{n_{22}} & 0 & J_{n_{22}} \\ J_{n_{33}} & -J_{n_{33}} & -J_{n_{33}} \end{bmatrix}. \tag{6.248}$$

The usual computations yield the residual sum of squares Q_{A_1} and, hence, $N_{A_1} = Q_{A_1} - Q(\tilde{\theta})$. The test statistic,

$$F = N_{A_1}/[Q(\tilde{\theta})/(N - 4)], \tag{6.249}$$

with one and $N - 4$ degrees of freedom, is seen to be appropriate for the hypothesis

$$H_{A_1}\colon \mu_{11} = \mu_{21}. \tag{6.250}$$

This is the effective hypothesis for this model.

An alternative procedure that is often used is to achieve full rank by setting $\beta_2 = 0$ as in (6.247) with resulting residual sum of squares. Then, to test for row effects, we set $\alpha_1 = \alpha_2 = 0$ in (6.247) to obtain

$$X_m^{***} = \begin{bmatrix} J_{n_{11}} & J_{n_{11}} \\ J_{n_{12}} & 0 \\ J_{n_{21}} & J_{n_{21}} \\ J_{n_{22}} & 0 \\ J_{n_{33}} & -J_{n_{33}} \end{bmatrix}. \tag{6.251}$$

From this, we obtain the residual sum of squares, Q_{A_2} and, hence, $N_{A_2} = Q_{A_2} - Q(\tilde{\theta})$. The test statistic, with two and $N - 4$ degrees of freedom, is

$$F = (N_{A_2}/2)/[Q(\tilde{\theta})/(N - 4)]. \tag{6.252}$$

This test statistic is appropriate for testing the two degrees of freedom hypothesis

$$\begin{aligned} H_{A_2}\colon \mu_{11} &= \mu_{21} \\ 2\mu_{12} &= \mu_{21} + \mu_{33}. \end{aligned} \tag{6.253}$$

The hypotheses, H_{A_1} and H_{A_2}, are readily determined by transforming back to the reduced model, W_R, and noting the consequences.

This second procedure tests a rather unusual measure of row effects for the second degree of freedom. In this case, the effective hypothesis seems preferable to the hypothesis H_{A_2}, which is determined by the computational procedure and not specified in advance. Thus, it is convenient that the first computing method yields the effective hypothesis.

■

We shall not attempt to develop general expressions for the hypotheses tested by this second computational procedure in the disconnected case. The essential points to make with missing cells in the no-interaction model are (a) if the design is connected, the standard hypotheses can be tested and (b) if the design is not connected, the analyst should develop hypotheses that are meaningful in the analysis and not resort to automated procedures. The effective hypotheses appear to be a reasonable default.

The confidence interval methods in the connected case are as given in Section 6.3.2.3 for the Scheffé procedure. The inconsistencies noted for the generalized Tukey procedure apply here as well. In the disconnected case, the problems noted with estimation in Section 6.2.2.3 suggest that we resort to the effective model.

6.3.3. A Constrained Model: The Zero-Level Problem

The theory developed in Chapter 3 applies to any linear constraints, but so far we have considered only the constraint of no interaction. We shall now discuss an interesting example illustrating a different kind of constraint. The problem has been discussed in the literature and has a long history (see Addelman, 1974). We shall see that the cell means formulation is particularly simple.

To illustrate, suppose that we are interested in the response to three fertilizers and that we apply each of these at four levels. Letting μ_{ij} denote the mean response to the jth level of the ith fertilizer, an appropriate model is

$$y_{ijk} = \mu_{ij} + e_{ijk} \qquad \begin{aligned} i &= 1, \ldots, 3 \\ j &= 1, \ldots, 4 \\ k &= 1, \ldots, n_{ij}. \end{aligned} \tag{6.254}$$

The methods developed in this chapter would usually be applied directly to analyze the resulting data.

The special case of interest here is that in which the lowest level of application is zero, as is often done to provide information over the full range of application. It is clear that, since it does not make any difference which fertilizer we do not apply, the cell means satisfy the constraint

$$\mu_{11} = \mu_{21} = \mu_{31}. \tag{6.255}$$

This constraint is easily incorporated into the model, using the general results of Chapter 3. The model reduction method is particularly simple.

It is of interest to examine some of the common hypotheses. For example, consider the hypothesis of no interaction between fertilizer type and level of application. In Figure 6.4 we show a typical plot of the cell means for the three types of fertilizer as a function of the level of application.

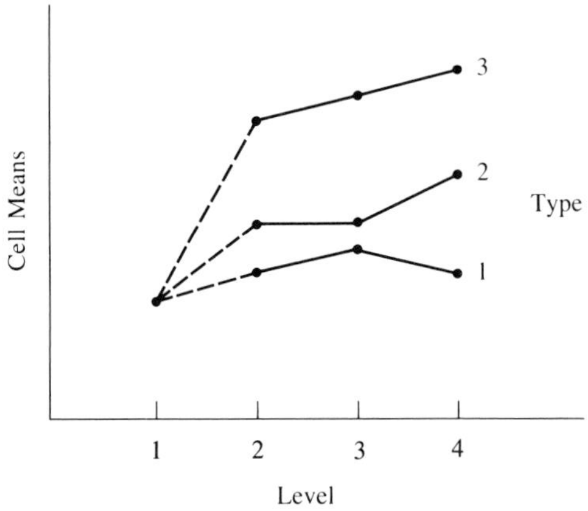

FIGURE 6.4
Cell means with a zero level

In general, the hypothesis of no interaction is that the line segments in each of the intervals are parallel. It is clear from this figure that, if we include the first interval (indicated by the dashed lines), the hypothesis is really that of identical response for the three fertilizer types. Thus, the appropriate interaction hypothesis in this case should be

$$H_{AB}: \mu_{ij} - \mu_{i4} - \mu_{3j} + \mu_{34} = 0 \qquad \begin{array}{l} i = 1, 2 \\ j = 2, 3. \end{array} \tag{6.256}$$

Clearly, if we ignore the constraint, (6.255), and test H_{AB} for $j = 1, 2, 3$, we could well get a significant interaction when in fact (6.256) is true.

The usual test for fertilizer differences is also less appropriate here; that is, although comparing the average response for all levels might provide an indication of fertilizer differences, it is more likely that we wish to examine the nature of the response as a function of the level of application.

To emphasize the types of analyses that might be appropriate, we note that we have already seen a zero-level problem but did not identify it as such. In Section 5.5, we examined the results of the application of five treatments in a one-way classification. The treatments consisted of plain water and two concentrations of two electrolytes that were mixed with water. Alternatively, we could have viewed the experiment as a two-way classification—type of electrolyte by concentration—with the first level of concentration being zero. The experiment is laid out symbolically in Table 6.16.

TABLE 6.16 Lactic acid example as a two-way classification, zero-level model.

		Concentration		
		Zero	Low	High
Electrolyte	A	$\mu_{11}(\mu_1)$	$\mu_{12}(\mu_2)$	$\mu_{13}(\mu_3)$
	B	$\mu_{21}(\mu_1)$	$\mu_{22}(\mu_4)$	$\mu_{23}(\mu_5)$

In Table 6.16, we indicate in parentheses the notation used for the cell means in Section 5.5. The assumption $\mu_{11} = \mu_{21}$ is implicit in the assumption that all runners who received only water (zero concentration) have the same mean response, μ_1. The comparisons C1 through C8 could have just as well been described in terms of the μ_{ij}. In particular, note that C8 is the appropriate contrast for testing the electrolyte-by-concentration interaction.

CHAPTER 6 EXERCISES

1. Establish the equivalence of the two expressions for the interaction constraints in (6.6) and (6.9). Verify that they may be written in matrix form as in (6.7) and (6.11).
2. Show that the reduced model given by (6.27) yields the estimates in (6.28).
3. Show that the canonical models in (6.30) and (6.47) may be written in matrix form as (6.32) and (6.48), respectively.
4. For the model given in Example 6.1, verify that the parameters are estimated by (6.63). Show that this expression generalizes for $a = 2$ and any b.
5. Show that the row–column definition of connectedness for the two-factor model without interaction in (6.73) is equivalent to Definition 4.3.
6. Verify that (6.84) gives the correct estimates for estimable parameters and parameter functions.
7. Verify that the parameter estimates given by (6.116) and (6.118) are correct for the model constrained by the hypothesis H_A.
8. Apply the Sweep Operator to determine the test statistics for the hypotheses H_A, H_B, and H_{AB} in the unconstrained, two-factor model with $n_{ij} = n$. Use the canonical form described in Section 6.2.3.
9. Verify that the Sequential Method summarized in Table 6.6 actually tests the hypotheses H_A^{**} and H_B^{*}.
10. Show that, with missing cells in the two-factor model with interaction, the Sweep Operator described in Section 6.3.1.9 yields the sums of squares for the effective hypotheses.
11. Write out the hypotheses $H_A(1)$, H_A^{*}, and H_A^{**} for the incidence matrix in Table 6.14(b).

12. Verify that the two procedures described in Example 6.7 yield the sums of squares for testing the hypotheses H_{A_1} and H_{A_2} in (6.250) and (6.253).

13. Consider the two-factor model with interaction in which the number of observations per cell is given by $n_{ij} = r_i c_j$ where r_i, $i = 1, \ldots, a$ and c_j, $j = 1, \ldots, b$ are known row and column constants. This is known as the *proportional frequencies model.*

(a) Write out the hypotheses corresponding to H_A, H_A^*, and H_A^{**} for this special case. Show that H_A^* and H_A^{**} are equivalent but that H_A is different.

(b) Develop formulas for the sums of squares for the three analyses described in Tables 6.5 through 6.7. (Hint: The sums of squares in Table 6.6 add up to the corrected total sum of squares, T, from Table 5.1.)

14. The data below represent the yield (bushels/acre) of three varieties of oats (Factor A) and three different fertilizers (Factor B).

		Factor B		
		B1	B2	B3
Factor A	A1	50, 51, 52 56, 60, 55	42, 40 38, 38	55, 56 56, 58
	A2	65, 69, 67	50, 50	62, 62
	A3	67, 67, 69	48, 50	65, 67

Compute the two AOM tables corresponding to the results of Exercise 13.

15. For the model in Example 6.3,

(a) write a reduced model.

(b) estimate the cell means, assuming all other cell frequencies are n, using the reduced model from part (a).

(c) write the effective model.

(d) use the effective model to estimate the cell means.

(e) suppose that n_{13} is also zero as in Example 6.4. Use (i) the reduced model, (ii) the canonical model, and (iii) the effective model to determine the estimates of estimable cell means.

16. Consider the two-factor model with $a = b = 2$ and all $n_{ij} \neq 0$.

(a) Develop the noncentrality parameters for the hypotheses H_A and H_A^*.

(b) Compare the power of these tests if $\mu_{11} + \mu_{22} = \mu_{12} + \mu_{21}$.

(c) Compare the power of the two tests in the general case where the μ_{ij} are not related.

17. For the unconstrained, two-factor model with $a = b = 2$, and all $n_{ij} \neq 0$,

(a) Develop the test for the hypothesis

$$H_0\colon \mu_{11} = \mu_{12}$$
$$\mu_{21} = \mu_{22}\,.$$

Can the numerator sum of squares for this hypothesis be obtained from Tables 6.5 through 6.7?

(b) Answer the same questions in part (a) for the no-interaction model.

(c) Illustrate parts (a) and (b) for the following data, and state whether the no-interaction assumption is reasonable for this data.

		Factor B	
		1	2
	1	9, 10, 11	13, 14
Factor A	2	10, 11 13, 14	18, 20 19, 21 20, 22

18. Using the data from Exercise 17, develop a set of 95 percent simultaneous confidence intervals for the cell means and the differences implied in H_0 of part (a) for both the constrained and the unconstrained models. Use the Scheffé method with different bases and the Tukey method, where possible, and compare the interval widths.

CHAPTER SEVEN
MORE COMPLEX MODELS

7.1. *K*-FACTOR MODELS

We have noted that the one-way classification, or single-factor model, naturally extends to situations in which the treatment combinations can be described in terms of several factors. Notationally, we saw that it was more convenient to describe the cell means in terms of two or more subscripts, but otherwise the basic model was unchanged. In Chapter 6, we saw that there were additional complexities introduced in the analysis by the extension to two factors. The concept of interaction introduced problems with the interpretation of effects of the treatments. The imbalance caused by unequal cell frequencies and missing cells led to a lengthy discussion of the appropriate analysis. The concept of constrained models was seen to affect both the estimation and testing problems. Finally, it was noted that much of the confusion in the analysis of the two-factor model is caused by a misunderstanding of the canonical model and the desire to apply standard analyses with unbalanced or incomplete data.

In this section, we extend the discussion to the three-factor model. Fortunately, the number of new concepts introduced here is small. The basic ideas noted for the two-factor model will apply, and thus the extension to three or more factors is relatively easy. The concepts will be described using the obvious extension of the canonical model from Section 6.2.3 and the *R*-notation from Section 6.3.1.3.

7.1.1. The Unconstrained, Three-Factor Model

The cell means model is written as

$$y_{ijkr} = \mu_{ijk} + e_{ijkr} \qquad \begin{aligned} i &= 1, \ldots, a \\ j &= 1, \ldots, b \\ k &= 1, \ldots, c \\ r &= 0, 1, \ldots, n_{ijk} \end{aligned} \tag{7.1}$$

where as usual the e_{ijk} are assumed to be independent $N(0, \sigma^2)$. The parameter estimates and their properties follow immediately from the general theory as

$$\hat{\mu}_{ijk} = \bar{y}_{ijk\cdot} \sim N(\mu_{ijk}, \sigma^2/n_{ijk}). \tag{7.2}$$

Here the $\hat{\mu}_{ijk}$ are independent, and they are independent of the estimate of σ^2 given by

$$s^2 = Q(\hat{\mu})/(N - abc) \tag{7.3}$$

where $N = n_{\dots}$ and

$$Q(\hat{\mu}) = \sum_i \sum_j \sum_k \sum_r (Y_{ijkr} - \bar{y}_{ijk\cdot})^2 \sim \sigma^2\chi^2(N - abc). \tag{7.4}$$

The canonical model, described in Section 6.2.3, may be extended by letting

$$\begin{aligned} \mu_{ijk} = \mu + \alpha_i + \beta_j + (\alpha\beta)_{ij} + \gamma_k + (\alpha\gamma)_{ik} \qquad & i = 1, \dots, a \\ + (\beta\gamma)_{jk} + (\alpha\beta\gamma)_{ijk} \qquad & j = 1, \dots, b \\ & k = 1, \dots, c. \end{aligned} \tag{7.5}$$

Here, we use the *marginal means* parameters

$$\begin{aligned}
\mu &= \bar{\mu}_{\dots} \\
\alpha_i &= \bar{\mu}_{i\cdot\cdot} - \bar{\mu}_{\dots} && i = 1, \dots, a-1 \\
\beta_j &= \bar{\mu}_{\cdot j\cdot} - \bar{\mu}_{\dots} && j = 1, \dots, b-1 \\
(\alpha\beta)_{ij} &= \bar{\mu}_{ij\cdot} - \bar{\mu}_{i\cdot\cdot} - \bar{\mu}_{\cdot j\cdot} + \bar{\mu}_{\dots} && i = 1, \dots, a-1 \\
&&& j = 1, \dots, b-1 \\
\gamma_k &= \bar{\mu}_{\cdot\cdot k} - \bar{\mu}_{\dots} && k = 1, \dots, c-1 \\
(\alpha\gamma)_{ik} &= \bar{\mu}_{i\cdot k} - \bar{\mu}_{i\cdot\cdot} - \bar{\mu}_{\cdot\cdot k} + \bar{\mu}_{\dots} && i = 1, \dots, a-1 \\
&&& k = 1, \dots, c-1 \\
(\beta\gamma)_{jk} &= \bar{\mu}_{\cdot jk} - \bar{\mu}_{\cdot j\cdot} - \bar{\mu}_{\cdot\cdot k} + \bar{\mu}_{\dots} && j = 1, \dots, b-1 \\
&&& k = 1, \dots, c-1 \\
(\alpha\beta\gamma)_{ijk} &= \mu_{ijk} - \bar{\mu}_{ij\cdot} - \bar{\mu}_{i\cdot k} - \bar{\mu}_{\cdot jk} + \bar{\mu}_{i\cdot\cdot} && i = 1, \dots, a-1 \\
&\quad + \bar{\mu}_{\cdot j\cdot} + \bar{\mu}_{\cdot\cdot k} - \bar{\mu}_{\dots} && j = 1, \dots, b-1 \\
&&& k = 1, \dots, c-1.
\end{aligned} \tag{7.6}$$

The remaining parameters in (7.5) are defined by

$$\alpha_a = -\sum_{i=1}^{a-1} \alpha_i$$

$$\beta_b = -\sum_{j=1}^{b-1} \beta_j$$

$$\gamma_c = -\sum_{k=1}^{c-1} \gamma_k$$

$$
\begin{aligned}
(\alpha\beta)_{ib} &= -\sum_{j=1}^{b-1}(\alpha\beta)_{ij} && i = 1, \ldots, a-1 \\
(\alpha\beta)_{aj} &= -\sum_{i=1}^{a-1}(\alpha\beta)_{ij} && j = 1, \ldots, b \\
(\alpha\gamma)_{ic} &= -\sum_{k=1}^{c-1}(\alpha\gamma)_{ik} && i = 1, \ldots, a-1 \\
(\alpha\gamma)_{ak} &= -\sum_{i=1}^{a-1}(\alpha\gamma)_{ik} && k = 1, \ldots, c \qquad (7.7) \\
(\beta\gamma)_{jc} &= -\sum_{k=1}^{c-1}(\beta\gamma)_{jk} && j = 1, \ldots, b-1 \\
(\beta\gamma)_{bk} &= -\sum_{j=1}^{b-1}(\beta\gamma)_{jk} && k = 1, \ldots, c \\
(\alpha\beta\gamma)_{ijc} &= -\sum_{k=1}^{c-1}(\alpha\beta\gamma)_{ijk} && i = 1, \ldots, a-1 \\
& && j = 1, \ldots, b-1 \\
(\alpha\beta\gamma)_{ibk} &= -\sum_{j=1}^{b-1}(\alpha\beta\gamma)_{ijk} && i = 1, \ldots, a-1 \\
& && k = 1, \ldots, c \\
(\alpha\beta\gamma)_{ajk} &= -\sum_{i=1}^{a-1}(\alpha\beta\gamma)_{ijk} && j = 1, \ldots, b \\
& && k = 1, \ldots, c.
\end{aligned}
$$

Let θ denote the abc-vector of new parameters in (7.6) in the order presented there. Note that these relations uniquely define θ in terms of μ. The additional parameters defined in (7.7) are primarily for notational convenience as noted earlier in the two-factor model. Note, too, that these relations define the additional parameters according to the same relations as in (7.6); that is,

$$\begin{aligned}
\alpha_a &= -\sum_{i=1}^{a-1}(\bar{\mu}_{i\cdot\cdot} - \bar{\mu}_{\cdots}) \\
&= \bar{\mu}_{a\cdot\cdot} - \bar{\mu}_{\cdots} \\
(\alpha\beta)_{ib} &= -\sum_{j=1}^{b-1}(\bar{\mu}_{ij\cdot} - \bar{\mu}_{i\cdot\cdot} - \bar{\mu}_{\cdot j\cdot} + \bar{\mu}_{\cdots}) \\
&= \bar{\mu}_{ib\cdot} - \bar{\mu}_{i\cdot\cdot} - \bar{\mu}_{\cdot b\cdot} + \bar{\mu}_{\cdots}
\end{aligned} \tag{7.8}$$

and so on. Let the combined vector of parameters of length $(abc + ab + ac + bc + a + b + c + 1)$ be denoted by θ_c.

The important point to be emphasized here is that the transformation from μ to θ is nonsingular and hence causes no conceptual difficulties. The remaining parameters arise naturally as functions of θ, as in the two-factor model.

Next, define the design matrix of dimension $abc \times (1 + a + b + ab + c + ac + bc + abc)$ as

$$\begin{aligned}
X_0 = [J_{abc}|I_a \otimes J_b \otimes J_c|J_a \otimes I_b \otimes J_c|I_a \otimes I_b \otimes J_c|J_a \otimes J_b \otimes I_c| \\
I_a \otimes J_b \otimes I_c|J_a \otimes I_b \otimes I_c|I_{abc}]
\end{aligned} \tag{7.9}$$

and the parameter matrix

$$P = \text{Diag}(1, \Delta_a', \Delta_b', \Delta_a' \otimes \Delta_b', \Delta_c', \Delta_a' \otimes \Delta_c', \Delta_b' \otimes \Delta_c', \Delta_a' \otimes \Delta_b' \otimes \Delta_c'). \tag{7.10}$$

Here, we recall from (6.8) that $\Delta_a = (I_{a-1}|-J_{a-1})$. With this notation, we note that

$$\theta_c = P\theta \tag{7.11}$$

and

$$E(Y) = WX_0P\theta = X\theta \tag{7.12}$$

or

$$E(Y) = WX_0\theta_c = X_c\theta_c. \tag{7.13}$$

Thus,

$$\mu = X_0\theta_c \tag{7.14}$$

as given in (7.5).

The natural hypotheses to be tested in a preliminary examination of the data will be described in terms of both the cell means model and the canonical model. These hypotheses, with the added restriction for now of $n_{ijk} \neq 0$, are as follows.

Three-factor interaction. The hypothesis is described by

$$H_{ABC}\colon (\Delta_a \otimes \Delta_b \otimes \Delta_c)\mu = 0 \tag{7.15}$$

or, equivalently,

$$H_{ABC}\colon (\alpha\beta\gamma)_{ijk} = 0 \qquad \begin{aligned} i &= 1, \ldots, a-1 \\ j &= 1, \ldots, b-1 \\ k &= 1, \ldots, c-1. \end{aligned} \tag{7.16}$$

The numerator sum of squares for testing this hypothesis is conveniently described, using the R-notation as applied to the canonical model, as

$$N_{ABC} = R[(\alpha\beta\gamma) \,|\, \mu, \alpha, \beta, (\alpha\beta), \gamma, (\alpha\gamma), (\beta\gamma)]. \tag{7.17}$$

Note that this quantity is easily computed by fitting the transformed model with and without the parameters $(\alpha\beta\gamma)_{ijk}$, $i = 1, \ldots, a-1; j = 1, \ldots, b-1; k = 1, \ldots, c-1$.

The test statistic is given by

$$\begin{aligned} &N_{ABC}/[(a-1)(b-1)(c-1)s^2] \\ &\quad \sim F[(a-1)(b-1)(c-1), N - abc, \lambda_{ABC}]. \end{aligned} \tag{7.18}$$

The interpretation of this hypothesis deserves a comment. It is easily seen that H_{ABC} is equivalent to assuming that the usual two-factor interactions are the same at all levels of the third factor. Thus, for example,

$$\mu_{ijk} - \mu_{ibk} - \mu_{ajk} + \mu_{abk} = \mu_{ijt} - \mu_{ibt} - \mu_{ajt} + \mu_{abt} \tag{7.19}$$

for any choice of $k, t = 1, \ldots, c$. Monlezun (1979) provides an interesting interpretation of three-factor interactions.

Two-factor interactions. There are three such hypotheses, corresponding to the three choices of two factors. We shall describe just one of them, as the others follow directly. The AB-interaction hypothesis is

$$H_{AB}\colon (\Delta_a \otimes \Delta_b \otimes (1/c)J_c')\mu = 0 \tag{7.20}$$

or, equivalently,

$$H_{AB}\colon (\alpha\beta)_{ij} = 0 \qquad \begin{aligned} i &= 1, \ldots, a-1 \\ j &= 1, \ldots, b-1. \end{aligned} \tag{7.21}$$

The numerator sum of squares is given by

$$N_{AB} = R[(\alpha\beta) \mid \mu, \alpha, \beta, \gamma, (\alpha\gamma), (\beta\gamma), (\alpha\beta\gamma)] \tag{7.22}$$

and is computed from the model, (7.12), with and without the parameters $(\alpha\beta)_{ij}$, $i = 1, \ldots, a - 1; j = 1, \ldots, b - 1$. It is important to realize that (7.22) must be computed from (7.12) and not from (7.13). The essential point is that the matrix, WX_0P, has full column rank (assuming $n_{ijk} \neq 0$), and the required computations are performed using the Sweep Operator as described in Section 6.3.1.2. The matrix WX_0 does not have full column rank. The usual techniques for solving equations of less than full rank can be applied, with and without $(\alpha\beta)_{ij}$, $i = 1, \ldots, a; j = 1, \ldots, b$, but the quantity (7.22), as defined, will be identically zero. This confusion does not arise in practice, as the full rank model, (7.12), is always used in this computation. (For further comments on this topic, see Speed and Hocking, 1976).

The test statistic for the hypothesis, H_{AB}, is given by

$$N_{AB}/[(a - 1)(b - 1)s^2] \sim F[(a - 1)(b - 1), N - abc, \lambda_{AB}]. \tag{7.23}$$

The interpretation of this test is aided by writing the hypothesis (7.20) in terms of the cell means as

$$H_{AB}: \bar{\mu}_{ij\cdot} - \bar{\mu}_{aj\cdot} - \bar{\mu}_{ib\cdot} + \bar{\mu}_{ab\cdot} = 0 \qquad \begin{array}{l} i = 1, \ldots, a - 1 \\ j = 1, \ldots, b - 1. \end{array} \tag{7.24}$$

In this form, we see that the hypothesis has the same interpretation as in the two-factor model except that, in this case, the means are averaged over all levels of the third factor.

Main effects. Again, there are three main effect hypotheses corresponding to the three factors. We shall discuss one of them—namely,

$$H_A: [\Delta_a \otimes (1/b)J_b' \otimes (1/c)J_c']\mu = 0 \tag{7.25}$$

or, equivalently,

$$H_A: \alpha_i = 0 \qquad i = 1, \ldots, a - 1. \tag{7.26}$$

The numerator sum of squares is given by

$$N_A = R[\alpha \mid \mu, \beta, (\alpha\beta), \gamma, (\alpha\gamma), (\beta\gamma), (\alpha\beta\gamma)] \tag{7.27}$$

as computed from (7.12). The test statistic is

$$N_A/[(a - 1)s^2] \sim F(a - 1, N - abc, \lambda_A). \tag{7.28}$$

This hypothesis is the extension of the main effect hypothesis in the two-factor model in which the different levels of factor A are compared when averaged over all levels of the other factors; that is, hypothesis (7.25), or (7.26), is equivalent to

$$H_A\colon \bar{\mu}_{i\cdot\cdot} = \bar{\mu}_{a\cdot\cdot} \qquad i = 1, \ldots, a-1. \tag{7.29}$$

For reference, these results are summarized in the Analysis of Means Table 7.1. As in the two-factor model, we shall refer to this as the *marginal means analysis*. In the last column we show, for the record, the expressions for the sums of squares in the special case in which $n_{ijk} = n$ for all i, j, k. Matrix expressions for these quadratic forms, in the case of equal cell frequencies, may be obtained in the obvious way, following the two-factor results in (6.141). This notation is generalized in Chapter 9.

There are several alternative hypotheses that are considered for main effects and two-factor interactions in the three-factor model. In Tables 7.2 and 7.3 we show the sequential and partially sequential analyses. The expressions for some of the hypotheses are simplified by defining

$$\mu^*_{ij} = \sum_{k=1}^{c} n_{ijk}\mu_{ijk}/n_{ij\cdot}. \tag{7.30}$$

TABLE 7.1 Marginal means analysis for the three-factor model.

Description	*Hypothesis*	*df*	*SS*	*SS* ($n_{ijk} = n$)
Factor A	H_A	$(a-1)$	N_A	$bcn\Sigma(\bar{y}_{i\ldots} - \bar{y}_{\ldots\cdot})^2$
Factor B	H_B	$(b-1)$	N_B	$acn\Sigma(\bar{y}_{\cdot j\cdot\cdot} - \bar{y}_{\ldots\cdot})^2$
AB interaction	H_{AB}	$(a-1)(b-1)$	N_{AB}	$cn\Sigma(\bar{y}_{ij\cdot\cdot} - \bar{y}_{i\ldots} - \bar{y}_{\cdot j\cdot\cdot} + \bar{y}_{\ldots\cdot})^2$
Factor C	H_C	$(c-1)$	N_C	$abn\Sigma(\bar{y}_{\cdot\cdot k\cdot} - \bar{y}_{\ldots\cdot})^2$
AC interaction	H_{AC}	$(a-1)(c-1)$	N_{AC}	$bn\Sigma(\bar{y}_{i\cdot k\cdot} - \bar{y}_{i\ldots} - \bar{y}_{\cdot\cdot k\cdot} + \bar{y}_{\ldots\cdot})^2$
BC interaction	H_{BC}	$(b-1)(c-1)$	N_{BC}	$an\Sigma(\bar{y}_{\cdot jk\cdot} - \bar{y}_{\cdot j\cdot\cdot} - \bar{y}_{\cdot\cdot k\cdot} + \bar{y}_{\ldots\cdot})^2$
ABC interaction	H_{ABC}	$(a-1)(b-1)$ $(c-1)$	N_{ABC}	$n\Sigma(\bar{y}_{ijk\cdot} - \bar{y}_{i\cdot k\cdot} - \bar{y}_{\cdot jk\cdot} + \bar{y}_{\cdot\cdot k\cdot}$ $- \bar{y}_{ij\cdot\cdot} + \bar{y}_{i\ldots} + \bar{y}_{\cdot j\cdot\cdot} - \bar{y}_{\ldots\cdot})^2$
Residual	—	$N - abc$	$Q(\hat{\mu})$	$\Sigma(y_{ijkr} - \bar{y}_{ijk\cdot})^2$

The hypotheses in Table 7.2 are defined by

$$H^S_A\colon \sum_j n_{ij\cdot}\mu^*_{ij}/n_{i\cdot\cdot} = \sum_j n_{aj\cdot}\mu^*_{aj}/n_{a\cdot\cdot}$$

$$H^S_B\colon \sum_i n_{ij\cdot} \sum_t n_{it\cdot}(\mu^*_{ij} - \mu^*_{it})/n_{i\cdot\cdot} = 0$$

$$H^S_{AB}\colon \mu^*_{ij} - \bar{\mu}^*_{i\cdot} - \bar{\mu}^*_{\cdot j} + \bar{\mu}^*_{\cdot\cdot} = 0 \tag{7.31}$$

$$H^S_C\colon \sum_{ij} n_{ijk} \sum_t n_{ijt}(\mu_{ijk} - \mu_{ijt})/n_{ij\cdot} = 0.$$

TABLE 7.2 Analysis of means for the sequential sums of squares method for the three-factor model.

Description	*Hypothesis*	*df*	*SS*
Factor A	H_A^S	$(a-1)$	$R(\alpha \mid \mu)$
Factor B	H_B^S	$(b-1)$	$R(\beta \mid \mu, \alpha)$
AB interaction	H_{AB}^S	$(a-1)(b-1)$	$R[(\alpha\beta) \mid \mu, \alpha, \beta]$
Factor C	H_C^S	$(c-1)$	$R[\gamma \mid \mu, \alpha, \beta, (\alpha\beta)]$
AC interaction	H_{AC}^S	$(a-1)(c-1)$	$R[(\alpha\gamma) \mid \mu, \alpha, \beta, (\alpha\beta), \gamma]$
BC interaction	H_{BC}^S	$(b-1)(c-1)$	$R[(\beta\gamma) \mid \mu, \alpha, \beta, (\alpha\beta), \gamma, (\alpha\gamma)]$
ABC interaction	H_{ABC}^S	$(a-1)(b-1)(c-1)$	N_{ABC}
Residual	—	$N - abc$	$Q(\hat{\mu})$

TABLE 7.3 Analysis of means for the partially sequential method for the three-factor model.

Description	*Hypothesis*	*df*	*SS*
Factor A	H_A^P	$(a-1)$	$R[\alpha \mid \mu, \beta, \gamma, (\beta\gamma)]$
Factor B	H_B^P	$(b-1)$	$R[\beta \mid \mu, \alpha, \gamma, (\alpha\gamma)]$
AB interaction	H_{AB}^P	$(a-1)(b-1)$	$R[(\alpha\beta) \mid \mu, \alpha, \beta, \gamma, (\alpha\gamma), (\beta\gamma)]$
Factor C	H_C^P	$(c-1)$	$R[\gamma \mid \mu, \alpha, \beta, (\alpha\beta)]$
AC interaction	H_{AC}^P	$(a-1)(c-1)$	$R[(\alpha\gamma) \mid \mu, \alpha, \beta, (\alpha\beta), \gamma, (\beta\gamma)]$
BC interaction	H_{BC}^P	$(b-1)(c-1)$	$R[(\beta\gamma) \mid \mu, \alpha, \beta, (\alpha\beta), \gamma, (\alpha\gamma)]$
ABC interaction	H_{ABC}^P	$(a-1)(b-1)(c-1)$	N_{ABC}
Residual	—	$N - abc$	$Q(\hat{\mu})$

Simple expressions for H_{AC}^S and H_{BC}^S are not available. The three-factor interaction hypothesis, H_{ABC}, is given by (7.15).

The hypotheses in Table 7.3 are defined by

$$H_A^P\colon \sum_{jk} n_{ijk} \sum_t n_{tjk}(\mu_{ijk} - \mu_{tjk})/n_{\cdot jk} = 0$$

$$H_B^P\colon \sum_{ik} n_{ijk} \sum_t n_{itk}(\mu_{ijk} - \mu_{itk})/n_{i\cdot k} = 0 \tag{7.32}$$

$$H_C^P\colon \sum_{ij} n_{ijk} \sum_t n_{ijt}(\mu_{ijk} - \mu_{ijt})/n_{ij\cdot} = 0.$$

Simple expressions for the two-factor interaction hypotheses are not available. The three-factor interaction hypothesis is the same as in Table 7.1. From these definitions, some of the hypotheses are seen to be the direct analogs of the two-factor hypotheses with μ_{ij} replaced by μ_{ij}^*.

Clearly, Table 7.2 is not unique, in that there are six possible sequential tables depending on the labeling of the factors. Similarly, Table 7.3 represents one of several possible choices for a partially sequential table. This particular choice was made to illustrate the SAS-Type II hypotheses. We further note that the analysis given in Table 7.1 corresponds to the SAS-Type III analysis. In the special case of equal cell frequencies, all three tables are identical, but not otherwise.

The choice of analysis depends upon the needs of the researcher and, as noted in Section 6.3.1.6, on the philosophies of the analyst with regard to changing the model based on nonsignificant interaction tests. The same comments apply here, so they will not be repeated. It is sufficient to note that, as general measures of main effects and two-factor interactions when the higher ordered interactions are present, the hypotheses in Table 7.1 are quite reasonable.

EXAMPLE 7.1

To illustrate these analyses, we recall the protein study from Section 6.3.1.8. That example is extended to three factors by recording the sex for each of the animals. The researcher suspected that there would be a difference in weight gain for the two sexes and, although this fact was not the purpose of the study, it was thought that removing this possible difference might improve the precision of the source-and-amount study. For illustration, we shall consider the unconstrained model, although it might be argued, as in Section 7.1.2, that some of the interactions could be assumed to be zero.

The cell frequencies for the three factors, A (source of protein), B (amount of protein), and C (sex) are shown in Table 7.4, and the sample cell means, $\bar{y}_{ijk\cdot}$, are shown in Table 7.5. The three analyses corresponding to Tables 7.1, 7.2, and 7.3, respectively, are shown in Tables 7.6, 7.7, and 7.8.

We shall not provide a detailed analysis of this data but simply make some observations about these AOM tables. The reader is encouraged to pursue the analysis further.

TABLE 7.4 Cell frequencies for the three-factor protein study.

		Males					*Females*		
		Amount					*Amount*		
		Low	Medium	High			Low	Medium	High
	Beef	2	2	3		Beef	2	3	3
Source	Pork	2	2	2	*Source*	Pork	1	2	3
	Grain	2	3	1		Grain	2	3	5

TABLE 7.5 Cell mean estimates for the three-factor protein study.

		Males					*Females*		
		Amount					*Amount*		
		Low	Medium	High			Low	Medium	High
Source	Beef	71.0	85.0	98.7	*Source*	Beef	81.0	87.7	105.0
	Pork	82.0	86.5	102.0		Pork	86.0	92.5	95.7
	Grain	93.5	88.0	98.0		Grain	74.0	79.0	83.8

TABLE 7.6 Marginal means analysis for the three-factor protein study.

Description	*Hypothesis*	*df*	*SS*	*F*	$P > F$
Source	H_A	2	130	1.7	0.202
Amount	H_B	2	1604	21.0	0.000
Source × amount	H_{AB}	4	382	2.5	0.068
Sex	H_C	1	46	1.2	0.282
Source × sex	H_{AC}	2	747	9.8	0.001
Amount × sex	H_{BC}	2	37	0.5	0.618
Source × amount × sex	H_{ABC}	4	152	1.0	0.43
Residual	—	25	954		

TABLE 7.7 Sequential analysis of means for the three-factor protein study.

Description	*Hypothesis*	*df*	*SS*	*F*	$P > F$
Source	H_A^S	2	389	5.1	0.014
Amount	H_B^S	2	1469	19.3	0.000
Source × amount	H_{AB}^S	4	622	4.1	0.011
Sex	H_C^S	1	100	2.6	0.120
Source × sex	H_{AC}^S	2	698	9.1	0.001
Amount × sex	H_{BC}^S	2	39	0.5	0.608

The most striking feature of the three tables is that the source factor (A) appears to be insignificant in Table 7.6, while it is significant ($\alpha = 0.05$) in Tables 7.7 and 7.8. This is just a consequence of the definition of the source effect, as given in Tables 7.2 and 7.3, as a result of differing computing philosophies with unequal cell

TABLE 7.8 Partially sequential analysis of means for the three-factor protein study.

Description	*Hypothesis*	*df*	*SS*	*F*	*P > F*
Source	H_A^P	2	279	3.7	0.040
Amount	H_B^P	2	1853	24.3	0.000
Source × amount	H_{AB}^P	4	384	2.5	0.068
Sex	H_C^P	1	45	1.2	0.289
Source × sex	H_{AC}^P	2	730	9.6	0.001
Amount × sex	H_{BC}^P	2	39	0.5	0.612

frequencies. We also note a significant source-by-sex (*AC*) interaction in all three tables even though the three measures of the effect are different. This is worth pursuing, especially in Table 7.6, since the researcher is led to believe that neither factor is important in this study. A detailed analysis of the cell means, using graphical methods as in Section 6.3.1.8 and supporting confidence intervals, is in order.

■

In the event of missing cells (some $n_{ijk} = 0$), the comments made in Section 6.3.1.9 are applicable; that is, serious consideration must be given to hypotheses that will test something close to those that would have been tested if all cells had been filled. The concept of the effective hypothesis provides a solution that may be acceptable but certainly needs to be examined carefully. The computations are performed as described in Section 6.3.1.9. The modification of this procedure, in which we set $(\alpha\beta\gamma)_{ijk} = 0$ if $n_{ijk} = 0$, is often used. It is not clear what hypothesis is tested by this procedure, but it is of interest to note that it does not correspond to SAS-Type III as it did in the two-factor case. Finally, we could use either of the analyses in Tables 7.2 and 7.3. Again, we emphasize that we are talking about the formulation of hypotheses to give preliminary indications of the presence of certain effects.

If it is not apparent what these hypotheses should be, as is often the case with missing cells, it seems that we should proceed directly to the second stage of the analysis, analyzing subhypotheses or developing confidence intervals. (For further discussion of the missing cell problem with three factors, see Hocking, 1981, or Hocking, Speed, and Coleman, 1980.)

7.1.2. The Constrained, Three-Factor Model

In the three-way and higher-way classification models, we have many options for imposing no-interaction constraints; that is, we may assume no three-factor (*ABC*) interaction, or we may further constrain the model by assuming that the *ABC* interactions are zero and that some of the two-factor interactions are zero. As might be expected, the strength of our conclusions with regard to lower ordered effects—for example, main effects and two-factor interactions—increases as we make assumptions about the higher ordered effects. To illustrate, we consider the three-factor model and assume first that there is no ABC interaction; that is, (7.19) holds for all choices of i, j, k, and t, or, equivalently, from (7.6) we may write

$$ABC\colon \mu_{ijk} - \bar{\mu}_{i\cdot k} - \bar{\mu}_{\cdot jk} + \bar{\mu}_{\cdot\cdot k} = \bar{\mu}_{ij\cdot} - \bar{\mu}_{i\cdot\cdot} - \bar{\mu}_{\cdot j\cdot} + \bar{\mu}_{\cdot\cdot\cdot} \quad \text{for all } i, j, k. \tag{7.33}$$

From this, we see that the AB-interaction hypothesis, normally written as

$$H_{AB}\colon \bar{\mu}_{ij\cdot} - \bar{\mu}_{i\cdot\cdot} - \bar{\mu}_{\cdot j\cdot} + \bar{\mu}_{\cdot\cdot\cdot} = 0 \tag{7.34}$$

is equivalent to the hypothesis

$$H_{AB}\colon \mu_{ijk} - \bar{\mu}_{i\cdot k} - \bar{\mu}_{\cdot jk} + \bar{\mu}_{\cdot\cdot k} = 0 \tag{7.35}$$

for all levels of k. The impact of this should be emphasized. Whereas (7.34) hypothesizes no interaction between factors A and B when the means are averaged over all levels of factor C, (7.35) hypothesizes no AB-interaction at any level of factor C, a much stronger statement. The analogous statements hold for the other two-factor interactions.

Similarly, if (7.33) holds, the A-effect hypothesis

$$H_A\colon \bar{\mu}_{i\cdot\cdot} = \mu_{\cdot\cdot\cdot} \tag{7.36}$$

is equivalent to the hypothesis

$$H_A\colon \mu_{ijk} - \bar{\mu}_{\cdot jk} - (\bar{\mu}_{i\cdot k} - \bar{\mu}_{\cdot\cdot k}) - (\bar{\mu}_{ij\cdot} - \bar{\mu}_{\cdot j\cdot}) = 0. \tag{7.37}$$

This is not a particularly revealing main effect hypothesis, but it is a stronger statement about the behavior of factor A than that given by (7.36).

Suppose we now assume, in addition to (7.33), that there is no AB- or AC-interaction; that is,

$$AB\colon \bar{\mu}_{ij\cdot} - \bar{\mu}_{i\cdot\cdot} - \bar{\mu}_{\cdot j\cdot} + \bar{\mu}_{\cdot\cdot\cdot} = 0 \tag{7.38}$$

$$AC\colon \bar{\mu}_{i\cdot k} - \bar{\mu}_{i\cdot\cdot} - \bar{\mu}_{\cdot k\cdot} + \bar{\mu}_{\cdot\cdot\cdot} = 0. \tag{7.39}$$

Combining (7.33), (7.38), and (7.39), we see that H_A is equivalent to

$$H_A\colon \mu_{ijk} = \bar{\mu}_{\cdot jk} \tag{7.40}$$

for all j, k; that is, (7.40) hypothesizes the equality of the response to the different levels of factor A for all combinations of levels of factors B and C.

It is now clear why we would like to assume that there are no three-factor or two-factor interactions. In that case, we could draw much stronger conclusions about the main effects. The resulting model, sometimes called the *additive model*, is thus very appealing, but we must remember that the strength of the conclusions does not justify the assumptions.

The computations for the analyses of these constrained models are straightforward. Using the R-notation, the sums of squares are computed as in Tables 7.1, 7.2, and 7.3 with the $(\alpha\beta\gamma)$ terms deleted if no three-factor interaction is assumed and with other two-factor terms deleted if the corresponding interactions are assumed to be zero. In all cases, the residual sums of squares are computed in the usual way; that is, $Q(\tilde{\mu}) = Y'Y - R(\mu, \alpha, \beta, \gamma, \ldots)$ for whatever terms are included in the model.

As a point of interest, note that, if we assume all interactions are zero, the analyses corresponding to Tables 7.1 and 7.3 are identical.

With missing cells, we have seen in our general discussion in Chapter 4 that, if the model is connected, all cell means are estimable and any linear hypothesis on the cell means can be tested. The question of determining connectedness was shown in Theorem 4.4 to be equivalent to the condition that the coefficient matrix in the reduced model is of full column rank. Analogously, the matrix $X = WX_0P$ in (7.12) must have full column rank. In the unconstrained model, this requires that all cells be filled. For constrained models, the question of connectedness depends on the constraints imposed. For example, it may be that assuming no ABC interaction is not sufficient, but adjoining the assumption that certain two-factor interactions are zero might achieve connectedness. Again, the end result does not justify the assumption, but this does point out that, in higher-way classifications, there may be various levels of connectedness.

The situation with regard to the unconnected case is essentially the same as that described in Section 6.3.2.4. Working with the effective model is preferred to working with the canonical model of less than full rank. The effective hypotheses are candidates for general preliminary tests in such cases, but as usual we prefer a detailed study of specific comparisons, rather than such general, and possibly misleading, tests. We shall not attempt to develop general results, but we shall illustrate the ideas with the following example.

EXAMPLE 7.2

An experiment was conducted to study the effect of long-term freezing on the rising of bread dough (see Murray et al., 1984). The three factors considered were as follows.

Factor A
  Chemical Additives, 2 levels
Factor B
  Flour Type, 3 levels
Factor C
  Freeze Time, 4 levels

The response was the amount the volume had increased four hours after removal from the freezer. Because of the cost and complexity of the experiment and because of some errors along the way, not all treatment combinations were observed. The missing cell patterns are shown in Table 7.9, with m and o denoting missing and observed cells.

TABLE 7.9 Missing cell pattern for the bread-dough experiment.

	Additive	*Flour*		*Freeze time* k			
				1	2	3	4
			1	o	o	m	m
	1	j	2	o	m	m	o
			3	m	o	m	o
i				k			
				1	2	3	4
			1	o	m	o	o
	2	j	2	m	o	o	m
			3	o	m	m	m

Discussions with the researchers suggested that it is reasonable to assume

$$\mu_{ijk} - \mu_{itk} = \mu_{ijr} - \mu_{itr} \tag{7.41}$$

for all i, j, t, k, r. The assumption allows us to conclude that there is no BC or ABC interaction. Unfortunately, these assumptions are not sufficient to achieve connectedness. For illustration, suppose we adjoin the assumption that there is no AB interaction. We still find that the model is not connected. On the other hand, if, rather than AB, we are able to justify the assumption of no AC interaction, then the model *is* connected. This example serves to emphasize the point that there are varying degrees of connectedness. The physical location of the missing cells is not sufficient to determine connectedness. We must also consider any assumptions made on the cell means. Note that once we have sufficient constraints to yield connectedness, additional constraints serve only to improve the precision of the estimates. Thus, for example, if we assume no interactions in this model, it is connected.

Returning to the original set of assumptions of no ABC or BC interaction, we consider the effective model and the resulting effective hypotheses. Following the usual reduction procedures described in Sections 4.2.3 and 4.3.3, we find that the single effective constraint is

$$\mu_{111} - \mu_{112} - \mu_{121} + \mu_{124} + \mu_{132} - \mu_{134} = 0. \tag{7.42}$$

The relations between observed and missing cells are given by

$$
\begin{array}{ll}
\textit{Missing} & \textit{Observed} \\
\mu_{113} - \mu_{133} & = \mu_{112} - \mu_{132} \\
\mu_{114} & = \mu_{112} - \mu_{132} + \mu_{134} \\
\mu_{122} & = \mu_{124} + \mu_{132} - \mu_{134} \\
\mu_{123} - \mu_{133} & = \mu_{124} - \mu_{134} \\
\mu_{131} & = \mu_{121} - \mu_{124} + \mu_{134} \\
\mu_{212} & = \mu_{213} + \mu_{222} - \mu_{223} \\
\mu_{221} & = \mu_{211} - \mu_{213} + \mu_{223} \\
\mu_{224} & = -\mu_{213} + \mu_{214} + \mu_{223} \\
\mu_{232} & = -\mu_{211} + \mu_{213} + \mu_{222} - \mu_{223} + \mu_{231} \\
\mu_{233} & = -\mu_{211} + \mu_{213} + \mu_{231} \\
\mu_{234} & = -\mu_{211} + \mu_{214} + \mu_{231}.
\end{array}
\tag{7.43}
$$

Note that the only cell means that are not uniquely estimable are μ_{113}, μ_{133}, and μ_{123}; hence, even though the model is not connected, we have nearly complete information with only half of the cells observed.

Development of the effective hypotheses shows that we can test the *AB* interaction and the *B* main effect as originally described in Table 7.1. Both the *AC*-interaction and the *C* main effect tests lose one degree of freedom, and there is no effective hypothesis for the *A* main effect. Additional discussions with the researchers regarding specific comparisons are required if alternative measures of the *A* main effect are to be considered.

■

7.2. NESTED MODELS

To this point, we have considered situations in which the treatment combinations represented all possible combinations of the levels of one factor with the levels of the second factor in the two-factor model, with the obvious extension of this idea to more factors. We now turn to an important class of models in which the treatment combinations may be described by two or more criteria, but they are not in a factorial arrangement.

We shall illustrate this concept with a simple model to fix the ideas and then turn to a more complex model. These models are often referred to as *nested,* or *hierarchical, models* or, as in the second illustration, *partially nested,* or *partially hierarchical, models.*

7.2.1. The Two-Fold Nested Model

To introduce the concept of a nested model, consider the following example.

EXAMPLE 7.3

Suppose we are interested in a study of fish captured in streams in various parks in a particular state. We have data on three parks. The first two parks each have two streams, while the third park has three streams. The data were obtained by interviewing fishermen at each stream, and the response is the number of fish captured on a given day. The data are shown in Table 7.10.

TABLE 7.10 Fish-capture data.

		Park						
		1		2		3		
		1	2	1	2	1	2	3
	1	6	5	4	2	3	2	4
	2	3	7	3	5	3	3	5
Stream	3	5	6		3	5	4	
	4		8		4		6	
	5				1		5	
	n_{ij}	3	4	2	5	3	5	2

The important point to stress here is that the same streams do not flow through each park. For example, stream one in park one is not the same as stream one in park two. This is the essential feature that distinguishes this situation from the two-factor factorial arrangement.

To model this data, let μ_{ij} denote the mean response from the jth stream in the ith park, and let y_{ijk} denote the kth response at that stream. Making the usual assumption about the error term, the model is

$$y_{ijk} = \mu_{ij} + e_{ijk} \qquad \begin{aligned} &i = 1, 2, \ldots, a = 3 \\ &j = 1, \ldots, b_i \; (b_1 = b_2 = 2, b_3 = 3) \\ &k = 1, \ldots, n_{ij}. \end{aligned} \qquad (7.44)$$

where the n_{ij} are shown on Table 7.10. We see, apart from the fact that the range on j is allowed to depend on i, that this model is identical to the one used for the two-way classification model in Chapter 6. In fact, the range on j could really be the same in all cases if we set $n_{13} = n_{23} = 0$. The important distinction is in the interpretation and the analysis. We shall see this as we consider meaningful hypotheses and confidence intervals on linear functions of the μ_{ij}.

The first question we might consider is a comparison of the three parks. If we consider the average of the mean responses, $\bar{\mu}_{i\cdot}$, as a measure of the park, then a reasonable hypothesis for comparing the parks is

$$H_P\colon \bar{\mu}_{1\cdot} = \bar{\mu}_{2\cdot} = \bar{\mu}_{3\cdot}. \tag{7.45}$$

Note that this hypothesis is identical to the main effect A hypothesis (6.101) in the two-factor model.

We next consider a comparison of streams. Now, however, it does not make sense to compare the quantities $\bar{\mu}_{\cdot j}$ since stream j is not the same in all parks. What is reasonable is a comparison of the mean responses of streams within a park; that is, μ_{ij} is compared with μ_{it} for $j \neq t$. If we make all such comparisons over all parks, the hypothesis is

$$H_{S/P}\colon \mu_{ij} = \mu_{it} \qquad \begin{matrix} i = 1, \ldots, a \\ j, t = 1, \ldots, b_i. \end{matrix} \tag{7.46}$$

Of course, other hypotheses might be more reasonable. For example, it might be argued that the number of fishermen, in the sample at a particular stream, is a reflection of the size, yield, or accessibility of the stream, and hence the appropriate measure of the park might be a weighted average of the cell means in that park. The hypothesis for comparing parks might then be,

$$H_P^*\colon \sum_j n_{ij}\mu_{ij}/n_{i\cdot} = \sum_j n_{tj}\mu_{tj}/n_{t\cdot} \tag{7.47}$$

for all $i, t = 1, \ldots, a$.

The analysis of this model involves no new concepts. For parameter estimates, the unconstrained analysis yields

$$\hat{\mu}_{ij} = \bar{y}_{ij\cdot} \sim N(\mu_{ij}, \sigma^2/n_{ij}) \tag{7.48}$$

for all i and j. These estimates are independent, and they are independent of the unbiased estimate of σ^2, given by

$$s^2 = Q(\hat{\mu})/(N - ab) \tag{7.49}$$

where $N = n_{\cdot\cdot}$ and

$$Q(\hat{\mu}) = \sum_{ijk}(y_{ijk} - \bar{y}_{ij\cdot})^2 \sim \sigma^2\chi^2(N - ab). \tag{7.50}$$

The development of the test statistics for the three hypotheses described above is also quite simple using previously developed algebra. The hypothesis, $H_{S/P}$, referred to as the *streams within parks hypothesis* may be written as

$$H_{S/P}\colon \mu_{ij} = \bar{\mu}_{i\cdot} \qquad \begin{array}{l} i = 1, \ldots, a \\ j = 1, \ldots, b_i - 1. \end{array} \tag{7.51}$$

Under this hypothesis, the reduced model is just the one-way classification model,

$$y_{ijk} = \bar{\mu}_{i\cdot} + e_{ijk} \tag{7.52}$$

with estimate

$$\hat{\bar{\mu}}_{i\cdot} = \bar{y}_{i\cdot\cdot} = \sum_j \sum_k y_{ijk}/n_{i\cdot} \tag{7.53}$$

and residual sum of squares

$$Q_{S/P} = \sum_{ijk}(y_{ijk} - \bar{y}_{i\cdot\cdot})^2. \tag{7.54}$$

Thus, the numerator sum of squares for this hypothesis with $\Sigma(b_i - 1) = (b. - a)$ degrees of freedom is

$$N_{S/P} = \sum_{ijk}(\bar{y}_{ij\cdot} - \bar{y}_{i\cdot\cdot})^2 \tag{7.55}$$

and the test statistic is

$$N_{S/P}/[(b. - a)s^2] \sim F(b. - a, N - ab, \lambda_{S/P}). \tag{7.56}$$

The hypothesis H_P is precisely that examined in Section 6.3.1.1 for the two-factor model under H_A. The constrained estimate of μ_{ij} under this hypothesis is developed in the same way, recognizing the variable range on the second subscript. We obtain,

$$\tilde{\mu}_{ij} = \bar{y}_{ij\cdot} - (h_{r_i}/n_{ij})(\bar{\bar{y}}_{i\cdot\cdot} - \bar{y}^*_r) \tag{7.57}$$

where

$$1/h_{r_i} = (1/b_i)\sum_j(1/n_{ij}) \tag{7.58}$$

$$\bar{\bar{y}}_{i\cdot\cdot} = (1/b_i)\sum_j \bar{y}_{ij\cdot}. \tag{7.59}$$

and

$$\overline{y}_r^* = \sum b_i h_{r_i} \bar{\bar{y}}_{i\cdot\cdot} \Big/ \sum b_i h_{r_i}. \tag{7.60}$$

The numerator sum of squares, with $a - 1$ degrees of freedom, is

$$N_P = \sum b_i h_{r_i} (\bar{\bar{y}}_{i\cdot\cdot} - \overline{y}_r^*)^2 \tag{7.61}$$

and the test statistic is

$$N_P/[(a - 1)s^2] \sim F(a - 1, N - ab, \lambda_P). \tag{7.62}$$

Reference to Section 6.3.1.4 shows that the hypothesis H_P^* is analogous to the hypothesis H_A^{**}, whose numerator sum of squares was denoted by $R(\alpha|\mu)$. Using this fact, we may immediately write the numerator sum of squares for the hypothesis H_P^* as

$$N_P^* = \sum n_{i\cdot}(\overline{y}_{i\cdot\cdot} - \overline{y}_{\cdot\cdot\cdot})^2 \tag{7.63}$$

with test statistic

$$N_P^*/[(a - 1)s^2] \sim F(a - 1, N - ab, \lambda_P^*). \tag{7.64}$$

Alternatively, we might appeal to the general results in Chapter 6, which yield the constrained estimator

$$\tilde{\mu}_{ij} = \overline{y}_{ij\cdot} - \overline{y}_{i\cdot\cdot} + \overline{y}_{\cdot\cdot\cdot} \tag{7.65}$$

and N_P^* is obtained directly.

A typical analysis for this model displays the results for H_P^* and $H_{S/P}$ in an AOM as shown in Table 7.11. The analysis is appealing in that we have simple expressions for the sums of squares. As usual, however, we must recognize that this table is just a convenient summary of the information for testing these hypotheses and that its simplicity does not justify the choice of hypotheses. If, for example, H_P is believed to be a more reasonable measure of park differences, then the test statistic (7.62) should be used. It is interesting to note that this latter test is rarely presented in textbooks or computer programs even though H_P is at least as reasonable as H_P^*.

TABLE 7.11 AOM table for the two-fold nested model.

Description	*Hypothesis*	*df*	*SS*
Between parks	H_P^*	$a - 1$	N_P^*
Streams/parks	$H_{S/P}$	$b. - a$	$N_{S/P}$
Residual	——	$N - ab$	$Q(\hat{\mu})$

The development of confidence intervals in this model is identical to that for the two-way classification. The only difference is in the types of contrasts that are of interest.

For automating the computations for this model it is common to express the model in a canonical form, as we did with the two-way classification model. We shall see that, with minor modifications, a single computer program could be used for the analysis of either model.

Motivated by the hypotheses H_P and $H_{S/P}$ and the development in Section 6.2.3, we define

$$\begin{aligned} \mu &= \sum \bar{\mu}_{i\cdot}/a \\ \alpha_i &= \bar{\mu}_{i\cdot} - \mu \qquad & i &= 1, \ldots, a-1 \\ \gamma_{ij} &= \mu_{ij} - \bar{\mu}_{i\cdot} \qquad & i &= 1, \ldots, a \\ & & j &= 1, \ldots, b_i - 1. \end{aligned} \tag{7.66}$$

[Note that the notation $(\beta/\alpha)_{ij}$, rather than γ_{ij}, is often used to suggest the hierarchical structure of the model.] Further, let the design matrix be given by

$$X_0 = [J_{b\cdot} | D_a(J_{b_i}) | I_{b\cdot}] \tag{7.67}$$

and the parameter matrix by

$$P_h = \text{Diag}[1, \Delta_a', D_a(\Delta_{b_i}')]. \tag{7.68}$$

Letting θ_h denote the vector of parameters in (7.66), we may now write the model in canonical form as

$$E(Y) = WX_0P_h\theta_h. \tag{7.69}$$

Here $W = D_a(D_{b_i}(J_{n_{ij}}))$ is the counting matrix of dimension $N \times \Sigma b_i$. From (7.69) it is suggested that we define

$$\begin{aligned} \alpha_a &= -\sum_{i=1}^{a-1} \alpha_i = \bar{\mu}_{a\cdot} - \mu \\ \gamma_{ib_i} &= -\sum_{j=1}^{b_i-1} \gamma_{ij} = \mu_{ib_i} - \bar{\mu}_{i\cdot} \qquad i = 1, \ldots, a. \end{aligned} \tag{7.70}$$

Combining (7.66) and (7.70) into a single vector θ_c, we have $\theta_c = P_h\theta_h$, and hence the canonical model in matrix form is given by

$$E(Y) = WX_0\theta_c = X_c\theta_c \tag{7.71}$$

where $X_c = WX_0$. Alternatively, we may write the canonical model algebraically as

$$E(y_{ijk}) = \mu + \alpha_i + \gamma_{ij}. \tag{7.72}$$

In terms of the model (7.69), the sums of squares for the three hypotheses are easily expressed using the R-notation from Section 6.3.1.3. In particular, we have

$$\begin{aligned} N_P^* &= R(\alpha \mid \mu) \\ N_P &= R(\alpha \mid \mu, \gamma) \\ N_{S/P} &= R(\gamma \mid \mu, \alpha). \end{aligned} \tag{7.73}$$

It is now clear that Table 7.11 corresponds to a sequential analysis of this canonical model.

Inspection of the design matrix (7.67) shows, in the case $b_i = b$, $i = 1, \ldots, a$, that

$$X_0 = (J_a \otimes J_b \mid I_a \otimes J_b \mid I_{ab}) \tag{7.74}$$

which is obtained from (6.94) by simply deleting the β columns from the two-way classification design matrix. Similarly, in this case the parameter matrix is given by

$$P_h = \text{Diag}(1, \Delta_a', I_a \otimes \Delta_b'). \tag{7.75}$$

The parameter vector θ_h in this case is

$$\theta_h' = \begin{bmatrix} \mu & i = 1, \ldots, a - 1 \\ \alpha_i & i = 1, \ldots, a \\ \gamma_{ij} & j = 1, \ldots, b - 1) \end{bmatrix}. \tag{7.76}$$

The notation is cumbersome for the design and parameter matrices in (7.67) and (7.68), for which the range on j is variable, and the notation will be even worse when we consider a more complex model in the next section. A simple device to simplify the notation is to use (7.74) and (7.75) as the design and parameter matrices for the general case and to modify the counting matrix W to account for the fact that the range on j depends on i. To do so, we use the same counting matrix as in the two-factor model—that is, $W = D_a(D_b(J_{n_{ij}}))$, if we let $n_{13} = n_{23} = 0$ and set $\gamma_{12} = \gamma_{22} = 0$. The obvious similarities between the two models clearly suggest the development of a single computer program that would accommodate both models.

Returning to the fish-capture data in Table 7.10, we find that the residual sum of squares and the numerator sum of squares described in (7.73) are

$$Q(\hat{\mu}) = 33.3 \qquad N_P^* = 24.2$$

$$N_P = 18.1 \qquad N_{S/P} = 6.9.$$

The test statistics are

$$F_P^* = 6.17 \qquad F_P = 4.61 \qquad F_{S/P} = 0.89.$$

The probabilities of greater F-ratios under the null hypotheses in these three cases are, respectively, Prob(P^*) = 0.01, Prob(P) = 0.025, and Prob(S/P) = 0.89. Thus, our preliminary indications are that, within parks, there are no differences between the streams but that there appear to be differences between the parks. Note that both H_P and H_P^* would indicate significant differences ($\alpha = 0.05$), but the sense in which this is interpreted is given by the precise statements of these hypotheses. It is clear that the parks might be judged different in one sense and not in the other if the cell frequencies are sufficiently unbalanced. We shall not pursue the analysis of this data, but in a typical analysis we would attempt to isolate the differences using simultaneous confidence intervals.

■

7.2.2. A Partially Nested Model

The concept of nesting can be extended to more complex models. The following example will illustrate the general ideas.

EXAMPLE 7.4

Suppose we are interested in the pollution of water emitted from laundromats as a function of the type of detergent and the type of washing machine used. Assume that there are three types of machines primarily used in laundromats and that four detergents are being studied. For the experiment, we have the cooperation of seven laundromats, but, unfortunately, each laundromat uses exclusively one type of machine. Of the seven, two laundromats use machine-types one and three, and three laundromats use machine-type two. Also, the laundromats have differing numbers of machines. For the experiment, the detergents are allocated at random to the machines within a laundromat. The response is the average amount of phosphorus in grams per liter from daily one-hour samples over a seven-day period.

Our model for this experiment is

$$y_{ijkr} = \mu_{ijk} + e_{ijkr} \qquad \begin{aligned} i &= 1, 2, \ldots, a = 4 \\ j &= 1, 2, \ldots, b = 3 \\ k &= 1, 2, \ldots, c_j \; (c_1 = c_3 = 2, c_2 = 3) \\ r &= 1, 2, \ldots, n_{ijk}. \end{aligned} \tag{7.77}$$

Here, μ_{ijk} is the mean response (grams/liter of phosphorus) from the ith detergent in the kth laundromat using machine type j. To fix the ideas, the model is shown schematically in Table 7.12.

TABLE 7.12 Cell means for the detergent study.

		Machine type						
		1		2			3	
		Laundromats within machine types						
		1	2	1	2	3	1	2
Detergent	1	μ_{111}	μ_{112}	μ_{121}	μ_{122}	μ_{123}	μ_{131}	μ_{132}
	2	μ_{211}	μ_{212}	μ_{221}	μ_{222}	μ_{223}	μ_{231}	μ_{232}
	3	μ_{311}	μ_{312}	μ_{321}	μ_{322}	μ_{323}	μ_{331}	μ_{332}
	4	μ_{411}	μ_{412}	μ_{421}	μ_{422}	μ_{423}	μ_{431}	μ_{432}

We shall see in Chapter 9 that this model may not be appropriate for this experiment and that a mixed model (nonscalar covariance matrix) is suggested. For now, we shall assume the fixed effect model for illustration. Also, we shall see in Chapter 10 that a common approach to analyzing mixed models is to first consider an analogous fixed effects model. The numerator sums of squares for the fixed effects analog will prove useful for the mixed model analysis.

The model in (7.77), apart from the variable range on the third subscript, looks exactly like the three-factor model, (7.1). (Even this difference is easily removed by letting $k = 1, \ldots, c = 3$ and defining $n_{i13} = n_{i33} = 0$ for $i = 1, \ldots, 4$.) Note that the two factors, detergent and machine type, occur in a factorial arrangement; however, this is not true for laundromats. While this does not affect the model statement, it does affect the analysis when we attempt to describe meaningful linear functions of the cell means. To illustrate, we shall describe a particular set of hypotheses that might be of interest.

As a measure of the detergent effect, we might consider the average over machines and laundromats in the usual way and test the hypothesis

$$H_D\colon \bar{\mu}_{i..} = \bar{\mu}_{t..} \qquad i \neq t = 1, \ldots, a. \tag{7.78}$$

As a measure of the machine effect, we consider the analogous hypothesis

$$H_M\colon \bar{\mu}_{.j.} = \bar{\mu}_{.t.} \qquad j \neq t = 1, \ldots, b. \tag{7.79}$$

The question of whether the detergent effect is the same for all machines may be posed as usual in terms of the detergent-by-machine interaction hypothesis

$$H_{DM}\colon \bar{\mu}_{ij\cdot} - \bar{\mu}_{iv\cdot} - \bar{\mu}_{tj\cdot} + \bar{\mu}_{tv\cdot} = 0 \qquad \begin{array}{l} i \neq t = 1, \ldots, a \\ j \neq v = 1, \ldots, b. \end{array} \tag{7.80}$$

These hypotheses emphasize the factorial arrangement of these two factors. Of course, different measures of these effects as described in Chapter 6 and in Section 7.1.1 might be considered.

Continuing, we see that the analog of (7.78) and (7.79) for laundromats does not make sense, since, for example, laundromat one using the first type of machine may be unrelated to laundromat one using the second type of machine. We may, however, wish to compare laundromats for a given machine type averaged over detergents. If we make such comparisons for all machine types, the hypothesis is

$$H_{L/M}\colon \bar{\mu}_{\cdot jk} = \bar{\mu}_{\cdot jt} \qquad \begin{array}{l} j = 1, \ldots, b \\ k \neq t = 1, \ldots, c_j. \end{array} \tag{7.81}$$

Also, we may ask whether the detergent effect is the same for all laundromats within a machine type—that is, the question of a detergent-by-laundromat interaction within machine type. Considering this question over all machine types leads us to the hypothesis

$$H_{DL/M}\colon \mu_{ijk} - \mu_{ijv} - \mu_{tjk} + \mu_{tjv} = 0 \qquad \begin{array}{l} i \neq t = 1, \ldots, a \\ j = 1, \ldots, b \\ k \neq v = 1, \ldots, c_j. \end{array} \tag{7.82}$$

Clearly, many other hypotheses could be formulated, but these will suffice for demonstration. The estimation problem here is simple. Point estimates of the parameters are the same as for the three-factor model and are given by (7.2) and (7.3), with appropriate changes in the residual degrees of freedom. Similarly, tests of the hypotheses (7.78) through (7.82) may be developed in the usual way using the results of Chapter 3. Rather than developing detailed expressions for the test statistics, we shall go directly to a canonical form for computations. This canonical form is motivated by our previous developments and is the one that is commonly used in computer programs. We shall develop the canonical form for the particular unbalanced model (7.77) and simply note the modifications necessary when the range on k is constant. This model is unbalanced in the sense that the cell frequencies n_{ijk} may differ. Also, since the c_j are not constant, we have, effectively, a missing cell problem. The consequences of this imbalance will be noted in our discussion of a numerical illustration of this example.

To develop the canonical form, we define

$$
\begin{aligned}
\mu &= (1/ab)\sum_{ij} \bar{\mu}_{ij\cdot} \\
\alpha_i &= (1/b)\sum_{j} \bar{\mu}_{ij\cdot} - \mu && i = 1, \ldots, a-1 \\
\beta_j &= \bar{\mu}_{\cdot j\cdot} - \mu && j = 1, \ldots, b-1 \\
(\alpha\beta)_{ij} &= \bar{\mu}_{ij\cdot} - \alpha_i - \beta_j - \mu && i = 1, \ldots, a-1 \\
& && j = 1, \ldots, b-1 \\
(\gamma/\beta)_{jk} &= \bar{\mu}_{\cdot jk} - \beta_j - \mu && j = 1, \ldots, b \\
& && k = 1, \ldots, c_j - 1 \\
(\alpha\gamma/\beta)_{ijk} &= \mu_{ijk} - \bar{\mu}_{ij\cdot} - (\gamma/\beta)_{jk} && i = 1, \ldots, a-1 \\
& && j = 1, \ldots, b \\
& && k = 1, \ldots, c_j - 1.
\end{aligned}
\tag{7.83}
$$

We note in the special case, $c_j = c$, that μ is simply the average of the cell means, $\bar{\mu}_{\cdots}$ and $\alpha_i = \bar{\mu}_{i\cdot\cdot} - \bar{\mu}_{\cdots}$. The fact that we have not used these definitions in this unbalanced case is motivated by our desire to define the canonical model to be consistent with that which is commonly used. The consequences of this will be noted.

To write the model in matrix form, we proceed as in Example 7.3, writing the design and parameter matrices in a general form and then using the counting matrix to adjust for the variable range on k. The design matrix is given as follows, with $c = \max c_j$. In this case, $c = 3$.

$$X_0 = (J_{abc} | I_a \otimes J_{bc} | J_a \otimes I_b \otimes J_c | I_{ab} \otimes J_c | J_a \otimes I_{bc} | I_{abc}). \tag{7.84}$$

Note that the design matrix is obtained from that for the three-factor model in (7.9) by deleting the columns corresponding to γ and $\alpha\gamma$. The appropriate parameter matrix is

$$P_h = \text{Diag}(1, \Delta_a', \Delta_b', \Delta_a' \otimes \Delta_b', I_b \otimes \Delta_c', \Delta_a' \otimes I_b \otimes \Delta_c'). \tag{7.85}$$

As described, X_0 and P_h are appropriate for the case in which the ranges on i, j, and k are constant. The vector of parameters, θ_h, would include

$$
\begin{aligned}
&(\gamma/\beta)_{jk} && j = 1, \ldots, b \\
& && k = c_j, \ldots, c - 1 \\
&(\alpha\gamma/\beta)_{ijk} && i = 1, \ldots, a - 1, \\
& && j = 1, \ldots, b, \\
& && k = c_j, \ldots, c - 1.
\end{aligned} \tag{7.86}
$$

To apply this notation to the model with variable range on the subscript j, we simply set these additional parameters to zero and write $W = D_a\{D_b[D_c(J_{n_{ijk}})]\}$ with $n_{ijk} = 0$ for $k = c_j + 1, \ldots, c$. Again, this illustrates how a single computer program could be developed to do the canonical analysis on a general class of factorial, nested, or partially nested models.

With these definitions and with θ_h denoting the parameters in (7.83) and the extraneous parameters in (7.86), we have the canonical form

$$E(Y) = WX_0P_h\theta_h. \tag{7.87}$$

As usual, it is suggested that we define, for convenience, the vector $\theta_c = P_h\theta_h$ where θ_c is the vector of parameters in (7.83) and (7.86) augmented by

$$
\begin{aligned}
\alpha_a &= -\sum_{i=1}^{a-1} \alpha_i & \beta_b &= -\sum_{j=1}^{b-1} \beta_j \\
(\alpha\beta)_{aj} &= -\sum_{i=1}^{a-1} (\alpha\beta)_{ij} & & j = 1, \ldots, b - 1 \\
(\alpha\beta)_{ib} &= -\sum_{j=1}^{b-1} (\alpha\beta)_{ij} & & i = 1, \ldots, a \\
(\gamma/\beta)_{jc} &= -\sum_{k=1}^{c-1} (\gamma/\beta)_{jk} & & j = 1, \ldots, b \\
(\alpha\gamma/\beta)_{ajk} &= -\sum_{i=1}^{a-1} (\alpha\gamma/\beta)_{ijk} & & j = 1, \ldots, b, \\
& & & k = 1, \ldots, c - 1 \\
(\alpha\gamma/\beta)_{ijc} &= -\sum_{k=1}^{c-1} (\alpha\gamma/\beta)_{ijk} & & i = 1, \ldots, a, \\
& & & j = 1, \ldots, b.
\end{aligned} \tag{7.88}
$$

This enables us to write the canonical model as

$$E(Y) = WX_0\theta_c = X_c\theta_c \tag{7.89}$$

where $X_c = WX_0$. Equivalently, we may write

$$\begin{aligned} \mu_{ijk} = \mu + \alpha_i + \beta_j + (\alpha\beta)_{ij} \qquad & i = 1, \ldots, a \\ + (\gamma/\beta)_{jk} + (\alpha\gamma/\beta)_{ijk} \qquad & j = 1, \ldots, b \\ & k = 1, \ldots, c. \end{aligned} \tag{7.90}$$

The model, in the form (7.89), must be used with care, as X_c does not have full column rank. While it may be convenient to use (7.90) as a descriptive model, the original cell means model should be used for interpretation. The canonical form, (7.87), should be used for computations, if appropriate. As noted, this canonical form was motivated by earlier forms of the model but should be analyzed carefully before being used to generate test statistics. We shall describe five distinct analyses for this example and illustrate the differences numerically. Where possible, we shall describe the sum of squares for the various hypotheses in terms of the R-notation as applied to the canonical model.

In Table 7.13 we show the sample cell means and cell frequencies for the detergent-study example. This data will be subjected to several of the standard analyses.

TABLE 7.13 Sample cell means for the detergent study. (Cell frequencies in parentheses.)

		Machine type						
		1		2			3	
		Laundromats within machine types						
		1	2	1	2	3	1	2
Detergents	1	6.0	5.5	2.5	3.5	4.0	4.0	4.5
		(3)	(2)	(2)	(2)	(3)	(2)	(2)
	2	4.5	4.0	3.0	4.0	4.0	3.0	3.0
		(2)	(3)	(2)	(2)	(2)	(3)	(2)
	3	4.5	4.0	3.7	4.5	4.0	1.5	2.0
		(4)	(2)	(3)	(2)	(2)	(2)	(2)
	4	2.0	4.0	2.3	2.0	3.0	1.0	1.3
		(3)	(3)	(3)	(2)	(3)	(2)	(3)

First, assuming we are interested in hypotheses (7.78) through (7.82), we ask whether the usual reduced model computations on the canonical model will yield the appropriate sums of squares. The answer is in the affirmative if we are in the special case $c_j = c$, but not otherwise. The only difference occurs in the A-effect—in this case, the detergent effect—where the hypothesis corresponding to $R[\alpha|\mu, \beta, (\alpha\beta), (\gamma/\beta), (\alpha\gamma/\beta)]$ is

$$H_D^*: \sum_j \bar{\mu}_{ij\cdot} = \sum_j \bar{\mu}_{aj\cdot} \qquad i = 1, \ldots, a - 1. \tag{7.91}$$

The corresponding Analysis of Means is shown in Table 7.14. (Note: For example, $R(\alpha|\text{all})$ means the reduction in the regression sum of squares obtained by deleting α, $i = 1, \ldots, a - 1$, from the canonical form but keeping all of the remaining parameters.)

TABLE 7.14 Analysis of means for the detergent study using the canonical model.

Description	*Hypothesis*	*df*	*SS*	*SS for example*	$P > F$
Detergent	H_D^*	$a - 1 = 3$	$R(\alpha\|\text{all})$	42.5	0.00
Machine type	H_M	$b - 1 = 2$	$R(\beta\|\text{all})$	30.0	0.00
Detergent × machine type	H_{DM}	$(a - 1)(b - 1) = 6$	$R[(\alpha\beta)\|\text{all}]$	17.0	0.01
Laundromat within machine type	$H_{L/M}$	$c. - b = 4$	$R[(\gamma/\beta)\|\text{all}]$	4.4	0.23
Detergent × laundromat within machine type	$H_{DL/M}$	$(a - 1)(c. - b) = 12$	$R[(\alpha\gamma/\beta)\|\text{all}]$	9.3	0.45
Residual	—	$N - ac. = 27$	$Q(\mu)$	30.5	

We have noted, in the two-way classification with missing data, that this regression analysis of the canonical model is equivalent to the Type III analysis of SAS. This is not true for more complex models such as the three-way classification nor for this partially nested model. In this case, the only difference is in the sum of squares for the detergent effect. The Type III sum of squares for our example is

$$N_{D_{\text{III}}} = 41.4, \qquad P > F = 0.00.$$

The hypothesis being tested by this sum of squares is

$$H_{D_{\text{III}}}: \sum_j \bar{\mu}_{ij\cdot}/(c_j + 1) = \sum_{ij} \bar{\mu}_{ij\cdot}/(c_j + 1) \qquad i = 1, 2, 3. \tag{7.92}$$

All other sums of squares for SAS-Type III agree with those given in Table 7.14.

This rather unusual hypothesis is a consequence of the computational procedure used by SAS but has no other justification.

If we wish to test the hypothesis (7.78) as a measure of the detergent effect, we obtain

$$N_D = 37.76, \qquad P > F = 0.00.$$

It happens in this case that this is exactly the sum of squares computed under the SAS-Type IV analysis. Otherwise, Type IV agrees with Table 7.14. The sum of squares, N_D, cannot be easily obtained from this canonical form, but it is easily obtained by working directly with the cell means model.

As a fourth analysis, we might consider the sequential analysis shown in Table 7.15. The hypotheses for the factorial effects, detergents and machines and their interaction, are the same as the three-factor analysis in Table 7.2. The laundromat-within-machine-type hypothesis, $H^*_{L/M}$, is a complex function of the cell frequencies.

TABLE 7.15 Sequential analysis of means for the detergent study.

Description	*Hypothesis*	*df*	*SS*	*SS for example*	$P > F$
Detergent	H^S_D	3	$R(\alpha \mid \mu)$	38.9	0.00
Machine type	H^S_M	2	$R(\beta \mid \mu, \alpha)$	31.1	0.00
Detergent × machine type	H^S_{DM}	6	$R[(\alpha\beta) \mid \mu, \alpha, \beta]$	16.3	0.01
Laundromat within machine type	$H^*_{L/M}$	4	$R[(\gamma/\beta) \mid \mu, \alpha, \beta, (\alpha\beta)]$	4.5	0.22
Detergent × laundromat within machine type	$H_{DL/M}$	12	$R[(\alpha\gamma/\beta) \mid \text{all})$	9.3	0.46

Finally, we present a partially sequential analysis in Table 7.16. The hypothesis for the first three rows of the table are the same as in Table 7.3 for the three-factor analysis, and the remaining two are the same as in Table 7.15.

TABLE 7.16 Partially sequential analysis of means for the detergent study.

Description	*Hypothesis*	*df*	*SS*	*SS for example*	$P > F$
Detergent	H^P_D	3	$R(\alpha \mid \mu, \beta)$	38.3	0.00
Machine type	H^P_M	2	$R(\beta \mid \mu, \alpha)$	31.1	0.00
Detergent × machine type	H^P_{DM}	6	$R[(\alpha\beta) \mid \mu, \alpha, \beta]$	16.3	0.1

The analysis in Table 7.16 differs from the SAS-Type II analysis in that, for Type II, the sum of squares for detergents is computed by $R[\alpha \mid \mu, \beta, (\gamma/\beta)] = 38.3$ and the detergent-by-machine interaction sum of squares is computed as $R[(\alpha\beta) \mid \mu, \alpha, \beta, (\alpha\gamma/\beta)] = 17.7$. These unusual sums of squares are a consequence of doing the computations as in the three-factor model analysis. The hypothesis associated with these sums of squares are, again, complex functions of the cell frequencies.

In this particular problem, the conclusions on the presence of effects from the preliminary analysis would be the same for all five analyses, but it is easy to construct cases where they would differ. Table 7.16 is often presented as the analysis for this model. It would appear that, lacking specific hypotheses of interest, Table 7.14 would be a more natural default hypothesis with, perhaps, H_D as the detergent hypothesis.

■

Nested and partially nested models of greater complexity frequently arise. Based on this discussion, a general approach to the analysis is to postulate a cell means model, which may be constrained, and then spell out reasonable hypotheses as general indicators of effects. These hypotheses may suggest a transformation to a canonical form for which the computations may be performed as indicated by the R-notation. It is clear that a standard canonical form should not be used without reference to the meaning of the parameter vector implicit in the definition of the parameter matrix. The hypotheses being tested are completely determined by that definition.

Finally, we note that nested and partially nested models are often viewed as mixed models. We shall return to a study of the detergent example in Section 9.3.1, where a mixed model analysis is given.

7.3. BALANCED INCOMPLETE BLOCK DESIGNS

We have seen, for the class of connected models, that the problem of missing cells is not a serious one in the sense that we can estimate all parameters and test any hypothesis about the cell means just as if all cells were observed. This fact raises the possibility of intentionally designing experiments so that we can perform the desired analyses without observing all cells. In many cases, the cost and/or difficulty of performing the experiments may dictate such a design. In other cases, it may be impossible to observe some of the combinations; hence, the missing cells are unavoidable. To illustrate the ideas, we consider a simple example.

EXAMPLE 7.5

Consider an experiment in which we wish to compare the effectiveness of three suntan lotions. We have three volunteers for the experiment, but, to avoid confounding the differences in the individuals with the differences in the lotions, we cannot assign one lotion to each individual. Instead, we decide to apply the lotions to the volunteers' arms, with one lotion to each arm. Since we have three lotions and only two arms per individual, we cannot compare all lotions on each individual, but we can compare two lotions per individual. The lotions are allocated to the individuals as shown in Table 7.17. Thus, the first individual receives lotions one and two, the second receives lotions one and three, and the third receives lotions two and three.

TABLE 7.17 Incidence matrix for the suntan-lotion study.

		Individual		
		1	2	3
Lotion	1	1	1	0
	2	1	0	1
	3	0	1	1

Note that the incidence matrix in Table 7.17 is identical to that in Table 6.3, with $n_{ij} = 1$ for the observed cells. In Example 6.5 we noted that, if we assumed the two-factor model without interaction, the model is connected and, hence, any hypothesis on the cell means can be tested. In particular, the hypothesis of no difference in lotions would be tested. Technically, we could test for equality of individuals, but this is not likely to be of interest in this example. In other applications, both factors might be of interest.

■

In this section, we shall discuss a class of two-factor experiments that have missing cells in a particular pattern. The experimental layouts are known as *Balanced Incomplete Block* (BIB) *designs* and are generally discussed as special cases of the randomized-block concept introduced in Section 6.2.2.4. We have used the term *balance* to refer to the situation in which we have the same number of observations on each population. For the BIB design this is clearly not the case, since certain treatment combinations are not observed. In this case, the term reflects the fact that the missing cells occur in a particular pattern. A consequence of this is that all treatment differences, $\mu_{ij} - \mu_{vj}$, are estimated with the same precision. For our purposes, we shall assume the two-factor model without interaction with at most one observation per cell, and we shall develop the analysis as in Chapter 6. In the case where one of the factors is a blocking variable, as it is in the suntan-lotion example, the model is generally considered to be a mixed model, and we shall consider that analysis in Chapter 9. Our detailed discussion of this fixed effects, two-factor model is of interest in its own right and also will be useful in the discussion in Chapter 9.

We shall not provide a thorough discussion of BIB designs but instead will use this simple example to motivate the ideas. [For more details, see, for example, Kempthorne (1952), Graybill (1961, 1976), or John (1971).]

From Table 7.17, we note the following properties, which we define in general terms.

1. There are b columns (blocks), and each column contains k observed cells with one observation.
2. There are t rows (treatments) with $t > k$ and $b \geq t$.
3. There are r filled cells in each row.
4. The filled cells for any pair of rows occur together in λ columns.

In Example 7.5, $b = 3$, $t = 3$, $k = 2$, $r = 2$, and $\lambda = 1$. These five parameters cannot vary independently but must satisfy the following relations ($n..$ is the total number of observations with $n_{ij} = 0$ or 1).

$$\begin{aligned} n = n.. &= bk = rt \\ \lambda(t - 1) &= r(k - 1). \end{aligned} \tag{7.93}$$

These relations serve to define the balanced incomplete block design.

The analysis of such experiments introduces no new concepts. The results of Chapter 6 on the no-interaction model apply directly. Either the reduced model or the canonical model may be used. In this case, since the missing cells occur in a particular pattern, it is possible to obtain algebraic expressions for the estimates of the cell means and for the test statistics for certain hypotheses. We shall indicate these results, recognizing that, in practice, the computations could be performed using a standard program as indicated earlier.

Following the development in Section 6.2.2.2, we may write the model as

$$E(Y) = X_c\theta_c \tag{7.94}$$

where

$$X_c = WX_0 \tag{7.95}$$

W is a diagonal matrix of ones and zeros denoting observed and unobserved cells, and

$$X_0 = (J_{tb} | I_t \otimes J_b | J_t \otimes I_b). \tag{7.96}$$

Here, $\theta_c = P_m\theta_m$ where P_m is the parameter matrix for the second canonical form, given by (6.50). Let θ be written in partitioned form as

$$\theta_c' = (1, \tau', \beta') \tag{7.97}$$

where $\tau' = (\tau_1, \ldots, \tau_t)$ and $\beta' = (\beta_1, \ldots, \beta_b)$. (Note that, to conform to standard BIB notation, we are using τ_i, $i = 1, \ldots, t$ to correspond to the first factor, rather than α_i, $i = 1, \ldots, a$, as in Chapter 6.)

It may be shown that $X_m = WX_0P_m$ has full column rank, $b + t - 1$; hence, by Theorem 6.1, all cell means μ_{ij}, $i = 1, \ldots, t, j = 1, \ldots, b$ can be estimated, and any linear hypothesis on these cell means can be tested.

For reference, we let

$$X_t = W(I_t \otimes J_b) \tag{7.98}$$

and

$$X_b = W(J_t \otimes I_b). \tag{7.99}$$

The incidence matrix, $N = (n_{ij})$, is given by

$$N = X_t' X_b. \tag{7.100}$$

Further, we define the matrix

$$A = X_t - X_b N'/k. \tag{7.101}$$

We may then establish the following properties that are useful in the analysis (see Weeks and Graybill, 1961, for details):

$$\begin{aligned} X_t J_t &= J_n & X_b J_b &= J_n \\ J_n' X_t &= r J_t' & J_n' X_b &= k J_b' \\ X_t' X_t &= r I_t & X_b' X_b &= k I_b \\ NN' &= (r - \lambda) I_t + \lambda J_t J_t'. \end{aligned} \tag{7.102}$$

The equations for estimating θ_m are defined as usual with

$$\begin{aligned} X_m' X_m &= P_m' X_c' X_c P_m \\ X_m' Y &= P_m' X_c' Y \end{aligned} \tag{7.103}$$

where

$$X_c' X_c = \begin{vmatrix} n.. & rJ_t' & kJ_b' \\ rJ_t & rI_t & N \\ kJ_b & N' & kI_b \end{vmatrix}. \tag{7.104}$$

It is easily established that θ_c is estimated by

$$\hat{\mu} = J'Y/n.. \tag{7.105}$$

$$\hat{\tau} = (k/\lambda t) A'Y \tag{7.106}$$

with A given by (7.101), and

$$\hat{\beta} = (X_b' Y - N' \hat{\tau} - k \hat{\mu} J_b)/k. \tag{7.107}$$

Here, we recall that τ_t and β_b are defined as in (6.45).

Utilizing the properties of X_t and X_b, we may show that

$$\hat{\tau}_i = (k/\lambda t)(y_{i\cdot} - \sum_j n_{ij}y_{\cdot j}/k) \qquad i = 1, \ldots, t \tag{7.108}$$

and

$$\hat{\beta}_j = (y_{\cdot j} - \sum_i n_{ij}\hat{\tau}_i - k\hat{\mu})/k \qquad j = 1, \ldots, b. \tag{7.109}$$

The estimates of the cell means are then given by

$$\hat{\mu}_{ij} = \hat{\mu} + \hat{\tau}_i + \hat{\beta}_j. \tag{7.110}$$

A nice feature of the BIB design is that each treatment difference, $\mu_{ij} - \mu_{vj}$, is estimated with the same precision. To establish this, we shall develop the general expression for the variance of the estimate of a linear contrast of the cell means in a particular column. Let p denote a vector of constants, p_i, $i = 1, \ldots, t$, with $J'p = 0$. Then,

$$\begin{aligned}\mathrm{Var}\Big(\sum_i p_i\hat{\mu}_{ij}\Big) &= \mathrm{Var}\Big(\sum_i p_i\hat{\tau}_i\Big)\\ &= (k/\lambda t)^2 p'A'Ap\sigma^2 \\ &= (k/\lambda t)p'p\sigma^2\end{aligned} \tag{7.111}$$

since, by (7.101) and (7.102), we have

$$A'A = (\lambda t/k)I_t - (\lambda/k)J_tJ_t'. \tag{7.112}$$

Thus,

$$\mathrm{Var}(\hat{\mu}_{ij} - \hat{\mu}_{vj}) = (2k/\lambda t)\sigma^2. \tag{7.113}$$

The estimate of σ^2 is obtained as usual as

$$s^2 = Q(\hat{\mu})/(n_{\cdot\cdot} - b - t + 1). \tag{7.114}$$

(Here, $\hat{\mu}$ denotes the vector of estimates of the cell means, μ_{ij}, from (7.110). It should be clear from the context when we are referring to this vector or to the scalar, μ, in the canonical model.) To spell out $Q(\hat{\mu})$, we write the estimated mean vector as

$$\begin{aligned}W\hat{\mu} &= \hat{\mu}J_n + X_t\hat{\tau} + X_b\hat{\beta}\\ &= A\hat{\tau} + X_bX_b'Y/k.\end{aligned} \tag{7.115}$$

We then have

$$\begin{aligned} Q(\hat{\mu}) &= \|Y - A\hat{\tau} - X_bX_b'Y/k\|^2 \\ &= Y'(I - X_bX_b'/k)Y - (\lambda t/k)\hat{\tau}'\hat{\tau}. \end{aligned} \tag{7.116}$$

The test of the hypothesis of equal row (treatment) effects is easily established using the model reduction approach. This hypothesis is

$$H_T\colon \mu_{ij} = \mu_{vj} \qquad \begin{aligned} i \neq v &= 1, \ldots, t \\ j &= 1, \ldots, b. \end{aligned} \tag{7.117}$$

If we let γ_j, $j = 1, \ldots, b$ denote the common value of the means in the jth column under this hypothesis, the reduced model is just the one-way classification model. In matrix form, this model is given by

$$E(Y) = X_b\gamma \tag{7.118}$$

with estimate

$$\hat{\gamma} = (X_b'Y)/k \tag{7.119}$$

and residual sum of squares

$$Q_T = Y'(I - X_bX_b'/k)Y. \tag{7.120}$$

From (7.116), we immediately have the numerator sum of squares for testing H_T, that is,

$$\begin{aligned} N_T &= R(\tau|\mu, \beta) \\ &= Q_T - Q(\hat{\mu}) \\ &= (\lambda t/k)\hat{\tau}'\hat{\tau}. \end{aligned} \tag{7.121}$$

Similarly, for testing the column hypothesis

$$H_B\colon \mu_{ij} = \mu_{iv} \qquad \begin{aligned} i &= 1, \ldots, t \\ j \neq v &= 1, \ldots, b, \end{aligned} \tag{7.122}$$

the vector of estimates of the common row means is $X_t'Y/r$, and the residual sum of squares is

$$Q_B = Y'(I - X_tX_t'/r)Y. \tag{7.123}$$

The numerator sum of squares for testing H_B is

$$\begin{aligned} N_B &= R(\beta | \mu, \tau) \\ &= Q_B - Q(\hat{\mu}) \\ &= (\lambda t/k)\hat{\tau}'\hat{\tau} + Y'X_bX_b'Y/k - Y'X_tX_t'Y/r. \end{aligned} \tag{7.124}$$

These results are displayed, for reference, as a standard two-factor AOM in Table 7.18.

TABLE 7.18 Analysis of means for BIB.

Description	*Hypothesis*	*df*	*SS*
Column effect	H_B	$b - 1$	$R(\beta \| \mu, t)$
Row effect	H_T	$t - 1$	$R(\tau \| \mu, \beta)$
Residual	—	$n_{..} - b - t + 1$	$Q(\hat{\mu})$

Alternatively, we might display the sequential AOM table using $R(\beta | \mu)$ for the column effect. We recall from our earlier discussions that this sum of squares is appropriate for testing the hypothesis

$$H_B^*\colon \sum_i n_{ij}\mu_{ij} = \sum_i n_{ib}\mu_{ib} \qquad j = 1, \ldots, b - 1 \tag{7.125}$$

subject to the constraint of no interaction. The corresponding sum of squares is given by

$$R(\beta | \mu) = Y'X_bX_b'Y/k - Y'JJ'Y/bk \tag{7.126}$$

and the AOM is displayed in Table 7.19.

TABLE 7.19 Sequential analysis of means for BIB.

Description	*Hypothesis*	*df*	*SS*
Column effect	H_B^*	$b - 1$	$R(\beta \| \mu)$
Row effect	H_T	$t - 1$	$R(\tau \| \mu, \beta)$

Other aspects of the analysis of this model, such as the construction of simultaneous confidence intervals, follow directly from the results of Chapter 6.

7.4. THE ANALYSIS OF MEANS AND CONCOMITANT VARIABLES

Linear statistical models are roughly dichotomized into regression models and means models. The latter are characterized by a design matrix that consists of zeros and ones indicating the presence or absence of an effect. In the regression situation the input variables are typically quantitative variables. It is not unusual that both types of variables are present, as in Example 1.8, and the analyst may be interested in making inferences on the coefficients associated with either or both. To emphasize the ideas, we present two additional examples.

EXAMPLE 7.6

Recall Example 4.2 in which the response was the test score on an exam given to incoming freshmen. The subjects consisted of n_1 males and $n_2 = N - n_1$ females. The analysis suggested was a comparison of mean responses assuming equal variability in the two groups. Now assume that we also have each student's high school grade-point average (GPA). It seems reasonable to expect that the test scores will be related to the GPA. In this situation it may well be that the primary purpose of the study is to try to establish a relation between test score and GPA. The reason for recording sex may be based on the concern that the relation may be obscured by differences in the sexes. Inclusion of the sex variable may allow us to do a better job of determining the relation by removing the differences due to sex.

■

EXAMPLE 7.7

In Section 5.5, we discussed a study of lactic acid build-up as a function of type of fluid used by runners. A natural concomitant variable for this study is the age of the runner. In this case, the relation between lactic acid and age may not be of importance. The primary objective is to study the various electrolytic drinks. The reason for including age in the study is to remove the effect of this variable, hence allowing a better measure of the effect of the fluid types.

■

We shall see that the models for these two situations are the same. In the literature, these are often referred to as *Analysis of Covariance models*. We might argue that this is a misnomer in the same sense that the Analysis of Variance is really an Analysis of Means. What we really mean is that we are analyzing the effect of the covariate, or concomitant, variables as well as the treatment effects. The term *Analysis of Covariance* is generally reserved for a special case of the model we shall consider and, more particularly, for a computational procedure designed for hand computations. In our development, we shall stress a more general model and recognize that the basic analysis techniques developed in Chapter 3 apply. General computer programs for the Analysis of Means or for regression analysis are easily extended to include these models. (For a recent discussion of this topic, see Cox and McCullagh, 1982.)

To develop the model, note that, in either Example 7.6 or Example 7.7, we can divide the data into groups according to the qualitative variable, sex, or type of electrolyte. A model for the two-group case was described in Section 1.1.4. To generalize to several groups, we postulate within each group a relation between the response, y, and the concomitant variable, z, such as

$$y_{ij} = \mu_i + \gamma_i z_{ij} + e_{ij} \qquad \begin{array}{l} i = 1, \ldots, p \\ j = 1, \ldots, n_i. \end{array} \tag{7.127}$$

That is, assuming p groups and n_i observations (y_{ij}, z_{ij}) in the ith group, we consider for each group a simple linear regression model with errors, e_{ij}, independently distributed $N(0, \sigma_i^2)$. The estimates of the parameters for each of these models may be obtained by applying the methods of Chapter 3 to each of the individual regression models. For reference, the model for each group is written in matrix form as

$$Y_i = \mu_i J_{n_i} + \gamma_i Z_i + e_i \tag{7.128}$$

where Y_i, Z_i, and e_i are the n_i-vectors of variables y_{ij}, z_{ij}, and e_{ij}, $j = 1, \ldots, n_i$.

We now proceed to develop the simultaneous analysis of the p groups. The essential assumption for this analysis is that the error variances are the same in each group; that is, we assume

$$\sigma_i^2 = \sigma^2 \qquad i = 1, \ldots, p. \tag{7.129}$$

If (7.129) is valid, we can combine the p models (7.128) into a single model written in matrix form as

$$Y = W\mu + Z\gamma + e \tag{7.130}$$

where Y and e are the concatenated vectors of responses and errors of length $N = n.$, given by

$$Y = \begin{bmatrix} Y_1 \\ Y_2 \\ \vdots \\ Y_p \end{bmatrix} \qquad e = \begin{bmatrix} e_1 \\ e_2 \\ \vdots \\ e_p \end{bmatrix}. \tag{7.131}$$

$W = \text{Diag}(J_{n_i})$ is the counting matrix from the one-way classification model, $Z = \text{Diag}(Z_i)$, $\mu = (\mu_1, \ldots, \mu_p)'$, and $\gamma = (\gamma_1, \ldots, \gamma_p)'$. It is easily seen that the estimates of μ_i and γ_i for the general model, (7.130), are the same as those obtained for the individual models in (7.128). The estimate, s^2, of σ^2 is a weighted average of the individual s_i^2 given by

$$s^2 = \sum(n_i - 1)s_i^2 \Big/ \sum(n_i - 1). \tag{7.132}$$

The obvious advantage of combining the data into a single model is that we can use the standard, maximum likelihood analysis developed in Chapter 3 to make inferences about the parameters. In particular, we might consider the hypotheses

$$H_1\colon \gamma_i = \gamma_j \qquad i, j = 1, \ldots, p \tag{7.133}$$

$$H_2\colon \gamma_i = 0 \qquad i = 1, \ldots, p \tag{7.134}$$

$$H_3\colon \mu_i + \gamma_i z_0 = \mu_j + \gamma_j z_0 \qquad i, j = 1, \ldots, p. \tag{7.135}$$

The hypothesis H_1, equality of slopes, states that the effect of the concomitant variable is the same for each group. H_2 states that the concomitant variable has no effect on the response for any group. H_3 states that the mean response at z_0 is the same for all groups. The special case $z_0 = 0$ corresponds to the hypothesis of equal intercepts.

The model (7.130) represents the general model to be considered in this section. The mean vector $E(Y)$ is the sum of two parts. The first part, $W\mu$, may represent any of the cell means models we have discussed. If the model is constrained, we shall assume that $W\mu$ represents the reduced model or the full rank canonical model. The second term, $Z\gamma$, represents the regression part of the model. Here we allow any of the common regression models to hold in each group; that is, there may be several concomitant variables, polynomial terms, and so on. The mean vector represents p regression models or surfaces (lines, planes, and so forth). These may be parallel (for example, if (7.133) is assumed), but that is not necessary. The regression part, $Z\gamma$, determines the nature of these surfaces. The design portion, $W\mu$, determines the intercepts of these surfaces. The hypothesis of equal intercepts may not be of interest unless the surfaces are assumed parallel. In that case, differences in the intercepts indicate treatment or group differences.

Before proceeding with the analysis, we return to the model, (7.130), and consider two reparameterizations that are the analogs of the two canonical forms used in the two-way classification model. (In fact, this model can be viewed as a two-way classification in which one of the classification criteria is continuous.)

As in the first canonical form, we write

$$Y = J_N\mu + X_a\alpha + X_0\beta_0 + X_b\beta + e \tag{7.136}$$

where

$$\begin{aligned} \mu &= \mu_p \\ \alpha_i &= \mu_i - \mu_p \qquad i = 1, \ldots, p-1 \\ \beta_0 &= \gamma_p \\ \beta_j &= \gamma_j - \gamma_p \qquad j = 1, \ldots, p-1 \end{aligned} \tag{7.137}$$

and

$$\begin{aligned} X_a &= W(I_{p-1}|0)' \\ X_0 &= (Z_1'Z_2', \ldots, Z_p')' \\ X_b &= Z(I_{p-1}|0)'. \end{aligned} \tag{7.138}$$

The relation between this form of the model and the first canonical form should be evident.

Alternatively, we might use the concepts from the second canonical form and write the model as in (7.136), where we define

$$\begin{aligned} \mu &= \bar{\mu}. \\ \alpha_i &= \mu_i - \bar{\mu}. \qquad i = 1, \ldots, p-1 \\ \beta_0 &= \bar{\gamma}. \\ \beta_i &= \gamma_i - \bar{\gamma}. \qquad i = 1, \ldots, p-1 \end{aligned} \tag{7.139}$$

and

$$\begin{aligned} X_a &= W\Delta_p' \\ X_b &= Z\Delta_p'. \end{aligned} \tag{7.140}$$

Here, $\Delta_p = (I_{p-1}|-J_{p-1})$ as used in Chapters 5 and 6.

For computational purposes, the nice feature of either of these forms is that the test statistics for any of the hypotheses (7.133) through (7.135) are easily developed. For example, the hypothesis of equal slopes, H_1, is equivalent in either case to $\beta = 0$.

In elementary treatments of the Analysis of Covariance, for this model, it is usually assumed that H_1 is true—that is, $\beta = 0$—and hence the effects of the covariates are the same for each group. This hypothesis is easily tested, and the condition should not be assumed, *a priori,* without justification. (Note that the term $X_b\beta$ is analogous to the interaction terms in the two-way classification model. The assumption $\beta = 0$ forces the difference in response to two treatments to be the same for all levels of the continuous factor.) The extensions of these canonical forms to more complex models should be apparent.

The analysis of the general model (7.130) is achieved using the techniques of Chapter 3. There are certain features of the model that are of interest, so we shall provide a brief discussion of the analysis in analytic form even though in practice the numerical computations would typically be performed using the concepts in Section 6.3.1.2. In our analysis, we shall not assume any particular structure on W except that it has full column rank. Thus, W may represent an unconstrained cell means model, a reduced cell means model, or a canonical form. Similarly, we shall assume only that Z has full column rank but emphasize that, in general, the structure of Z is similar to that of W.

Rewriting (7.130) in partitioned form as

$$Y = (W|Z)\begin{bmatrix} \mu \\ \gamma \end{bmatrix} + e \tag{7.141}$$

we immediately obtain the likelihood equations for estimating μ and γ as

$$\begin{bmatrix} W'W & W'Z \\ Z'W & Z'Z \end{bmatrix}\begin{bmatrix} \mu \\ \gamma \end{bmatrix} = \begin{bmatrix} W'Y \\ Z'Y \end{bmatrix}. \tag{7.142}$$

Eliminating μ from the second equation in (7.142), we obtain the equation

$$Z'UZ\gamma = Z'UY \tag{7.143}$$

where U is the symmetric, idempotent matrix

$$U = I - W(W'W)^{-1}W'. \tag{7.144}$$

The equation (7.143) has an interesting structure. Note that UY is the vector of residuals obtained by fitting only the design portion of the model—that is,

$$E(Y) = W\mu. \tag{7.145}$$

Similarly, UZ is the matrix whose columns are the residuals obtained by fitting the columns of Z to the model (7.145). The equation (7.143) for γ thus arises by fitting the model

$$E(UY) = UZ\gamma. \tag{7.146}$$

This model is often referred to as the *error model,* reflecting the fact that we are regressing the vector of residuals UY on the matrix of residuals UZ. We note that the residual sum of squares for fitting this model is

$$Q_e = Y'[U - UZ(Z'UZ)^{-1}Z'U]Y. \tag{7.147}$$

Having estimated γ from (7.143), we return to the first equation in (7.142) to obtain the estimate of μ as

$$\hat{\mu} = \tilde{\mu} - (W'W)^{-1}W'Z\hat{\gamma} \tag{7.148}$$

where $\tilde{\mu}$ is the estimate of μ for the design model (7.145). The residual vector for the original model (7.141) is then given by

$$\begin{aligned} Y - W\hat{\mu} - Z\hat{\gamma} &= Y - W\tilde{\mu} - UZ\hat{\gamma} \\ &= UY - UZ\hat{\gamma} \\ &= [U - UZ(Z'UZ)^{-1}Z'U]Y. \end{aligned} \tag{7.149}$$

We may now write the residual sum of squares for the original model (7.141) in two ways using the last two expressions for the residual vector. Thus,

$$Q = Y'UY - \hat{\gamma}'Z'UZ\hat{\gamma} \tag{7.150}$$

or

$$Q = Y'[U - UZ(Z'UZ)^{-1}Z'U]Y. \tag{7.151}$$

We note first that (7.151) is identical to (7.147); that is, the residual sum of squares computed for the error model, (7.146), is actually the correct residual sum of squares for the original model.

To further illustrate the role of the error model, consider testing the hypothesis, (7.134), that the concomitant variables have no effect; that is, the hypothesis

$$H_\gamma\colon \gamma = 0. \tag{7.152}$$

Using the model reduction approach, we fit the reduced model (7.145), obtaining the residual sum of squares,

$$Q_\gamma = Y'UY. \tag{7.153}$$

Comparing (7.153) to the expression (7.150) for the residual sum of squares for the original model, we obtain the numerator sum of squares,

$$N_\gamma = \hat{\gamma}'Z'UZ\hat{\gamma}. \tag{7.154}$$

Note that (7.154) is just the regression sum of squares for the error model. Assuming W and Z have p_1 and p_2 columns, respectively, the test statistic for H_γ is

$$F = (N_\gamma/p_2)/[Q/(N - p_1 - p_2)]. \tag{7.155}$$

We note that the elements of this test statistic were computed from the regression and residual sums of squares for the error model. This observation led to the form of the Analysis of Covariance presented in many introductory texts. (See Exercise 7 at the end of this chapter.)

In practice, we may be interested in other hypotheses on γ, such as H_1 given in (7.133). Test statistics for such hypotheses can be developed from the error model, but there is no particular advantage in doing so and we will generally do the analysis in the original model.

To illustrate one more result in which the test statistic is simply described, we consider the hypothesis of equal intercepts—that is, the hypothesis that all of the cell means are equal, say, $\mu_i = \mu^*$, where μ^* is not specified. Thus, we may write

$$H_\mu\colon \mu = \mu^*J. \tag{7.156}$$

In this case, the reduced model is

$$E(Y) = \mu^*J + Z\gamma + e \tag{7.157}$$

and the residual sum of squares is given by

$$Q_\mu = Y'MY - Y'MZ(Z'MZ)^{-1}Z'MY \tag{7.158}$$

where

$$M = I - JJ'/N. \tag{7.159}$$

The numerator sum of squares for testing H_μ is then

$$N_\mu = Q_\mu - Q \tag{7.160}$$

with Q given by (7.151). We note that Q_μ has the same structure as Q with U replaced by M. As noted earlier, this observation leads to simple hand computations for simple models of the form (7.130), but, in general, there is no practical advantage to this approach if the analysis is done by computer. If we express the design portion of the model in canonical form as in (7.136), the reduced model for testing equality of intercepts is obtained by setting $\alpha = 0$.

Whether we view models of this type as design models with concomitant variables or as regression models with indicator variables for treatment groups is of no consequence. The emphasis depends on the desires of the experimenter. The relation to regression is important, as it reminds us of the hazards involved. The possibility exists that certain observations may be highly influential. When more than one concomitant variable is involved, the potential for collinear predictors must be considered. Belsley, Kuh, and Welsch (1980) and Cook and Weisberg (1982) provide thorough discussions of these topics. Of course, as in any of our analyses, we should be aware of the possibility of unreasonable values for the observations. These are often caused by errors in recording the data but may be the result of other problems in conducting the experiments. The discussion in Section 3.5 applies directly to this type of model.

To conclude this section, we present a simple example to illustrate the relation of our general development to the elementary discussions of the Analysis of Covariance.

EXAMPLE 7.8

Referring to the lactic acid example, suppose we are willing to assume that there is no interaction between age and type of fluid; that is, the five regressions in (7.127) are assumed to be parallel. Assuming (7.129) allows us to write the single model,

$$y_{ij} = \mu_i + \gamma z_{ij} + e_{ij}. \tag{7.161}$$

The design model (7.145), in this case, is

$$y_{ij} = \mu_i + e_{ij} \tag{7.162}$$

and, hence, the vector of residuals for this model is given by

$$UY = (y_{ij} - \bar{y}_{i\cdot}). \tag{7.163}$$

Similarly,

$$UZ = (z_{ij} - \bar{z}_{i\cdot}) \tag{7.164}$$

and we obtain from (7.142)

$$Q_e = \sum\sum(y_{ij} - \bar{y}_{i\cdot})^2 - \hat{\gamma}\sum\sum(y_{ij} - \bar{y}_{i\cdot})(z_{ij} - \bar{z}_{i\cdot}) \tag{7.165}$$

where

$$\hat{\gamma} = \sum\sum(y_{ij} - \bar{y}_{i\cdot})(z_{ij} - \bar{z}_{i\cdot})\Big/\sum\sum(z_{ij} - \bar{z}_{i\cdot})^2. \tag{7.166}$$

The numerator sum of squares, (7.154), for testing the hypothesis H_γ: $\gamma = 0$ is

$$N_\gamma = \hat{\gamma}^2\sum\sum(z_{ij} - \bar{z}_{i\cdot})^2. \tag{7.167}$$

The residual sum of squares, (7.158), for the reduced model under the hypothesis H_μ: $\mu_i = \mu^*$ is given by

$$Q(\hat{\mu}) = \sum\sum(y_{ij} - \bar{y}_{\cdot\cdot})^2 - \tilde{\gamma}\sum\sum(y_{ij} - \bar{y}_{\cdot\cdot})(z_{ij} - \bar{z}_{\cdot\cdot}) \tag{7.168}$$

where

$$\tilde{\gamma} = \sum\sum(y_{ij} - \bar{y}_{\cdot\cdot})(z_{ij} - \bar{z}_{\cdot\cdot})\Big/\sum\sum(z_{ij} - \bar{z}_{\cdot\cdot})^2. \tag{7.169}$$

These computations are typically described in an Analysis of Covariance Table (see Exercise 7). Similar expressions are easily obtained for situations in which we have a more complex design matrix, say a two-way classification without interaction when we have equal cell frequencies.

Situations involving unequal cell frequencies, with K-factor models, including two or more concomitant variables, are generally intractable for hand computations. Allowing for interaction between treatments and concomitant variables increases the computational burden and allows for more general hypotheses to be considered. In such cases, the general techniques developed in Chapter 3 may be applied.

7.5. SUMMARY

The role of the cell means model in the analysis of fixed effects linear models should now be clear. Any of these models can be expressed in the cell means format. The interpretation of results, formulation of hypotheses, and specification of linear functions for confidence intervals are best done in terms of the cell means. In the case of models constrained to have certain interactions equal to zero, the computations are more efficient if the model is transformed to a canonical form. With regard to developing test statistics, the canonical form may be more efficient for unconstrained models as well. The dangers inherent in using a canonical model lie in the lack of understanding of the definition of the terms in that model. This is most evident in the case of unbalanced data.

The concept of connectedness in situations involving missing cells is clarified in terms of the cell means model. In the disconnected case, the specification of preliminary hypotheses for the Analysis of Means is problem specific.

Controversies with regard to the preliminary analysis of unbalanced or incomplete models are a consequence of various attempts at trying to extend techniques that agree for balanced data. The only justifiable approach is a precise specification of the hypothesis of interest. This is most easily done in terms of the cell means model.

This concludes our discussion of the special case of the cell means model, in which we assumed that the covariance structure on the errors was described as $V = \sigma^2 I$. In the remaining chapters, we shall consider models in which more general covariance structures are allowed.

CHAPTER 7 EXERCISES

1. Consider the three-factor model

$$y_{ijkr} = \mu_{ijk} + e_{ijkr} \qquad i = 1, 2; \quad j = 1, 2, 3; \quad k = 1, 2, 3; \quad r = 0, 1, \ldots, n_{ijk}$$

constrained to have no three-factor interaction; that is, (7.19) holds. Suppose $n_{112} = n_{121} = n_{123} = n_{221} = n_{223} = 0$.

(a) Determine the effective constraints.

(b) Is the model connected?

(c) Determine the effective hypothesis based on the column hypothesis

$$H_B\colon \bar{\mu}_{\cdot j \cdot} = \bar{\mu}_{\cdot 3 \cdot} \qquad j = 1, 2.$$

(d) Determine the effective hypothesis based on the BC-interaction hypothesis

$$H_{BC}\colon\ \bar{\mu}_{\cdot jk} - \bar{\mu}_{\cdot 3k} - \bar{\mu}_{\cdot j3} + \bar{\mu}_{\cdot 33} = 0 \qquad j = 1, 2; \quad k = 1, 2.$$

(e) Determine the column effect hypotheses for the sequential and partially sequential methods—that is, H_B^S and H_B^P.

(f) Suggest other measures of column effects and the BC-interaction.

2. For Example 7.2,

(a) Verify the relations in (7.43).

(b) Describe the effective hypotheses for the AB and AC interactions and the A, B, and C main effects.

3. Consider the partially nested model defined by equations (7.89) and (7.90), but assume the balanced case in which $i = 1, \ldots, a$; $j = 1, \ldots, b$; $k = 1, \ldots, c$; and $r = 1, \ldots, n$. Develop expressions for the sums of squares in Tables 7.14 through 7.16.

4. For the incidence matrices shown below, observed cells are denoted by X. Verify that each incidence matrix describes a BIB, and determine the parameters for each design.

(a)

		Block			
		1	2	3	4
Treatment	1	X	X	X	
	2		X	X	X
	3	X		X	X
	4	X	X		X

(b)

		Block									
		1	2	3	4	5	6	7	8	9	10
Treatment	1	X	X	X	X	X	X				
	2	X	X	X				X	X	X	
	3	X			X	X		X	X		X
	4		X		X		X	X		X	X
	5			X		X	X		X	X	X

(c)

		Block						
		1	2	3	4	5	6	7
	1	X	X	X				
	2	X			X	X		
	3	X					X	X
Treatment	4		X		X		X	
	5		X			X		X
	6			X	X			X
	7			X		X	X	

(d)

		Block						
		1	2	3	4	5	6	7
	1		X	X	X		X	
	2			X	X	X		X
	3	X			X	X	X	
Treatment	4		X			X	X	X
	5	X		X			X	X
	6	X	X		X			X
	7	X	X	X		X		

5. For design (d) in Exercise 4, write out the model in matrix form and verify the properties of the matrices given in (7.100) and (7.102).
6. For the data shown below corresponding to design (d) in Exercise 4, determine the relevant quantities for the analysis in Tables 7.18 and 7.19.

		Block						
		1	2	3	4	5	6	7
	1		68	73	120		112	
	2			73	89	79		107
	3	70			94	93	92	
Treatment	4		89			78	74	121
	5	92		86			84	88
	6	43	69		81			111
	7	57	113	76		91		

7. In texts on statistical methods, the data for an Analysis of Covariance are often presented in a table as shown below for the one-way classification model with one covariate.

$$\text{Model: } Y_{ij} = \mu_i + \gamma z_{ij} + e_{ij} \qquad i = 1, \ldots, p, \quad j = 1, \ldots, n_i$$

Description	*df*	*SS and SP*
Treatment	$p - 1$	T_{zz} T_{zy} T_{yy}
Error	$n. - p$	E_{zz} E_{zy} E_{yy}
Total	$n. - 1$	S_{zz} S_{zy} S_{yy}

To relate to the development in Section 7.4, note that the sums of squares and sums of products (SS and SP) are defined by the following quadratic and bilinear forms.

(a) From the design model, (7.145), we have

$$\text{Residual SS} = Y'UY = E_{yy}$$

$$\text{Reg SS} = Y'(I - U)Y = R_{yy}$$

$$\text{Treatment SS} = T_{yy} = R_{yy} - Y'JJ'Y/n.$$

$$S_{yy} = T_{yy} + E_{yy}$$

(b) By fitting the covariate to the design model, we obtain the analogous quantities

$$E_{zz} = Z'UZ$$

$$R_{zz} = Z'(I - U)Z$$

$$T_{zz} = R_{zz} - Z'JJ'Z/n.$$

$$S_{zz} = T_{zz} + E_{zz}.$$

(c) By analogy, we define

$$E_{zy} = Z'UY$$

$$R_{zy} = Z'(I - U)Y$$

$$T_{zy} = R_{zy} - Z'JJ'Y/n.$$

$$S_{zy} = T_{zy} + E_{zy}.$$

With this notation, verify that

(i) The residual sum of squares, Q_e, is given by

$$Q_e = E_{yy} - E_{yz}E_{zz}^{-1}E_{zy}.$$

(ii) The numerator sum of squares for testing the hypothesis

$$H_\gamma: \gamma = 0$$

is given by

$$N_\gamma = E_{yz}E_{zz}^{-1}E_{zy}.$$

(iii) The numerator sum of squares for testing the hypothesis

$$H_\mu: \mu_i = \mu_j \text{ for all } i \text{ and } j$$

is given by

$$N_\mu = S_{yy} - S_{yz}S_{zz}^{-1}S_{zy} - Q_e.$$

8. Consider the two-factor model without interaction with one observation per cell, except that $n_{11} = 0$. Suppose we add a covariate defined as $z_{11} = -1$, and $z_{ij} = 0$, $i, j \neq (1, 1)$; and let $y_{11} = 0$.

(a) Write this covariance model in canonical form, and show that the estimates of μ, α_i, and β_j are identical to those obtained for the canonical form of the two-factor model with $n_{11} = 0$.

(b) Let γ denote the coefficient of the covariate, and show that $E(\hat{\gamma}) = \mu_{11}$ and hence that $\hat{\gamma}$ can be used as an estimate of this parameter.

(c) Determine the estimate of γ, and compare it with the estimate of μ_{11} obtained in Exercise 6 of Chapter 4.

CHAPTER EIGHT
ESTIMATION AND HYPOTHESIS TESTING FOR THE GENERAL LINEAR MODEL: GENERAL THEORY

8.1. INTRODUCTION

We now turn to the analysis of linear models in which we allow for a more general covariance structure than $\sigma^2 I$, as was assumed in Chapters 3 through 7. Our primary interest will be in the general cell means model, but the results are quite general and can be applied to other mean structures such as regression and analysis of covariance models.

For reference, we define the general cell means model as follows.

DEFINITION 8.1
The general cell means model is given by

$$\begin{aligned} Y &= W\mu + e \\ \text{subject to} \quad G\mu &= g. \end{aligned} \tag{8.1}$$

Here, Y, W, μ, G, and g are as defined in Definition 4.1, and we now assume $e \sim N(0, V)$ where V is a symmetric, positive definite, covariance matrix, which may be written in the form

$$V = \sum_{t=0}^{T} \phi_t V_t. \tag{8.2}$$

The known matrices V_t are assumed to be symmetric, and the unknown parameters, ϕ_t, are constrained to satisfy the requirement that V is positive definite. In special cases, we may adjoin additional constraints on the parameters.

■

To simplify our discussion in the general development, we shall assume either that the means are not constrained or, if constraints are present, that they have been imposed to reduce the model. Also, we may choose to write the mean structure in a canonical form as in Chapters 6 and 7. For this reason, we shall write the mean vector as $E(Y) = X\theta$, recognizing that X could be W, W_R, or any canonical form matrix and that θ is the appropriate vector of parameters. The essential point is that X has full column rank, say r. In fact, X could just as well correspond to a regression model or a combined model as in Section 7.4. Our examples will focus on means models.

The structure assumed on V in (8.2) is not restrictive, since any matrix may be written in this form. Our interest will focus on cases in which the number of parameters, ϕ_t, is relatively small. We shall also specify, without loss of generality, that $V_0 = I$ and $\phi_0 \geq 0$, although our general treatment will not utilize this information.

The analysis will consist of determining point estimates of the r-vector, θ, and the $(T + 1)$-vector, $\phi = (\phi_t, t = 0, 1, \ldots, T)$; testing linear hypotheses about these parameters; and developing confidence regions, or intervals. The general theory will be developed in this chapter, and then specific models will be considered in the following chapters. Unfortunately, the general theory is not as complete as that developed in Chapter 3 for the case $V = \sigma^2 I$. Closed form solutions and exact distributional results are available in special cases, but typically we resort to iterative numerical solutions and approximate distributional results. The special models discussed in the following chapters will serve to clarify the concepts.

It will again be convenient to display test statistics in tabular form. We shall use the term *Analysis of Variance* (*ANOVA*) to refer to these tables, recognizing that we may be interested in testing hypotheses about θ or ϕ. Typical ANOVA tables will be illustrated in Chapter 9.

To motivate the discussion and model formulation, we present several examples.

EXAMPLE 8.1

A researcher is concerned with estimating the yield of a new strain of cotton. Several experimental plots, consisting of a large number of subplots of equal size, have been planted. The data consist of the yield of bc subplots where c subplots have been selected at random from each of b whole plots. Our primary interest is in the estimation of the mean subplot yield, μ, but the problem is complicated by the fact that it is not reasonable to assume a scalar covariance matrix.

To develop the model, let y_{jk}, $j = 1, \ldots, b$, $k = 1, \ldots, c$, denote the yield from the kth subplot of the jth whole plot. It is reasonable to assume that responses from different whole plots are uncorrelated but that the observations on subplots within a whole plot may be correlated. Such a covariance structure may be described as follows:

$$
\begin{aligned}
\operatorname{Cov}(y_{jk}, y_{j^*k^*}) &= \phi_0 + \phi_1 \qquad && j = j^*, k = k^* \\
&= \phi_1 && j = j^*, k \neq k^* \\
&= 0 && j \neq j^*.
\end{aligned}
\tag{8.3}
$$

From (8.3) we see that the variance of an individual observation is given by $\phi_0 + \phi_1$ and that the covariance between observations in the same plot is ϕ_1. This notation may seem curious, since we use two parameters to describe the variance. The primary reason for this parameterization is for notational convenience in describing the covariance matrix. This in no way restricts the allowable values for the variance and covariance. The only requirement is that the covariance matrix, V, as defined by (8.3) is positive definite. In this example, the conditions are that $\phi_0 \geq 0$ and $\phi_0 + c\phi_1 \geq 0$. Note that these conditions ensure that the variance, $\phi_0 + \phi_1$, is positive but allows ϕ_1 to be negative; that is, this model allows for the possibility of a negative covariance between subplot responses.

The specification of the model is completed by requiring

$$E(y_{jk}) = \mu \qquad \begin{array}{l} j = 1, \ldots, b, \\ k = 1, \ldots, c \end{array} \tag{8.4}$$

and assuming normality. Letting Y denote the vector of responses, we write

$$E(Y) = \mu J_{bc}. \tag{8.5}$$

To develop the expression for V in the form of (8.2), let Y_j denote the c-vector of observations on the jth plot. Then the covariance matrix for Y_j may be determined from (8.3) as

$$\text{Var}(Y_j) = \phi_0 I_c + \phi_1 J_c J_c'. \tag{8.6}$$

Noting that $\text{Cov}(Y_j, Y_{j^*}) = 0$ for $j \neq j^*$, we may write

$$\begin{aligned} V = \text{Var}(Y) &= I_b \otimes (\phi_0 I_c + \phi_1 J_c J_c') \\ &= \phi_0 I_{bc} + \phi_1 (I_b \otimes J_c J_c'). \end{aligned} \tag{8.7}$$

Thus, V has the linear structure (8.2), where we define $V_0 = I_{bc}$ and $V_1 = (I_b \otimes J_c J_c')$.

It should be noted that, in many texts, this model is described as a random, one-way classification model. (See, for example, Graybill, 1976, Scheffé, 1959, or Searle, 1971a.) In these references, ϕ_0 and ϕ_1 are both assumed positive and are called *variance components*. Note that this requirement will assure that V is positive definite, but it is stronger than necessary. The relation of this model to the one-way classification model of Chapter 5 will be established in Chapter 10.

■

EXAMPLE 8.2

We extend Example 8.1 by assuming that we wish to study yield as a function of a treatment—say, different fertilizers. In this case, assume that we have ab experimental plots and that each of a treatments are applied to b plots. Again, c observations are made on each plot.

It seems reasonable to make the same covariance assumptions as in (8.3); that is, observations within a plot are correlated, but observations between plots are uncorrelated. To develop the model, let y_{ijk}, $i = 1, \ldots, a$, $j = 1, \ldots, b$, $k = 1$,

. . . , c denote the response from the kth subplot of the jth plot receiving the ith treatment. Let Y_{ij} denote the c-vector of responses on the (i, j)th plot, let $Y_i = C_b(Y_{ij})$ denote the bc-vector of responses for the ith treatment, and let $Y = C_a(Y_i)$ denote the abc-vector of responses. Note that $\text{Var}(Y_{ij})$ is given by (8.6), $\text{Var}(Y_i)$ is given by (8.7), and hence

$$\begin{aligned} V = \text{Var}(Y) &= I_a \otimes I_b \otimes (\phi_0 I_c + \phi_1 J_c J_c') \\ &= \phi_0 I_{abc} + \phi_1 (I_{ab} \otimes J_c J_c'). \end{aligned} \tag{8.8}$$

Also, if we let μ denote the a-vector of means for the treatments, we have

$$E(Y) = W\mu$$

where (8.9)

$$W = I_a \otimes J_{bc}.$$

The covariance matrix V given by (8.8) has the linear structure of Definition 8.1. The condition for positive definiteness is the same as in Example 8.1. Note that the mean structure (8.9) is that of the one-way classification model, but in this case we have a more general covariance structure. In the literature, this model is referred to as a *mixed, two-fold nested model*. The relation to the analogous model in Section 7.2.1 will be made in Chapter 10.

■

EXAMPLE 8.3

A random, two-fold nested model will arise under a different modification of Example 8.1. In this case, we do not examine the entire subplot; rather, we sample from it. For example, we may observe the yield on a random sample of n plants in each subplot. Again, we assume that observations in a plot are correlated, and in addition we assume that observations on plants within a subplot are correlated but with a different correlation than observations on plants in different subplots of the same experimental plot. If we let y_{jkr}, $j = 1, \ldots, b$, $k = 1, \ldots, c$, $r = 1, \ldots, n$ denote the yield of the rth plant in the kth subplot of the jth plot, we shall assume $E(Y_{jkr}) = \mu$ for all j, k, r and

$$\begin{aligned} \text{Cov}(y_{jkr}, y_{j^*k^*r^*}) &= \phi_0 + \phi_2 + \phi_3 && j = j^*, k = k^*, r = r^* \\ &= \phi_2 + \phi_3 && j = j^*, k = k^*, r \neq r^* \\ &= \phi_2 && j = j^*, k \neq k^* \\ &= 0 && j \neq j^*. \end{aligned} \tag{8.10}$$

From (8.10) we see that the covariance between observations on plants within a subplot is given by $\phi_2 + \phi_3$ and that the covariance between observations within the same whole plot but in different subplots is given by ϕ_2. Let Y_{jk} denote the n-vector of responses in the (j, k)th subplot, Y_j denote the cn-vector of responses in the jth whole plot, and Y denote the bcn-vector of observations. Then

$$\begin{aligned} \text{Var}(Y_{jk}) &= \phi_0 I_n + (\phi_2 + \phi_3) J_n J_n' \\ \text{Var}(Y_j) &= \phi_0 I_{cn} + \phi_2 J_{cn} J_{cn}' + \phi_3 (I_c \otimes J_n J_n') \\ V = \text{Var}(Y) &= \phi_0 I_{bcn} + \phi_2 (I_b \otimes J_{cn} J_{cn}') + \phi_3 (I_{bc} \otimes J_n J_n'). \end{aligned} \tag{8.11}$$

The mean vector is

$$E(Y) = \mu J_{bcn}. \tag{8.12}$$

Again, we see that V, as given by (8.11), has the linear structure of our general model. The condition for positive definiteness of V in this case is

$$\phi_0 > 0,\ \phi_0 + n\phi_3 > 0,\ \phi_0 + n\phi_3 + cn\phi_2 > 0. \tag{8.13}$$

The relations between these last two examples and the two-fold nested example from Section 7.2.1 will be examined later. We do note certain similarities in the covariance matrices in these two examples that will be seen to be a part of a general scheme.

■

EXAMPLE 8.4

In this example, due to Scheffé (1959), we are interested in the daily production of a particular type of machine and the variability of the responses among machines of this type and among the operators of the machines. An experiment is run using a machines and b operators. Each operator uses each machine on n different days. On a given day, operators are randomly assigned to machines.

Let y_{ijk}, $i = 1, \ldots, a$, $j = 1, \ldots, b$, and $k = 1, \ldots, n$ denote the production by the jth operator on the kth day using the ith machine. Let $E(y_{ijk}) = \mu$. To develop the covariance structure, we first assume that the responses due to a given operator on the same machine are correlated, with covariance $\phi_1 + \phi_2 + \phi_3$. Responses of an operator on different machines also will be assumed to be correlated, with covariance ϕ_2, and, finally, the responses for different operators on the same machine will have covariance ϕ_1. Again, the curious notation for these three covariances is for convenience in developing the covariance matrix in linear form and in no way restricts the magnitudes of the correlations. Of course, we shall require V to be positive definite. To summarize the covariance structure, we write

$$\begin{aligned} \text{Cov}(y_{ijk}, y_{i^*j^*k^*}) &= \phi_0 + \phi_1 + \phi_2 + \phi_3 && i = i^*, j = j^*, k = k^* \\ &= \phi_1 + \phi_2 + \phi_3 && i = i^*, j = j^*, k \neq k^* \\ &= \phi_1 && i = i^*, j \neq j^* \\ &= \phi_2 && i \neq i^*, j = j^* \\ &= 0 && i \neq i^*, j \neq j^*. \end{aligned} \tag{8.14}$$

Let Y_{ij} be the n-vector of responses for the machine/operator combination (i, j), let $Y_i = C_b(Y_{ij})$ be the bn-vector of responses for the ith machine, let $Y_j = C_a(Y_{ij})$ be the an-vector of responses for the jth operator, and let Y be the abn-vector of responses. Then we may write (8.14) in matrix form as follows.

$$\begin{aligned}
\mathrm{Var}(Y_{ij}) &= \phi_0 I_n + (\phi_1 + \phi_2 + \phi_3) J_n J_n' \\
\mathrm{Var}(Y_i) &= \phi_0 I_{bn} + \phi_1 J_{bn} J_{bn}' + (\phi_2 + \phi_3)(I_b \otimes J_n J_n') \\
\mathrm{Var}(Y_j) &= \phi_0 I_{an} + \phi_2 J_{an} J_{an}' + (\phi_1 + \phi_3)(I_a \otimes J_n J_n').
\end{aligned} \tag{8.15}$$

Combining the information in (8.15) (or 8.14), we write

$$\begin{aligned}
V = \mathrm{Var}(Y) = \phi_0 I_{abn} &+ \phi_1 (I_a \otimes J_{bn} J_{bn}') + \phi_2 (J_a J_a' \otimes I_b \otimes J_n J_n') \\
&+ \phi_3 (I_{ab} \otimes J_n J_n').
\end{aligned} \tag{8.16}$$

If we assume $\phi_i > 0$, $i = 0, 1, 2, 3$, this formulation agrees with that of Scheffé, the ϕ_i are called *variance components,* and the model is referred to as the *random, two-way classification model*. The conditions for positive definiteness in this case are

$$\begin{aligned}
\phi_0 &> 0 \\
\phi_0 + n\phi_3 &> 0 \\
\phi_0 + n\phi_3 + bn\phi_1 &> 0 \\
\phi_0 + n\phi_3 + an\phi_2 &> 0 \\
\phi_0 + n\phi_3 + an\phi_2 + bn\phi_1 &> 0.
\end{aligned} \tag{8.17}$$

■

EXAMPLE 8.5
Now suppose the machines are of different brands with potentially different production capabilities. This is reflected in our model by assuming $E(Y_{ijk}) = \mu_i$ or, equivalently,

$$E(Y) = (I_a \otimes J_{bn})\mu \tag{8.18}$$

where μ is the a-vector of μ_i. The covariance structure is similar to that in Example 8.4, except that now we assume the responses of different operators on the same machine are uncorrelated—that is, $\phi_1 = 0$. The covariance matrix is then given by (8.16) with $\phi_1 = 0$; that is,

$$V = \mathrm{Var}(Y) = \phi_0 I_{abn} + \phi_2 (J_a J_a' \otimes I_b \otimes J_n J_n') + \phi_3 (I_{ab} \otimes J_n J_n'). \tag{8.19}$$

This model, with ϕ_0, ϕ_2, and ϕ_3 assumed positive, is known as the *two-way classification, mixed model*. Note that the mean structure is that of the one-way classification model and that the covariance structure is, apart from subscripts, the same as the two-fold nested model in Example 8.3.

■

EXAMPLE 8.6

Continuing with Example 8.5, suppose each operator uses each machine only once; hence, $n = 1$. From (8.19), or equivalently (8.14) with $\phi_1 = 0$, we see that it does not make sense to include in the model the covariance between responses for a given operator on a given machine, since we have only one observation for each machine/operator combination. The only covariance is between responses for the same operator on different machines, given by ϕ_2. Further, it is sufficient to denote the variance of an observation by $\phi_0 + \phi_2$. This is equivalent to setting $\phi_3 = 0$ in (8.19).

The reason for pointing out this special case is that it was mentioned earlier in Section 6.2.2.5 under the heading of the Randomized Block Design. In this case, the different machines correspond to the different treatments, and the operators correspond to the blocks. It is important to emphasize the distinction between this model and the two-way classification model discussed in Chapter 6. In the latter, the experimental units were assumed to be homogeneous and the treatments were described by two criteria. We were forced to make the assumption of no interaction, since $n = 1$; otherwise, no inferences could be made.

In the present case, with operators as blocks, we assign each machine type to each operator, recognizing the potential for operator differences. Further, the operators are of no interest in the study. It is assumed that they are selected from a pool of potential operators and that their different contributions are random. We are not interested in these differences but do wish to control for them. We thus make the assumption that responses from a particular operator are correlated. Note, further, that there is no need to make any assumptions about interaction. If we have replicate observations, $n > 1$, we may wish to assume, as above, that the correlation between observations for a given operator on the same machine is different from that on different machines.

This important difference between the randomized block model and the two-way classification without interaction is strongly emphasized.

■

EXAMPLE 8.7

Returning to the cotton-yield study, suppose we are interested in b varieties of cotton and c different fertilizers. We have available a large fields for our experiment. Each field is divided into b whole plots, and the b varieties are assigned at random to these whole plots. Further, each whole plot is divided into c subplots, and to these the c fertilizers are assigned at random. The point to be emphasized here is that the bc combinations of varieties and fertilizers are not assigned at random to the subplots in a field.

Let y_{ijk}, $i = 1, \ldots, a$, $j = 1, \ldots, b$, $k = 1, \ldots, c$ denote the responses, and assume $E(y_{ijk}) = \mu_{jk}$, the mean response of a variety/fertilizer combination. (In some cases, it may be reasonable to assume constraints on the μ_{jk}—say, the no-interaction constraints.) To develop the covariance structure, we assume that observations in the same field are correlated with covariance ϕ_1 if they are in different whole plots and $\phi_1 + \phi_2$ if they are in the same whole plot. Responses in different fields are assumed to be uncorrelated. Thus, we write

$$\begin{aligned} \operatorname{Cov}(y_{ijk}, y_{i^*j^*k^*}) &= \phi_0 + \phi_1 + \phi_2 && i = i^*, j = j^*, k = k^* \\ &= \phi_1 + \phi_2 && i = i^*, j = j^*, k \neq k^* \\ &= \phi_1 && i = i^*, j \neq j^* \\ &= 0 && i \neq i^*. \end{aligned} \tag{8.20}$$

To write this in matrix form, let Y_{ij} be the c-vector of responses on whole plot j in field i, let $Y_i = C_b(Y_{ij})$ be the bc-vector of responses on field i, and let $Y = C_a(Y_i)$. Then

$$\begin{aligned} \operatorname{Var}(Y_{ij}) &= \phi_0 I_c + (\phi_1 + \phi_2) J_c J_c' \\ \operatorname{Var}(Y_i) &= \phi_0 I_{bc} + \phi_1 J_{bc} J_{bc}' + \phi_2 (I_b \otimes J_c J_c') \\ \operatorname{Var}(Y) &= \phi_0 I_{abc} + \phi_1 (I_a \otimes J_{bc} J_{bc}') + \phi_2 (I_{ab} \otimes J_c J_c'). \end{aligned} \tag{8.21}$$

The mean structure is given by

$$E(Y) = (J_a \otimes I_{bc})\mu \tag{8.22}$$

where μ is the bc-vector of μ_{jk}.

This model is known as the *split-plot model* in which it is generally assumed that ϕ_0, ϕ_1, and ϕ_2 are positive. The condition for positive definiteness is

$$\phi_0 > 0 \qquad \phi_0 + c\phi_2 > 0 \qquad \phi_0 + c\phi_2 + bc\phi_1 > 0. \tag{8.23}$$

■

These examples will serve to motivate the general discussion in this chapter. In Chapter 9, we shall examine a single structure that includes Examples 8.1 through 8.7 as special cases.

A point that should be emphasized about these examples is that they have a balanced structure; that is, the range on any subscript is not affected by the value of other subscripts. Thus, in Example 8.1 we had the same number of subplots per whole plot, and in Example 8.4 each operator spent n days on each machine. In the analysis of simple linear models, lack of such balance did not seriously complicate the theory. In the general linear model, we shall see that such imbalance causes serious theoretical and computational difficulties.

8.2. ESTIMATION OF PARAMETERS

Following our general philosophy on parameter estimation, we shall first consider maximum likelihood (ML) estimates of the parameter vectors θ and ϕ. Examination of the likelihood equations will reveal some of the important features of the problem even when the equations are not easily solved. Harville (1977) suggests that, with improved computer techniques and equipment, the computational problems we shall encounter may no longer be a limiting factor.

Many other estimation procedures have been suggested, reflecting different estimation philosophies and/or simplified computing techniques. We shall not discuss all of these, but we refer you to Harville (1977) or Searle (1971a or 1971b).

An interesting modification of maximum likelihood estimation is the so-called *restricted maximum likelihood* (REML) *procedure*. We shall discuss it, as well as the general class of what might be called *moment estimators*. These latter estimators arise by equating sets of quadratic forms to their expected values, and they were popularized by the methods of Henderson (1953).

In Chapter 3, we were able to make precise statements about the distributional properties of the estimators. In the present situation, no such general results are available. In special cases we will be able to be more precise, but, since the general maximum likelihood results are iterative, our usual recourse is to appeal to large sample properties.

8.2.1. Maximum Likelihood Estimation (ML)

The model to be examined states that the N-vector, Y, is normally distributed with mean $E(Y) = X\theta$ and covariance matrix $V = \Sigma\phi_t V_t$, where V is positive definite. Thus, the log-likelihood function is given by

$$\text{Ln}L(\theta, \phi) = -(N/2)\text{Ln}(2\pi) - (\text{Ln}|V|)/2 - [(Y - X\theta)'V^{-1}(Y - X\theta)]/2. \quad (8.24)$$

The requirement that V is positive definite places constraints on ϕ as noted in the examples in Section 8.1. For now we shall simply write

$$\phi \in \Phi \quad (8.25)$$

to define the constraint space. The problem then is to determine vectors θ and ϕ that maximize (8.24) subject to (8.25). In the development that follows, we shall temporarily ignore the constraints on ϕ. Recall that, in Chapter 3, we actually had a constrained problem—that is, $\sigma^2 > 0$—but it caused no difficulty, as the unconstrained maximizer, s^2, automatically satisfied the constraint. It is interesting to speculate that, if the model is properly stated, the positive definiteness of the estimate of V will be automatic. We will return to this point later when we discuss specific models.

The stationary equations for the unconstrained maximization of (8.24) are easily obtained by applying the results given in Appendix A.II. In particular, we note that

$$\partial(\text{Ln}|V|)/\partial\phi_t = tr(V^{-1}V_t) \quad (8.26)$$

and

$$\partial[(Y - X\theta)'V^{-1}(Y - X\theta)]/\partial\phi_t = -(Y - X\theta)'V^{-1}V_tV^{-1}(Y - X\theta). \quad (8.27)$$

Thus, differentiating $\mathrm{Ln}L(\theta, \phi)$ with respect to ϕ_t and setting the derivatives to zero yields the likelihood equations

$$tr(V^{-1}V_t) = (Y - X\theta)'V^{-1}V_tV^{-1}(Y - X\theta) \qquad t = 0, 1, \ldots, T. \quad (8.28)$$

Differentiation with respect to θ yields the weighted least squares equations

$$X'V^{-1}X\theta = X'V^{-1}Y. \quad (8.29)$$

Simultaneous solution of (8.28) and (8.29) for θ and ϕ yields the stationary points of the likelihood function, one of which will be the unconstrained, maximum likelihood estimator.

The linear structure of V enables us to write

$$\begin{aligned} tr(V^{-1}V_i) &= tr(V^{-1}V_iV^{-1}V) \\ &= tr\left(V^{-1}V_iV^{-1}\sum_{j=0}^{T} \phi_jV_j\right) \\ &= \sum_{j=0}^{T} \phi_j tr(V^{-1}V_iV^{-1}V_j) \\ &= \sum_{j=0}^{T} \phi_j\omega_{ij}. \end{aligned} \quad (8.30)$$

where we define

$$\omega_{ij} = tr(V^{-1}V_iV^{-1}V_j). \quad (8.31)$$

Let $\Omega = (\omega_{ij}, i, j = 0, \ldots, T)$, and let $\rho(\theta)$ denote the column vector whose elements are

$$\rho_i(\theta) = (Y - X\theta)'V^{-1}V_iV^{-1}(Y - X\theta) \qquad i = 0, \ldots, T. \quad (8.32)$$

Then we may write the equations (8.28) as

$$\Omega\phi = \rho(\theta). \quad (8.33)$$

Note that in (8.33) both Ω and $\rho(\theta)$ may depend on ϕ; hence, the equations are not linear in ϕ. It is natural to ask whether there are special cases in which these nonlinear equations have a closed form solution. In Chapter 9, we shall examine a special class of covariance matrices and see that, within this class, certain covariance structures yield simple expressions for the estimators. One such example is illustrated in Exercise 2. The first problem encountered when examining these likelihood equations is that they involve the inverse of V. The key to the results in Chapter 9 is that, for the special covariance matrices considered, we can write a simple expression for V^{-1}. The models for which this is possible are the balanced data cases analogous to those encountered in the fixed effects models.

The pseudo-linear structure of (8.29) and (8.33) does suggest an obvious computational procedure in the general case. For example, with an initial estimate of ϕ, we could use (8.29) to obtain an estimate of θ. Using these estimates to compute Ω and $\rho(\theta)$ in (8.33), we could then solve the resulting linear equations to obtain a new estimate of ϕ. Iterating on this procedure until suitable convergence is obtained will provide a solution to the likelihood equations.

Pursuing the analysis of the likelihood function, we now consider the matrix of second derivatives of (8.24). Differentiating twice with respect to θ yields

$$[\partial^2 \mathrm{Ln}L(\theta, \phi)]/\partial\theta^2 = -X'V^{-1}X. \tag{8.34}$$

The matrix of second derivatives with respect to the elements of ϕ has components

$$\begin{aligned}[\partial^2 \mathrm{Ln}L(\theta, \phi)]/(\partial\phi_i\partial\phi_j) = \omega_{ij}/2 - (Y - X\theta)'(V^{-1}V_jV^{-1}V_i \\ + V^{-1}V_iV^{-1}V_j)(Y - X\theta)/2.\end{aligned} \tag{8.35}$$

The matrix of mixed derivatives consists of vectors defined by

$$[\partial^2 \mathrm{Ln}L(\theta, \phi)]/(\partial\phi_i\partial\theta) = X'V^{-1}V_iV^{-1}X\theta - X'V^{-1}V_iV^{-1}Y. \tag{8.36}$$

Taking the expected value of (8.35) yields

$$E[\partial^2 \mathrm{Ln}L(\theta, \phi)/(\partial\phi_i\partial\phi_j)] = -(\omega_{ij})/2. \tag{8.37}$$

Similarly,

$$E[\partial^2 \mathrm{Ln}L(\theta, \phi)/(\partial\phi_i\partial\theta)] = 0. \tag{8.38}$$

Combining the results of (8.34), (8.37), and (8.38), we see that the information matrix (see Appendix B.I) given by the expected value of the negative matrix of second derivatives is, in this case,

$$I(\theta, \phi) = \mathrm{Diag}(X'V^{-1}X, \Omega/2). \tag{8.39}$$

The inverse of this matrix gives the large sample covariance matrix of the unconstrained maximum likelihood estimator. Thus, as an approximation for the covariance matrix of our estimators, we might use

$$\mathrm{Var}(\hat{\theta}, \hat{\phi}) = \mathrm{Diag}[(X'\hat{V}^{-1}X)^{-1}, 2\hat{\Omega}^{-1}] \tag{8.40}$$

with $\hat{V}$ and $\hat{\Omega}$ denoting the value of V and Ω at $\phi = \hat{\phi}$.

Returning to the problem of computing the estimates, we now relate the computing procedure above to other methods. Consider first a quadratic approximation of the log-likelihood at an initial point (θ^*, ϕ^*), given by

$$\begin{aligned}\mathrm{Ln}L(\theta, \phi) \doteq \mathrm{Ln}L(\theta^*, \phi^*) + \nabla_\theta \mathrm{Ln}L(\theta^*, \phi^*)(\theta - \theta^*) \\ + \nabla_\phi \mathrm{Ln}L(\theta^*, \phi^*)(\phi - \phi^*) \\ - (1/2)\begin{bmatrix} \theta - \theta^* \\ \phi - \phi^* \end{bmatrix}' I(\theta^*, \phi^*)\begin{bmatrix} \theta - \theta^* \\ \phi - \phi^* \end{bmatrix}.\end{aligned} \tag{8.41}$$

(The gradient vectors ∇_θ and ∇_ϕ are defined in Appendix A.II.2.) The *method of scoring* suggested by Fisher (1925) consists of a sequence of maximizations of this quadratic, using the current solution as the next initial point for the approximation. Note that at each iteration this requires the solution of the systems of equations

$$\begin{aligned} X'V^{-1}X(\theta - \theta^*) &= \nabla_\theta Ln(\theta^*, \phi^*) \\ &= -X'V^{-1}X\theta^* + X'V^{-1}Y. \end{aligned} \tag{8.42}$$

and

$$\begin{aligned} (1/2)\Omega(\phi - \phi^*) &= \nabla_\phi Ln(\theta^*, \phi^*) \\ &= -(1/2)\Omega\phi^* + (1/2)\rho(\theta) \end{aligned} \tag{8.43}$$

where V, Ω, and $\rho(\theta)$ are evaluated at (θ^*, ϕ^*). Note that this procedure is identical to that described above for the solution of the likelihood equations. Vandaele and Chowdhury (1971) suggested a revised method of scoring that will ensure convergence to a local maximum of the likelihood function, but there is no assurance that the global maximum will be determined.

Closely related to the method of scoring is the Newton-Raphson method proposed by Hemmerle and Hartley (1973). The difference is that the information matrix in the quadratic approximation (8.41) is replaced by the negative of the matrix of second derivatives as defined by (8.34), (8.35), and (8.36). Both of these methods have quadratic convergence in the neighborhood of a local optimum and, hence, are generally more efficient than the steepest ascent method proposed by Hartley and Rao (1967). That paper also noted that, if the ϕ_i are constrained to be nonnegative, we can rewrite the likelihood function in terms of the parameters

$$\tau_i = \phi_i^{1/2}. \tag{8.44}$$

We then maximize with respect to θ and τ and use $\phi_i = \tau_i^2$. This will clearly ensure positive estimates for the ϕ_i, but there is no claim that they will give the constrained maximum. Little attention has been given in the literature to the less demanding requirement that V be positive definite.

Apart from the fact that none of these methods guarantees a global maximum, the biggest deterrent to their use would appear to be the magnitude of the computing effort. Hemmerle and Hartley (1973), however, made several observations that will greatly decrease this effort. To appreciate the magnitude of the computations, consider the simultaneous solution of (8.29) and (8.33) as in the method of scoring. (The same problems exist in the steepest ascent and Newton-Raphson methods.) We note that, at each iteration, the inverse of the $N \times N$ matrix V must be computed, and the $N \times N$ matrix computations required in the formation of ω_{ij} in (8.30) must be performed. The interesting result of the Hemmerle and Hartley (1973) paper is that the iterative computations can be performed using quantities that do not depend on N in any way.

To describe their procedure, we must assume that the matrices V_t in the covariance matrix can be easily written in the form

$$V_t = U_t U_t' \qquad t = 1, \ldots, T. \tag{8.45}$$

This is seen to be true for Examples 8.1 through 8.7 and is frequently true in practice. Indeed, for the special models considered in Chapter 10, the matrices U_t arise naturally.

The first observation we make is that the inversion of the $N \times N$ matrix, V, can be reduced to the inversion of an $m \times m$ matrix followed by appropriate matrix multiplications where $m = \Sigma_t m_t$ and m_t is the number of columns in U_t. To see this, let U denote the partitioned matrix

$$U = (U_1 | U_2 | \ldots | U_t) \tag{8.46}$$

and let D denote the block diagonal matrix

$$D = \text{Diag}(\phi_t I_{m_t}). \tag{8.47}$$

Assuming that $V_0 = I$, we may then write

$$V = \phi_0 I + UDU'. \tag{8.48}$$

Applying the results of Appendix A.I.10, we have

$$V^{-1} = (1/\phi_0)(I - UQ^{-1}U') \tag{8.49}$$

where

$$Q = \phi_0 D^{-1} + U'U. \tag{8.50}$$

Note that Q is $m \times m$—hence, the inversion is simpler—but U is $N \times m$, so we are still dealing with the dimension N. To avoid this, define the symmetric matrix

$$B = \begin{bmatrix} U' \\ X' \\ Y' \end{bmatrix} V^{-1}(U\ X\ Y). \tag{8.51}$$

[Note that Hemmerle and Hartley (1973) call this matrix W and refer to it as the W-*transform*.] This matrix contains the essential elements to describe the likelihood equations (8.29) and (8.33). To see this denote the submatrices of B by $B(i, j)$. Thus,

$$\begin{aligned} B(i, j) &= U_i' V^{-1} U_j \qquad i, j = 1, \ldots, T \\ B(T+1, j) &= X' V^{-1} U_j \\ B(T+2, j) &= Y' V^{-1} U_j \\ B(T+1, T+1) &= X' V^{-1} X \\ B(T+2, T+1) &= Y' V^{-1} X. \end{aligned} \tag{8.52}$$

We see that the equation (8.29) is given by

$$B(T+1, T+1)\theta = B(T+1, T+2). \tag{8.53}$$

From (8.32) note that, using the properties of the trace operator, we may write

$$\begin{aligned} \omega_{ij} &= tr(U_i'V^{-1}U_jU_j'V^{-1}U_i) \\ &= tr[B(i, j)B(j, i)]. \end{aligned} \tag{8.54}$$

Also, we may write

$$\begin{aligned} \delta_i' &= (Y - X\theta)'V^{-1}U_i \\ &= B(T+2, i) - \theta' B(T+1, i). \end{aligned} \tag{8.55}$$

Hence, from (8.32) we may write

$$\rho_i(\theta) = \delta_i'\delta_i. \tag{8.56}$$

The likelihood equation (8.33) is now written in terms of (8.54), (8.55), and (8.56). It remains to show that in the iterative computation we can efficiently evaluate the matrix B. To do so, let

$$B_0 = \begin{bmatrix} U' \\ X' \\ Y' \end{bmatrix} (U\ X\ Y) \tag{8.57}$$

and note that, using (8.49), we may write

$$B = (1/\phi_0)[B_0 - \begin{bmatrix} U'U \\ X'U \\ Y'U \end{bmatrix} Q^{-1}(U'U\ U'X\ U'Y)]. \tag{8.58}$$

The matrices in B_0 need only be computed once since they do not depend on the unknown parameters. Thus, in the evaluation of B, only the inverse of Q must be performed in each iteration, and, of course, the multiplications in (8.58) must be done. Hemmerle and Hartley (1973) showed that the number of computations to form B on each iteration is less than that required for the inversion of V alone, even in the form of (8.49); hence, the overall saving is substantial.

8.2.2. Restricted Maximum Likelihood Estimation (REML)

Patterson and Thompson (1971) suggested a modification of the maximum likelihood procedure in which the likelihood function is factored into two parts, one depending only on ϕ. This factor is then maximized with respect to ϕ, and the result is used as the estimator of ϕ. The maximization of the second factor with respect to

θ—using this estimate of ϕ—then yields an estimate of θ. Corbeil and Searle (1976) extended this concept and called the procedure *restricted maximum likelihood (REML) estimation*. Perhaps a better terminology would have been *modified maximum likelihood estimation,* since the word *restricted* appears to imply that the estimators are constrained. This is not the case, and, if constraints such as $\phi_i > 0$ are required, they must be imposed using standard techniques of constrained optimization.

The factorization described by Patterson and Thompson (1971) is achieved by a linear transformation on the original data. Recalling that $Y \sim N(X\theta, V)$, let

$$Z = PY \tag{8.59}$$

where P and Z are partitioned as

$$P = \begin{bmatrix} P_1 \\ P_2 \end{bmatrix} \qquad Z = \begin{bmatrix} Z_1 \\ Z_2 \end{bmatrix} \tag{8.60}$$

with

$$\begin{aligned} P_1 &= [I - X(X'X)^{-1}X'] \\ P_2 &= X'V^{-1}. \end{aligned} \tag{8.61}$$

It follows that $Z \sim N(PX\theta, PVP')$, where

$$PX\theta = \begin{bmatrix} 0 \\ X'V^{-1}X\theta \end{bmatrix} \tag{8.62}$$

and

$$PVP' = \text{Diag}(P_1VP_1', X'V^{-1}X). \tag{8.63}$$

Thus, Z_1 and Z_2 are independent, and the distribution of Z_1 does not depend on θ. Note that P_1 is idempotent of rank $N - r$ where r is the rank of X. It follows that Z_1 has a singular normal distribution of rank $N - r$. To eliminate this degeneracy, let P_{11} be any set of $N - r$ linearly independent rows of P, and let

$$Z_{11} = P_{11}Y. \tag{8.64}$$

It follows that $Z_{11} \sim N(0, P_{11}VP_{11}')$ and is independent of Z_2. The likelihood function is then the product of the marginal densities of Z_{11} and Z_2. The log likelihood is the sum of the following two terms:

$$\begin{aligned} \text{Ln}L_1(\phi) = &-[(N - r)/2]\text{Ln}(2\pi) - (1/2)\text{Ln}|P_{11}VP_{11}'| \\ &- (1/2)Z_{11}'(P_{11}VP_{11}')^{-1}Z_{11} \end{aligned} \tag{8.65}$$

and

$$\begin{aligned} \mathrm{Ln}L_2(\theta, \phi) = &-(r/2)\mathrm{Ln}(2\pi) - (1/2)\mathrm{Ln}|X'V^{-1}X| \\ &- (1/2)(Z_2 - X'V^{-1}X\theta)'(X'V^{-1}X)^{-1}(Z_2 - X'V^{-1}X\theta). \end{aligned} \tag{8.66}$$

The suggestion of Patterson and Thompson (1971) is that (8.65) be used to estimate ϕ and then, using this estimate of ϕ, (8.66) be used to estimate θ.

Looking first at (8.66), differentiation with respect to θ yields the stationary equation

$$X'V^{-1}X\theta = Z_2. \tag{8.67}$$

Noting that $Z_2 = X'V^{-1}Y$, we see that this equation is identical to the original likelihood equation (8.29). The difference between ML and REML thus lies in the estimation of ϕ.

Examination of (8.65) reveals that it has the same form as the original likelihood function (8.24) if we let $\theta = 0$ and define the covariance matrix

$$V^* = P_{11}VP'_{11}. \tag{8.68}$$

It follows that we can write the modified likelihood equations for ϕ in the same form as (8.33)—say,

$$\Omega^*\phi = \rho \tag{8.69}$$

where Ω^* has elements

$$\omega^*_{ij} = tr(V^{*-1}V^*_iV^{*-1}V^*_j) \tag{8.70}$$

and ρ has elements

$$\begin{aligned} \rho_i &= Z'_{11}V^{*-1}V^*_iV^{*-1}Z_{11} \\ &= Y'P'_{11}V^{*-1}V^*_iV^{*-1}P_{11}Y. \end{aligned} \tag{8.71}$$

Here, $V^*_i = P_{11}V_iP'_{11}$.

The iterative solution of (8.69) proceeds as usual, and the W-transform of Hemmerle and Hartley (1973) can be applied to reduce the computational effort.

This development illustrates quite simply the concept of the REML estimation method, but it is somewhat cumbersome in practice since it requires the specification of P_{11}. Following Patterson and Thompson (1971), we now present an alternative development that avoids the specification of P_{11}. This development also allows for a comparison of REML with other methods.

Using Appendix A.I.9 and the idempotency of P_1, we may write

$$P_1 = AA' \tag{8.72}$$

where A is $N \times (N - r)$ and

$$A'A = I. \tag{8.73}$$

The development then follows as above with P_{11} replaced by A'. That is, we consider the likelihood defined by

$$Z = \begin{bmatrix} A' \\ P_2 \end{bmatrix} Y \sim N\left\{\begin{bmatrix} 0 \\ X'V^{-1}X\theta \end{bmatrix}, \text{Diag}(A'VA, X'V^{-1}X)\right\}. \tag{8.74}$$

In (8.74) we have used the fact that

$$A'X = A'AA'X = A'P_1X = 0. \tag{8.75}$$

The modified likelihood equations for ϕ are then given by (8.69), (8.70), and (8.71) with

$$V^* = A'VA. \tag{8.76}$$

We now show that it is not necessary to explicitly specify the matrix A in (8.72). Observe that, under the transformation in (8.74), the exponent in the likelihood may be written as

$$\begin{aligned}(Y - X\theta)'V^{-1}(Y - X\theta) &= (Y - X\theta)'V^{-1}X(X'V^{-1}X)^{-1}X'V^{-1}(Y - X\theta) \\ &\quad + Y'A(A'VA)^{-1}A'Y.\end{aligned} \tag{8.77}$$

Hence,

$$\begin{aligned}A(A'VA)^{-1}A' &= V^{-1}[I - X(X'V^{-1}X)^{-1}X'V^{-1}] \\ &= V^{-1}M\end{aligned} \tag{8.78}$$

where

$$M = I - X(X'V^{-1}X)^{-1}X'V^{-1}. \tag{8.79}$$

We note three properties of M:

1. $V^{-1}M = M'V^{-1}$
2. $V^{-1}M = M'V^{-1}M = V^{-1}MVM'V^{-1}$
3. M is idempotent.

The REML equations, (8.69), for ϕ may now be written in terms of the matrix M without specifying the matrix A. In particular, from the definitions of the components of Ω^* and ρ in (8.70) and (8.71) and V^* given in (8.76), we have

$$\begin{aligned}\omega^*_{ij} &= tr[(A'VA)^{-1}A'V_iA(A'VA)^{-1}A'V_jA] \\ &= tr(V^{-1}MV_iM'V^{-1}V_j)\end{aligned} \tag{8.80}$$

and

$$\begin{aligned}\rho_i &= Y'A(A'VA)^{-1}A'V_iA(A'VA)^{-1}A'Y \\ &= Y'V^{-1}MV_iM'V^{-1}Y.\end{aligned} \tag{8.81}$$

To characterize the REML equations, define the matrix

$$Q_i = V^{-1}MV_iM'V^{-1}. \tag{8.82}$$

Then

$$\rho_i = Y'Q_iY \tag{8.83}$$

and

$$\begin{aligned} E(\rho_i) &= tr(Q_iV) + (X\theta)'Q_iX\theta \\ &= tr(Q_iV) \end{aligned} \tag{8.84}$$

since $MX = 0$. Further,

$$\begin{aligned} tr(Q_iV) &= \sum_j \phi_j tr(Q_iV_j) \\ &= \sum_j \phi_j tr(V^{-1}MV_iM'V^{-1}V_j). \end{aligned} \tag{8.85}$$

Combining (8.84) and (8.85), we see that the REML equations may be written as

$$E(\rho_i) \stackrel{S}{=} \rho_i \qquad i = 0, \ldots, T. \tag{8.86}$$

That is, we equate the quadratic forms, $Y'Q_iY$, to their expected values. (The symbol ($\stackrel{S}{=}$) is used to emphasize that (8.86) does not imply that ρ_i is equal to its expected value.) This concept will be explored further in the next section, where the quadratic forms will not depend on the unknown parameters as is the case here.

Comparing (8.80) and (8.81) with their analogs (8.31) and (8.32) in the ordinary likelihood equations for ϕ reveals that the key to the difference in the estimators is the matrix M. This matrix arises quite naturally if we use the likelihood equation for θ given by (8.29) to eliminate θ from $\rho(\theta)$. Note that

$$\begin{aligned} Y - X\hat{\theta} &= [I - X(X'V^{-1}X)^{-1}X'V^{-1}]Y \\ &= MY. \end{aligned} \tag{8.87}$$

We then note that, in (8.32),

$$\rho_i(\hat{\theta}) = Y'M'V^{-1}V_iV^{-1}MY = \rho_i. \tag{8.88}$$

To summarize, the ML equations for ϕ given by (8.33) may be written as

$$\Omega\phi = \rho \tag{8.89}$$

while the REML equations for ϕ are given by

$$\Omega^*\phi = \rho \tag{8.90}$$

where Ω and Ω^* are defined by (8.31) and (8.80), respectively, and ρ is defined by (8.81).

We thus see that the ML and REML equations differ only in the coefficient matrices Ω and Ω^*. To examine the relation between these matrices, note that the ML equation (8.28) with θ replaced by $\hat{\theta}$, as obtained from (8.29), is

$$tr(V^{-1}V_t) = \rho_t \qquad t = 0, \ldots, T. \tag{8.91}$$

The REML equations may be written, using (8.86), as

$$tr(Q_t V) = \rho_t \qquad t = 0, \ldots, T. \tag{8.92}$$

Note that

$$tr(Q_t V) = tr(M'V^{-1}MV_t) \tag{8.93}$$

and that, from the properties of M,

$$\begin{aligned} M'V^{-1}M &= V^{-1}M \\ &= V^{-1} - V^{-1}(I - M). \end{aligned} \tag{8.94}$$

We may then write

$$tr(V^{-1}V_t) = tr(Q_t V) + tr[V^{-1}(I - M)V_t]. \tag{8.95}$$

Returning to the matrix form of the ML equations, we may write

$$\Omega\phi = \Omega^*\phi + \Psi\phi = \rho \tag{8.96}$$

where the elements of Ψ are

$$\Psi_{tj} = tr[V^{-1}(I - M)V_t V^{-1}V_j]. \tag{8.97}$$

The term $\Psi\phi$ distinguishes between the ML and REML equations. At this point, we can go no further in the interpretation of this term, but we shall examine it in detail in Chapter 9.

Exercises 2 through 5 in Chapter 8 ask for the development of the ML and REML estimators for two simple cases where V^{-1} is known explicitly. In Example 8.1, discussed in Exercise 2 of this chapter, either system of equations can be solved in closed form. In Example 8.4, discussed in Exercise 4 of this chapter, we find that the REML equations are easily solved but that the ML equations are not. This issue is addressed in detail for the special class of problems discussed in Chapter 9.

8.2.3. Moment Estimators

The method of moments consists of equating statistics to their expected values and solving the resulting equations for the parameters, the solution being the moment estimators.

For the purpose of estimating ϕ, this suggests that we consider quadratic functions of Y, since the linear structure of V ensures that the expected value will be linear in ϕ. In particular, we consider the quadratic forms

$$Y'Q_iY \qquad i = 0, 1, \ldots, T \tag{8.98}$$

where

$$Q_iX = 0 \qquad i = 0, 1, \ldots, T. \tag{8.99}$$

From Theorem 2.2 we have

$$\begin{aligned} E(Y'Q_iY) &= tr(Q_iV) + \theta'X'Q_iX\theta \\ &= \sum_{j=0}^{T} \phi_j tr(Q_iV_j). \end{aligned} \tag{8.100}$$

Note that the requirement (8.99) removes θ from the expected value in (8.100). Equating (8.100) to (8.98), we have a system of linear equations in ϕ which we may now compare directly with the ML or REML equations (8.89) and (8.90). In particular, if we define Ω_Q to be the matrix with elements

$$\omega_{ij} = tr(Q_iV_j) \qquad i, j = 0, \ldots, T \tag{8.101}$$

and ρ_Q to be the vector with components

$$\rho_i = Y'Q_iY \qquad i = 0, \ldots, T \tag{8.102}$$

we may write the general moment equations as

$$\Omega_Q\phi = \rho_Q. \tag{8.103}$$

Assuming that we choose Q_i so that Ω_Q is nonsingular, it follows directly that the solution of (8.103) provides an unbiased estimate of ϕ. The problem then rests on the choice of the quadratic forms. In Chapter 9 we shall develop quadratic forms for specific models. For now, we observe two general approaches to the problem.

The first choice is a simple one suggested by LaMotte (1970). From the REML development, we consider the quadratic forms suggested by ρ_i in (8.81). Replacing V by—say, V^o—where this matrix denotes V evaluated at a particular vector—say, ϕ^o—we consider

$$Q_i = M^{o\prime}(V^o)^{-1}V_i(V^o)^{-1}M^o. \tag{8.104}$$

Here M^o denotes M evaluated at V^o. Since $MX = 0$, the requirement (8.99) is satisfied, and this matrix may be used as in (8.103) to develop unbiased estimators. Clearly, the choice of ϕ^o is important. Two simple suggestions are $\phi_t^o = 1$, $t = 0, \ldots, T$ and $\phi_0^o = 1$, $\phi_t^o = 0$, $t = 1, \ldots, T$, the latter yielding $V^o = I$. We observe that such moment estimators correspond to one iteration of the REML procedure.

As another alternative, Rao (1970) proposed a minimum norm, quadratic, unbiased estimator (MINQUE) as follows: For estimating the linear function $a'\phi$, we choose Q as the solution to the problem

$$\begin{aligned} &\text{minimize } tr(QV^o)^2 \\ &\text{subject to} \quad QX = 0 \\ &tr(QV_t) = a_t \qquad t = 0, \ldots, T. \end{aligned} \tag{8.105}$$

Here V^o again denotes V evaluated at a specified value of ϕ. Reference to (8.100) shows that the constraints in (8.105) are required so that $E(Y'QY) = a'\phi$. The motivation for the particular norm to be minimized in (8.105) is found in the expression for the variance of $Y'QY$ in Theorem 2.2; that is, with $QX = 0$, $\text{Var}(Y'QY) = 2tr(QV)^2$. Thus, with normally distributed data, the solution to this problem yields a minimum variance estimator of $a'\phi$ if $V = V^o$.

LaMotte (1970) established the relation between the quadratic forms arising from this procedure and those given by (8.104). He showed that the matrix Q in 8.105 has the form

$$Q = \sum_{t=0}^{T} \gamma_t (V^o)^{-1} M^o V_t M^{o\prime} (V^o)^{-1}. \tag{8.106}$$

Here γ_t are chosen so that $E(Y'QY) = a'\phi$. We thus see that this procedure is equivalent to the moment estimation procedure with Q_i defined by (8.104) and that it corresponds to the first iteration of REML with initial guess ϕ^o. Actually, Rao (1970) suggested $V^o = I$ for the MINQUE procedure. In the future, we shall refer to these procedures collectively as REML.

From this development, we note that, if we do only one iteration of REML, the resulting estimators are unbiased; that is, with Q_i given by (8.104), the solution, ϕ, of (8.103) has expected value

$$\begin{aligned} E(\hat{\phi}_Q) &= \Omega_Q^{-1} E(\rho_Q) \\ &= \phi. \end{aligned} \tag{8.107}$$

Also, from the MINQUE development, we note that, if the solution does not depend on V^o, then the REML estimators have minimum variance in the class of unbiased estimators. We shall see such cases in Chapter 9.

Henderson (1953) also suggested quadratic forms that satisfy (8.99), but they are generally different from those that arise in the REML method. We shall illustrate these in Chapter 10.

Smith (1971) noted that the inverse of V will have the general form

$$V^{-1} = \sum_{t=0}^{S} \delta_t(\phi) A_t \tag{8.108}$$

where $\delta_t(\phi)$ is a function of ϕ and the matrices A_t are symmetric. It can be shown that the statistics

$$Y'A_tY \qquad t = 0, \ldots, S \tag{8.109}$$

and

$$Y'A_tX \qquad t = 0, \ldots, S \tag{8.110}$$

are sufficient, but not complete sufficient, statistics for ϕ and θ. Thus, it is suggested that these quadratic forms, (8.109), be considered as candidates in the moment estimation procedure. However, this observation clearly depends on having a closed form expression for the inverse of V. We shall see, in Chapter 9, a class of problems for which this is readily available and for which optimal estimates will be obtained.

We have not mentioned estimation of θ in this section. The sufficient statistics above suggest that linear functions of the linear forms (8.110) be considered. Generally, given an estimate of ϕ and hence V, the likelihood equations, (8.29), are used to estimate θ.

We conclude this section on estimation by noting a special situation in which the estimation of θ is independent of ϕ. We contrast the likelihood equations for θ, sometimes called the *weighted least squares equations,* with the likelihood equations for θ if V is $\sigma^2 I$, called the *ordinary least squares equations*. The former are given by

$$\theta = (X'V^{-1}X)^{-1}X'V^{-1}Y \tag{8.111}$$

and the latter by

$$\theta_0 = (X'X)^{-1}X'Y. \tag{8.112}$$

The following theorem describes the requirement for these two vectors to be identical.

THEOREM 8.1

The vectors θ and θ_0 are identical if and only if there exists a nonsingular matrix B such that

$$VX = XB. \tag{8.113}$$

Proof: First, if B exists satisfying (8.113), then we may write

$$X'V^{-1} = (B')^{-1}X'. \tag{8.114}$$

Multiplying on the right by X, we have

$$X'V^{-1}X = (B')^{-1}X'X. \tag{8.115}$$

Inverting (8.115) and combining with (8.114), we have

$$(X'V^{-1}X)^{-1}X'V^{-1} = (X'X)^{-1}X' \tag{8.116}$$

hence, in view of (8.111) and (8.112), we have $\theta = \theta_0$.

Conversely, if we assume $\theta = \theta_0$, (8.116) holds. Multiplying on the right by V and on the left by $X'X$ and transposing, we obtain

$$\begin{aligned} VX &= X(X'V^{-1}X)^{-1}X'X \\ &= XB. \end{aligned} \tag{8.117}$$

■

We shall see that this relation does hold for some of the models to be discussed in the next chapter.

8.3. TESTS OF HYPOTHESES

As noted in the introduction to this chapter, the general results on hypothesis testing are limited. We shall develop the likelihood ratio test and a modified likelihood ratio test, but the distribution of the test statistics are not known in general. In Chapter 9, we shall see that, in some special cases, exact tests are available.

8.3.1. The Likelihood Ratio Test

The likelihood function as defined by (8.24) is written as

$$L(\theta, \phi) = (2\pi)^{-N/2}|V|^{-1/2} \exp[-Q(\theta, \phi)/2] \tag{8.118}$$

where

$$Q(\theta, \phi) = (Y - X\theta)'V^{-1}(Y - X\theta). \tag{8.119}$$

Consider first, the value of the likelihood function at the maximum likelihood solution. From Section 8.2.1, we recall that closed form expressions for the estimators were not generally available, but we do know that they satisfy the likelihood equations

$$X'V^{-1}X\theta = X'V^{-1}Y \tag{8.120}$$

and

$$tr(V^{-1}V_t) = \rho_t \qquad t = 0, \ldots, T \tag{8.121}$$

where

$$\begin{aligned} \rho_t &= Y'Q_tY \\ Q_t &= V^{-1}MV_tM'V^{-1} \\ M &= I - X(X'V^{-1}X)^{-1}X'V^{-1}. \end{aligned} \tag{8.122}$$

Recall also that

$$V^{-1}M = M'V^{-1} = M'V^{-1}M. \tag{8.123}$$

Denote the maximum likelihood estimators by $\hat{\theta}$ and $\hat{\phi}$, and let

$$\begin{aligned} \hat{V} &= \sum_t \hat{\phi}_t V_t \\ \hat{\rho}_t &= \rho_t(\hat{\phi}). \end{aligned} \tag{8.124}$$

Then

$$\begin{aligned} Q(\hat{\theta}, \hat{\phi}) &= (Y - X\hat{\theta})'\hat{V}^{-1}(Y - X\hat{\theta}) \\ &= Y'\hat{M}'\hat{V}^{-1}\hat{M}Y \\ &= Y'\hat{V}^{-1}\hat{M}\hat{V}\hat{M}'\hat{V}^{-1}Y \\ &= \sum_t \hat{\phi}_t \hat{\rho}_t \\ &= \sum_t \hat{\phi}_t tr(\hat{V}^{-1}V_t) \\ &= tr(\hat{V}^{-1}\hat{V}) \\ &= N. \end{aligned} \tag{8.125}$$

Thus, we have

$$L(\hat{\theta}, \hat{\phi}) = (2\pi)^{-N/2}|\hat{V}|^{-1/2} \exp(-N/2). \tag{8.126}$$

Further, letting λ_k, $k = 1, \ldots, s$ denote the distinct eigenvalues of V with multiplicities r_k, we may write

$$|V| = \prod_{k=1}^{s} \lambda_k^{r_k}. \tag{8.127}$$

The likelihood ratio test then requires that we evaluate the likelihood function at the maximum likelihood solution, constrained by the hypothesis. We consider, separately, hypotheses on θ and ϕ.

Linear hypotheses on $\boldsymbol{\theta}$. We have seen that for testing linear hypotheses on θ we can use the hypothesis to form a reduced model. Thus, the likelihood has the same functional form as (8.118) with X and θ replaced by their reduced counterparts. Evaluation of the likelihood at the solution $(\tilde{\theta}, \tilde{\phi})$ follows exactly the same pattern as above; hence, the likelihood ratio statistic is given by

$$\begin{aligned} \gamma &= L(\tilde{\theta}, \tilde{\phi})/L(\hat{\theta}, \hat{\phi}) \\ &= (|\hat{V}|/|\tilde{V}|)^{1/2}. \end{aligned} \tag{8.128}$$

In terms of the eigenvalues of V with $\hat{\lambda}_k$ and $\tilde{\lambda}_k$ denoting the eigenvalues of $\hat{V}$ and $\tilde{V}$, we have

$$\gamma^2 = \prod_{k=1}^{s} (\hat{\lambda}_k/\tilde{\lambda}_k)^{r_k}. \tag{8.129}$$

Linear hypotheses on ϕ. Linear constraints on ϕ will result in a reduced likelihood that has the same form as (8.118), except that V is replaced by

$$V^* = \sum_{i=0}^{T^*} \phi_i^* V_i^* \tag{8.130}$$

where ϕ_i^* and V_i^* are linear functions of ϕ and the V_i. The reduction of the likelihood follows as above. With $\tilde{V}^*$ denoting the estimate of V^* and $\tilde{\lambda}_k^*$, $k = 1, \ldots, s^*$ denoting the distinct eigenvalues of $\tilde{V}^*$ with multiplicities r_k^*, we have

$$\begin{aligned}\gamma^2 &= |\hat{V}|/|\tilde{V}^*| \\ &= \prod_{k=1}^{s} \hat{\lambda}_k^{r_k} \Big/ \prod_{k=1}^{s^*} \tilde{\lambda}_k^{*r_k^*}.\end{aligned} \tag{8.131}$$

In general, little can be said about the distribution of either (8.129) or (8.131), although we shall see some special cases in which the exact distribution is available. In general, we can appeal to the asymptotic result (Appendix B.II), which says that under the null hypothesis

$$-2\text{Ln}\gamma \sim \chi^2(v) \tag{8.132}$$

where v is the number of constraints in the hypothesis.

8.3.2. A Modified Likelihood Ratio Test

In Section 8.2.2, we described a modification of the maximum likelihood estimator (REML), which arose by factoring the original likelihood. We shall see that in some cases these estimators have more appeal than the ML estimators. We now describe a modified likelihood ratio test in which we use REML, rather than ML, estimators.

Linear hypotheses on θ. To evaluate the likelihood function at the REML estimators, recall from (8.77) and (8.78) that we may write

$$\begin{aligned}Q(\theta, \phi) &= (Y - X\theta)'V^{-1}X(X'V^{-1}X)^{-1}X'V^{-1}(Y - X\theta) \\ &\quad + Y'V^{-1}MVM'V^{-1}Y.\end{aligned} \tag{8.133}$$

Evaluating this expression at the REML estimators $(\hat{\theta}, \hat{\phi})$, with $\hat{\phi}$ defined by (8.92), we see that the first term is zero and that the second simplifies as follows:

$$
\begin{aligned}
Y'V^{-1}MVM'V^{-1}Y &= \sum_t \hat{\phi}_t \rho_t \\
&= \sum_t \hat{\phi}_t tr(Q_t V) \\
&= \sum_t \hat{\phi}_t tr(V^{-1}MV_t M') \\
&= \sum_t \hat{\phi}_t tr(M'V^{-1}V_t) \\
&= tr(M') \\
&= N - r.
\end{aligned}
\tag{8.134}
$$

(Here, V, M, Q, and ρ are all evaluated at $\hat{\phi}$.) Thus,

$$L(\hat{\theta}, \hat{\phi}) = (2\pi)^{-N/2}|\hat{V}|^{-1/2} \exp[-(N - r)/2]. \tag{8.135}$$

Noting the similarity between this expression with V evaluated at the REML estimator and (8.126) with V evaluated at the ML estimator, it follows that, for testing linear hypotheses on θ, the modified likelihood ratio statistic will have the form (8.129) using the REML estimates for λ_k.

Linear hypotheses on ϕ. In this case, we could use (8.135) and obtain the modified likelihood ratio statistic analogous to (8.131) using the REML estimates for λ_k. Alternatively, we might consider a further modification in which we use only the factor of the likelihood function, $L_1(\phi)$, used to obtain the REML estimates of ϕ. Using (8.134), we have

$$L_1(\hat{\phi}) = (2\pi)^{-(N-r)/2}|A'\hat{V}A|^{-1/2} \exp[-(N - r)/2]$$

where A is defined by (8.72). The squared, modified likelihood ratio statistic is then given by

$$\gamma^2 = |A'\hat{V}A|/|A'\tilde{V}^*A|. \tag{8.136}$$

We thus have the analog of (8.131), which may be expressed in terms of the REML estimates of the eigenvalues of $A'VA$ and $A'V^*A$. This test statistic would seem to require an explicit expression for A that we have avoided. We shall see, in Chapter 9, examples in which we may determine these eigenvalues without determining A. In fact, this will yield the classical analysis of variance test with balanced data.

While there is no general argument to support these modified test statistics, it seems reasonable to assume that their general properties are analogous to those of the ordinary likelihood ratio test.

8.3.3. Tests Based on Ratios of Quadratic Forms

Based on our experience with the special case $V = \sigma^2 I$, we might consider test statistics based on ratios of quadratic forms. Thus, for testing linear hypotheses on θ, we might seek independent chi-square statistics, one of which is a central chi-square and the other a noncentral chi-square whose noncentrality parameter is zero under the null hypothesis. The resulting F-ratio could then be used.

The analogous situation for testing hypotheses on ϕ suggests the ratio of two central chi-square statistics that is proportional to an F-statistic such that the proportionality factor does not depend on ϕ under the null hypothesis.

We shall see that special cases exist in which such test statistics do arise and, further, that the quadratic forms are suggested by the modified likelihood ratio test.

8.3.4. An Approximate Procedure for Hypotheses on θ

To motivate this approach, we consider the special case in which the covariance matrix has the form

$$V = \sigma^2 U \tag{8.137}$$

where U is a known matrix. For example, this might imply, from (8.2), that $\sigma^2 = \phi_0$ and that the ratios ϕ_t/ϕ_0 are known.

Since U is positive definite, we may write $C'UC = I$ for nonsingular C and transform as follows:

$$Z = C'Y \sim N(C'X\theta,\ \sigma^2 I).$$

Noting that we are now in the Chapter 3 structure, we have

$$\begin{aligned} \hat{\theta} &= (X'U^{-1}X)^{-1}X'U^{-1}Y \\ s^2 &= Y'[U^{-1} - U^{-1}X(X'U^{-1}X)^{-1}X'U^{-1}]Y/(N - r) \end{aligned} \tag{8.138}$$

as estimates of θ and σ^2. Further, for testing the hypothesis $H\theta = h$, we have the test statistic

$$(H\hat{\theta} - h)'[H(X'U^{-1}X)^{-1}H']^{-1}(H\hat{\theta} - h)/vs^2 \sim F(v, N - r, \lambda) \tag{8.139}$$

where H has row rank v.

The suggestion in the general case is to replace U by an estimator in which the ratios ϕ_t/ϕ_0 are replaced by their ML or REML estimates. (It can be shown that the ML or the REML estimator of ϕ_0 is s^2 in (8.138) if U is replaced by its estimator.)

While this procedure is intuitively appealing, its properties are not generally known, and thus it should be used with caution. Formally, if we use the REML estimator, this test corresponds to a modified likelihood ratio test in which we use the likelihood factor, L_2, in (8.66) as if the estimated ratios were known.

Finally, we note that this approach also will yield approximate confidence regions and simultaneous confidence intervals in the usual way, as described in Chapter 3.

8.4. CONFIDENCE REGIONS AND INTERVALS

The situation with regard to confidence regions is similar to that encountered in testing hypotheses; that is, there are special cases in which exact confidence intervals are available, but in general only approximate results are known.

8.4.1. Confidence Regions and Intervals for θ

As noted in Section 8.3.4, we might invert the test statistic (8.139) as if the covariance matrix were known to yield the usual elliptical confidence ellipsoid for linear functions of θ. Simultaneous confidence intervals can be constructed as usual using the methods of Chapter 3. Of course, these results are only approximate (see Harville, 1976).

In those special cases in which the test of the linear hypothesis on θ is exact, we may invert the test statistic to obtain exact confidence regions.

These ideas will be illustrated in Chapter 9.

8.4.2. Confidence Intervals for Functions of ϕ

We shall be interested in confidence regions and confidence intervals for the components of ϕ, linear combinations of these components, and ratios of linear combinations. While exact intervals do exist in special cases, there are no general results available. Confidence intervals and regions based on the inversion of test statistics are not generally available. Even when we have exact tests, they are generally limited to such hypotheses as $\phi_i = 0$ and do not extend to the general case $\phi_i = \phi_i^o$, which is required for the inversion.

An approximate approach that is useful is based on the fact that we often have independent quadratic forms q_i such that

$$q_i/g_i(\phi) \sim \chi^2(d_i) \qquad i = 0, \ldots, T \tag{8.140}$$

where the $g_i(\phi)$ are linear functions of ϕ. Now suppose we are interested in a confidence interval on a linear function of ϕ—say, $h(\phi)$. If $h(\phi)$ happens to coincide with $g_i(\phi)$ for some i, then we immediately have

$$\text{Prob}[\chi^2(1 - \alpha_1; d_i) \leq q_i/h(\phi) \leq \chi^2(\alpha_2; d_i)] = 1 - \alpha_1 - \alpha_2 \tag{8.141}$$

and hence an interval with confidence coefficient $(1 - \alpha_1 - \alpha_2)$ on $h(\phi)$ is given by

$$q_i/\chi^2(\alpha_2; d_i) \leq h(\phi) \leq q_i/\chi^2(1 - \alpha_1; d_i). \tag{8.142}$$

In other cases, there may exist constants a_i, $i = 0, \ldots, T$ such that

$$h(\phi) = \sum_i a_i g_i(\phi). \tag{8.143}$$

We are thus led to consider the statistic

$$q = \sum_i a_i q_i / d_i. \tag{8.144}$$

Note that q has mean $h(\phi)$ and

$$\mathrm{Var}(q) = \sum_i 2a_i^2 g_i^2(\phi)/d_i. \tag{8.145}$$

In general, the distribution of q is not known, but Satterthwaite (1946) and Welch (1956) suggested that the distribution of q is proportional to a chi-square variate; that is,

$$q \mathrel{\dot{\sim}} c\chi^2(d) \tag{8.146}$$

where c and d are determined so that the first two moments of (8.146) are given by (8.143) and (8.145). It follows that

$$c = \sum_i a_i^2 g_i^2(\phi)/h(\phi)d_i \tag{8.147}$$

and

$$\begin{aligned} d &= \sum_i a_i g_i(\phi)/c \\ &= h(\phi)/c. \end{aligned} \tag{8.148}$$

To the accuracy of this approximation, we may then write, in analogy with (8.141),

$$\mathrm{Prob}[\chi^2(1 - \alpha_1;\, d) \le q/c \le \chi^2(\alpha_2;\, d)] = 1 - \alpha_1 - \alpha_2. \tag{8.149}$$

Solving for $h(\phi)$, we obtain the approximate confidence interval

$$qd/\chi^2(\alpha_2;\, d) \le h(\phi) \le qd/\chi^2(1 - \alpha_1;\, d). \tag{8.150}$$

Since ϕ is unknown, the degrees of freedom d in (8.150) are approximated by replacing $g_i(\phi)$ by q_i/d_i in (8.148). Welch (1956) also suggested an improvement on this approximation.

We note that this same approximation may be used to test certain linear hypotheses. Thus, with q as defined in (8.144), we also define

$$\begin{aligned} p &= \sum_i b_i q_i/d_i \\ &\mathrel{\dot{\sim}} u\chi^2(v) \end{aligned} \tag{8.151}$$

where $E(p) = k(\phi)$ and u and v are defined as in the definition of c and d. It follows that

$$[k(\phi)/h(\phi)]q/p \mathrel{\dot\sim} F(d, v). \tag{8.152}$$

Thus, for testing the hypothesis H_0: $k(\phi) = h(\phi)$ against the alternative $h(\phi) > k(\phi)$ or $h(\phi) \neq k(\phi)$, we may use the fact that, under H_0, the statistic q/p is approximately $F(d, v)$.

CHAPTER 8 EXERCISES

1. In Example 8.1, we could have described the covariance structure by letting the variance of an observation be ζ_0 and the covariance between two observations in the same plot be ζ_1. Describe the covariance matrix in terms of these parameters, and note the relation to the matrix (8.7). Do the analogous exercise for Examples 8.2 through 8.7 to verify that the choice of parameters ϕ_t is primarily for notational convenience.
2. Develop the ML and REML equations for estimating ϕ_0 and ϕ_1 in Example 8.1. Solve the equations and determine the mean and variance of the estimators.
3. Verify that the inverse of the covariance matrix (8.16) for Example 8.4 is given by

$$\begin{aligned} V^{-1} &= [V_0 - (1/n)V_3]/\lambda_0 + [(1/bn)V_1 - A]/\lambda_1 + [(1/an)V_2 - A]/\lambda_2 \\ &\quad + [(1/n)V_3 - (1/bn)V_1 - (1/an)V_2 + A]/\lambda_3 + A/\lambda \end{aligned}$$

where V_0, V_1, V_2, and V_3 are the matrices associated with $\phi_0, \ldots, \phi_3$ in (8.16),

$$\begin{aligned} \lambda_0 &= \phi_0 \\ \lambda_1 &= \phi_0 + bn\phi_1 + n\phi_3 \\ \lambda_2 &= \phi_0 + an\phi_2 + n\phi_3 \\ \lambda_3 &= \phi_0 + n\phi_3 \\ \lambda &= \phi_0 + bn\phi_1 + an\phi_2 + n\phi_3 \end{aligned}$$

and

$$A = (1/N)J_N J_N' \qquad N = abn.$$

4. Develop the ML and REML equations for estimating ϕ_0, ϕ_1, and ϕ_2 in Example 8.4 for the special case $\phi_3 = 0$ and $n = 1$. Provide solutions for each if possible.
5. Develop the ML and REML estimates for μ, ϕ_0, and ϕ_1 in Example 8.6.

6. In Exercise 5, determine the mean, variances, and covariances of the estimates. Compare the variances and covariances with the large sample moments given by the inverse of the information matrix (8.39).

7. Spell out the details of the Hemmerle–Hartley W-transform for the unbalanced, one-way classification model defined as

$$E(Y) = \mu J$$

$$\text{Var}(Y) = \phi_0 I + \phi_1 UU'$$

with $U = D_a(J_{n_i})$.

8. For the one-way classification, random model in Example 8.1, determine the distributions of the sums of squares defined by

$$T = \sum_j \sum_k (\bar{y}_{j\cdot} - \bar{y}_{\cdot\cdot})^2$$

$$E = \sum_j \sum_k (y_{jk} - \bar{y}_{j\cdot})^2.$$

Show how T and E can be used in the method of moments to estimate ϕ_0 and ϕ_1. Contrast these results with those obtained in Exercise 2.

9. Show that the MINQUE estimator defined by (8.105) is given by (8.106).

10. Develop the modified likelihood ratio test for the hypothesis H_0: $\phi_1 = 0$ versus H_A: $\phi_1 \neq 0$ in Example 8.1. Compare this test with the test based on the ratio of the quadratic forms T and E in Exercise 8.

11. Consider the linear model with

$$E(Y) = X\beta \quad \text{and} \quad \text{Var}(Y) = \sigma^2 V$$

where X has full column rank. Show that the weighted least squares and the ordinary least squares estimators of β are identical if the columns of X are eigenvectors of V.

12. For the mixed model in Example 8.5, develop the modified likelihood ratio test for the hypothesis H_0: $\mu_i = \mu_{i*}$, $i \neq i* = 1, \ldots, a$. Compare this test with the approximate one suggested in Section 8.3.4.

13. Show that the REML estimate of V is positive definite for the covariance matrices in Exercises 2, 4, and 5.

CHAPTER NINE THE ANALYSIS OF MEANS AND VARIANCE: SPECIAL MODELS

9.1. INTRODUCTION

The general results developed in Chapter 8 are applicable to linear models with any structure on the mean vector and the covariance matrix. As noted, the results were limited. The parameter estimates required iterative solutions of nonlinear equations, and no exact tests or confidence intervals were obtained.

In this chapter, we shall examine a class of problems for which optimal point estimators, exact tests, and exact confidence intervals are obtainable. We shall see that the mean structure is that of the cell means model and that the covariance matrix is of a special form. A general requirement is that the data be balanced in the same sense as in the fixed effects model; that is, the range on each subscript is a constant not depending on other subscripts.

We begin, in Section 9.2, by assuming a simple mean structure and defining the special covariance matrix. In Section 9.3, the analysis is extended to include a general cell means structure on the mean vector.

The model is extended to include the unbalanced case in Section 9.5. A special case of unbalanced data—the mixed, Balanced Incomplete Block model—is examined to illustrate the difficulties.

9.2. A SPECIAL COVARIANCE STRUCTURE

9.2.1. Definition of V and Development of V^{-1}

In Chapter 8, when concerned with estimating the parameter vector ϕ that determines the covariance matrix, we saw that both the ML and the REML equations involved the inverse of V. It is clear that, if we hope to obtain closed form solutions for these estimates, we need an expression for V^{-1}. Similarly, when discussing moment estimators, the form (8.108) for the inverse matrix suggested matrices to be used in the quadratic forms. Motivated by our examples in Chapter 8, we now consider a special covariance matrix for which the inverse is readily obtained. We shall see that this matrix is appropriate for a broad class of models for which the data is balanced.

To motivate our covariance matrix, we recall the structure of the covariance matrices in Examples 8.1 through 8.7. We shall also associate these matrices with the names for these models that were implied in Chapter 8. (For convenience in developing the general model, we shall change the subscripts slightly from those used in Chapter 8.)

EXAMPLE 8.1 The Random, One-Way Classification Model

$$V = \phi_0 I_{an} + \phi_1(I_a \otimes J_n J_n') \tag{9.1}$$

EXAMPLE 8.2 The Mixed, Two-Fold Nested Model

$$V = \phi_0 I_{abn} + \phi_3(I_a \otimes I_b \otimes J_n J_n') \tag{9.2}$$

EXAMPLE 8.3 The Random, Two-Fold Nested Model

$$V = \phi_0 I_{abn} + \phi_1(I_a \otimes J_b J_b' \otimes J_n J_n') + \phi_3(I_a \otimes I_b \otimes J_n J_n') \tag{9.3}$$

EXAMPLE 8.4 The Random, Two-Way Classification Model

$$\begin{aligned} V = \phi_0 I_{abn} &+ \phi_1(I_a \otimes J_b J_b' \otimes J_n J_n') + \phi_2(J_a J_a' \otimes I_b \otimes J_n J_n') \\ &+ \phi_3(I_a \otimes I_b \otimes J_n J_n') \end{aligned} \tag{9.4}$$

EXAMPLE 8.5 The Mixed, Two-Way Classification Model

$$V = \phi_0 I_{abn} + \phi_2(J_a J_a' \otimes I_b \otimes J_n J_n') + \phi_3(I_a \otimes I_b \otimes J_n J_n') \tag{9.5}$$

EXAMPLE 8.6 The Randomized Block Model

$$V = \phi_0 I_{ab} + \phi_2(J_a J_a' \otimes I_b \otimes J_n J_n') \tag{9.6}$$

EXAMPLE 8.7 The Split-Plot Model

$$V = \phi_0 I_{abn} + \phi_1(I_a \otimes J_b J_b' \otimes J_n J_n') + \phi_3(I_a \otimes I_b \otimes J_n J_n') \tag{9.7}$$

■

Examination of (9.1) through (9.7) shows that all of these covariance matrices are special cases of (9.4). For reference, we rewrite (9.4) in a notation that will be more easily generalized.

We write

$$V = \phi_0 I + \phi_1 V_1 + \phi_2 V_2 + \phi_{12} V_{12} \tag{9.8}$$

where

$$
\begin{aligned}
V_1 &= I_{a_1} \otimes U_{a_2} \otimes U_n \\
V_2 &= U_{a_1} \otimes I_{a_2} \otimes U_n \\
V_{12} &= I_{a_1} \otimes I_{a_2} \otimes U_n
\end{aligned}
\tag{9.9}
$$

and

$$U_i = J_i J_i'. \tag{9.10}$$

Here a and b have been replaced by a_1 and a_2, and ϕ_3 has been replaced by ϕ_{12}. This notation will be used in the remainder of this book. Note that V has dimension na_1a_2.

We are now prepared to define the covariance structure to be considered in this section.

DEFINITION 9.1
Let T denote the collection of all $2^k - 1$ subsets of the integers $(1, 2, \ldots, k)$ excluding the empty set. ■

For example, with $k = 3$,

$$T = (1, 2, 3, 12, 13, 23, 123). \tag{9.11}$$

These subsets will be used to index parameters and matrices as in (9.8) and (9.9). It will be convenient to include the zero index, so we augment T with zero and denote the collection of indices

$$T_0 = (0, T). \tag{9.12}$$

DEFINITION 9.2
Let $a_1, a_2, \ldots, a_k$ and n be positive integers, and let

$$N = n\prod_{i=1}^{k} a_i. \tag{9.13}$$

Also, let

$$V_0 = I_N \tag{9.14}$$

and, for $t \in T$, define

$$V_t = (L_1 \otimes L_2 \otimes \ldots \otimes L_k \otimes U_n) \tag{9.15}$$

where

$$L_i = \begin{bmatrix} I_{a_i} & i \in t \\ U_{a_i} & i \notin t. \end{bmatrix} \tag{9.16}$$

Finally, with ϕ_t, $t \in T_0$ denoting a set of parameters, we define the covariance matrix

$$V = \sum_{t \in T_0} \phi_t V_t \tag{9.17}$$

where the sum in (9.17) ranges over all subsets in T_0.

■

To motivate the expression for the inverse of V, we first describe the inverse for the special case $k = 2$. Note that this matrix is given by (9.8) through (9.10) in accordance with Definition 9.2.

EXAMPLE 9.1

With V defined by (9.8), we first write V in an alternative form as

$$\begin{aligned} V &= \lambda_0 A_0 + \lambda_1 A_1 + \lambda_2 A_2 + \lambda_{12} A_{12} + \lambda_\mu A_\mu \\ &= \sum_{t \in T_0} \lambda_t A_t + \lambda_\mu A_\mu \end{aligned} \tag{9.18}$$

where

$$\begin{aligned} \lambda_0 &= \phi_0 \\ \lambda_1 &= \phi_0 + na_2\phi_1 + n\phi_{12} \\ \lambda_2 &= \phi_0 + na_1\phi_2 + n\phi_{12} \\ \lambda_{12} &= \phi_0 + n\phi_{12} \\ \lambda_\mu &= \phi_0 + na_2\phi_1 + na_1\phi_2 + n\phi_{12} \end{aligned} \tag{9.19}$$

and

$$\begin{aligned} A_0 &= V_0 - (1/n)V_{12} \\ A_1 &= (1/na_2)V_1 - A_\mu \\ A_2 &= (1/na_1)V_2 - A_\mu \\ A_{12} &= (1/n)V_{12} - (1/na_2)V_1 - (1/na_1)V_2 + A_\mu \\ A_\mu &= (1/N)J_N J_N'. \end{aligned} \tag{9.20}$$

In this notation, it is easily verified that

$$V^{-1} = \sum_{t \in T_0} (1/\lambda_t)A_t + (1/\lambda_\mu)A_\mu. \tag{9.21}$$

This result is a consequence of the idempotency and orthogonality of the matrices in (9.20), which are established below in Theorem 9.1.

From (9.20), we may verify that alternative expressions for A_t, $t \in T$ are given by

$$\begin{aligned} A_1 &= (1/na_2)[I_{a_1} - (1/a_1)U_{a_1}] \otimes U_{a_2} \otimes U_n \\ A_2 &= (1/na_1)U_{a_1} \otimes [I_{a_2} - (1/a_2)U_{a_2}] \otimes U_n \\ A_{12} &= (1/n)[I_{a_1} - (1/a_1)U_{a_1}] \otimes [I_{a_2} - (1/a_2)U_{a_2}] \otimes U_n. \end{aligned} \tag{9.22}$$

For future reference, it is convenient to express (9.19) in matrix notation. Thus,

$$\begin{bmatrix} \lambda_0 \\ \lambda_1 \\ \lambda_2 \\ \lambda_{12} \end{bmatrix} = \begin{bmatrix} 1 & 0 & 0 & 0 \\ 1 & na_2 & 0 & n \\ 1 & 0 & na_1 & n \\ 1 & 0 & 0 & n \end{bmatrix} \begin{bmatrix} \phi_0 \\ \phi_1 \\ \phi_2 \\ \phi_{12} \end{bmatrix}. \tag{9.23}$$

Let λ and ϕ denote the vectors in (9.23), and let P denote the coefficient matrix where

$$\begin{aligned} \lambda &= (\lambda_m,\ m \in T_0) \\ \phi &= (\phi_t,\ t \in T_0) \\ P &= (p_{mt},\ m,\ t \in T_0). \end{aligned} \tag{9.24}$$

It is easily verified that P is nonsingular. (Note that the indexing on λ, ϕ, and P is unusual in that the subscripts are the subsets of T_0 as given in Definition 9.1. Thus, for example, the element in the lower right corner of P is $p_{12,12} = n$.)

To define λ_μ, we write

$$\lambda_\mu = (1\ na_2\ na_1\ n)\phi. \tag{9.25}$$

The row vector in (9.25) is denoted by

$$P_\mu' = (p_t,\ t \in T_0). \tag{9.26}$$

The matrix P and the vector P_μ may be used to describe an alternative relation between the V and A matrices as follows:

$$V_t = \sum_{m \in T_0} p_{mt} A_m + p_t A_\mu \qquad t \in T_0. \tag{9.27}$$

We shall see that the relation in (9.20) is based on the elements of the inverse of P.

■

Example 9.1 will be used for illustration later in this chapter. For now, we use it to motivate our general result, which is based on the paper by Smith and Hocking (1978). (See also Khuri, 1983, and Wansbeck and Kapteyn, 1982.)

DEFINITION 9.3
For the index set T in Definition 9.1, define

$$A_t = (1/na_t^*)G_1 \otimes G_2 \otimes \ldots \otimes G_k \otimes U_n \tag{9.28}$$

where

$$a_t^* = \prod_{i \neq t} a_i \tag{9.29}$$

and

$$G_i = \begin{bmatrix} I_{a_i} - (1/a_i)U_{a_i} & i \in t \\ U_{a_i} & i \notin t. \end{bmatrix} \tag{9.30}$$

Also, define

$$\begin{aligned} A_\mu &= (1/N)U_N \\ A_0 &= V_0 - \sum_{t \in T} A_t - A_\mu. \end{aligned} \tag{9.31}$$

Further, for the parameters ϕ_t, $t \in T_0$ define

$$\begin{aligned} \lambda_0 &= \phi_0 \\ \lambda_m &= \phi_0 + n \sum_{\substack{t \in T \\ m \subseteq t}} a_t^* \phi_t \qquad m \in T \end{aligned} \tag{9.32}$$

and

$$\lambda_\mu = \phi_0 + n \sum_{t \in T} a_t^* \phi_t. \tag{9.33}$$

The sum in (9.32) ranges over all subsets $t \in T$ such that m is contained in t. We express (9.32) in matrix notation as

$$\lambda = P\phi \tag{9.34}$$

where the elements of P, p_{mt}, m, $t \in T_0$ are given by

$$\begin{aligned} p_{m0} &= 1 \qquad m \in T_0 \\ p_{0t} &= 0 \qquad t \in T \\ p_{mt} &= \begin{bmatrix} na_t^* & m \subseteq t \\ 0 & \text{otherwise.} \end{bmatrix} \end{aligned} \tag{9.35}$$

It is easily shown that P is nonsingular. The relation (9.33) is written as

$$\lambda_\mu = P'_\mu \phi \tag{9.36}$$

where

$$P'_\mu = (1, p_t = na_t^*, t \in T). \tag{9.37}$$

(Note: If $t = (1, 2, \ldots, k)$, then $a_t^* = 1$.)

■

We now state our general result as follows.

THEOREM 9.1
The covariance matrix

$$V = \sum_{t \in T_0} \phi_t V_t \tag{9.38}$$

as given in Definition 9.2 may be written as

$$V = \sum_{t \in T_0} \lambda_t A_t + \lambda_\mu A_\mu \tag{9.39}$$

with inverse matrix

$$V^{-1} = \sum_{t \in T_0} (1/\lambda_t) A_t + (1/\lambda_\mu) A_\mu \tag{9.40}$$

where V_t, A_t, and λ_t are defined as in Definitions 9.2 and 9.3 for the index sets T and T_0 in Definition 9.1. It is assumed that λ_t and λ_μ are nonzero.

Proof: From (9.28), it is easily shown that, for $t \in T$,

$$A_t = (1/na_t^*)V_t - \sum_{m \subset t} A_m - A_\mu. \tag{9.41}$$

The argument is by induction and simply involves the expansion of (9.28). From (9.41) we may write, for $t \in T$,

$$V_t = na_t^* \sum_{m \subseteq t} A_m + na_t^* A_\mu \tag{9.42}$$

or, using (9.35), (9.37), and (9.31), we may write, for $t \in T_0$,

$$V_t = \sum_{m \in T_0} p_{mt} A_m + p_t A_\mu. \tag{9.43}$$

To establish (9.39), write

$$
\begin{aligned}
V &= \sum_{t \epsilon T_0} \phi_t V_t \\
&= \sum_{t \epsilon T_0} \phi_t \Big(\sum_{m \epsilon T_0} p_{mt} A_m + p_t A_\mu \Big) \\
&= \sum_{m \epsilon T_0} \Big(\sum_{t \epsilon T_0} p_{mt} \phi_t \Big) A_m + A_\mu \sum_{t \epsilon T_0} p_t \phi_t \\
&= \sum_{m \epsilon T_0} \lambda_m A_m + \lambda_\mu A_\mu .
\end{aligned}
\tag{9.44}
$$

From (9.28) and (9.30) we immediately establish that

$$
A_t A_m = \begin{bmatrix} A_t & t = m \\ 0 & t \neq m \end{bmatrix} \qquad t \epsilon T. \tag{9.45}
$$

From (9.31), we see that

$$
\begin{aligned}
A_\mu A_\mu &= A_\mu \\
A_0 A_0 &= A_0 \\
A_\mu A_t &= 0 \qquad t \epsilon T_0 \\
A_0 A_t &= 0 \qquad t \epsilon T.
\end{aligned}
\tag{9.46}
$$

The matrices in (9.39) are thus idempotent and orthogonal; hence, from (9.39), (9.40), and (9.31) we have

$$
\begin{aligned}
VV^{-1} &= \sum_{t \epsilon T_0} A_t + A_\mu \\
&= V_0 = I
\end{aligned}
\tag{9.47}
$$

and the proof is complete.

■

The properties of the idempotent matrices in Theorem 9.1 immediately yield the following theorem.

THEOREM 9.2
The matrices A_μ and A_t, $t \epsilon T_0$, have ranks

$$
\begin{aligned}
r(A_\mu) &= 1 \\
r(A_0) &= (n - 1) \prod_{i=1}^{k} a_i \\
r(A_t) &= \prod_{i \epsilon t} (a_i - 1) \qquad t \epsilon T.
\end{aligned}
\tag{9.48}
$$

Further, the eigenvalues of V are

$$\begin{aligned} &\lambda_t \text{ with multiplicity } r(A_t) = r_t \qquad t \in T_0. \\ &\lambda_\mu \text{ with multiplicity } r(A_\mu) = 1 \end{aligned} \tag{9.49}$$

The associated eigenvectors are the linearly independent columns of A_t, $t \in T_0$, and A_μ. (Note that λ_μ is a linear combination of λ_t, $t \in T_0$.) Further,

$$|V| = \lambda_\mu \prod_{t \in T_0} \lambda_t^{r_t}. \tag{9.50}$$

■

EXAMPLE 9.2
To illustrate Theorem 9.2, we note that, for V given in Example 9.1, the eigenvalues are

$$\begin{aligned} &\lambda_0 \text{ with multiplicity } a_1 a_2 (n - 1) \\ &\lambda_1 \text{ with multiplicity } (a_1 - 1) \\ &\lambda_2 \text{ with multiplicity } (a_2 - 1) \\ &\lambda_{12} \text{ with multiplicity } (a_1 - 1)(a_2 - 1) \\ &\lambda_\mu \text{ with multiplicity } 1. \end{aligned} \tag{9.51}$$

■

Since we are assuming that V is positive definite, it follows that λ_μ and λ_t, $t \in T_0$, are positive. We may, however, allow some of the ϕ_t, $t \in T$, to be zero. This was the case in Examples 8.1 through 8.7, repeated at the beginning of the chapter. It follows that Theorem 9.1 gives the inverse matrix for any of these covariance matrices as long as $\lambda_t \neq 0$, $t \in T_0$. To illustrate, consider the covariance matrix associated with Example 8.3.

EXAMPLE 9.3
For the random, two-fold nested model in Example 8.3, we see that the covariance matrix is the special case of (9.8) with $\phi_2 = 0$. We may thus write V as

$$\begin{aligned} V &= \phi_0 I + \phi_1 V_1 + \phi_{12} V_{12} \\ &= \lambda_0 A_0 + \lambda_1 (A_1 + A_\mu) + \lambda_{12} (A_2 + A_{12}) \end{aligned} \tag{9.52}$$

since $\lambda_1 = \lambda_\mu$ and $\lambda_2 = \lambda_{12}$ when $\phi_2 = 0$. The inverse matrix is given by

$$V^{-1} = (1/\lambda_0) A_0 + (1/\lambda_1)(A_1 + A_\mu) + (1/\lambda_{12})(A_2 + A_{12}). \tag{9.53}$$

■

Rules for determining the relations between the parameters and for determining which matrices are combined are established in Theorem 9.3 and will be essential in the estimation problem.

THEOREM 9.3
For covariance matrices in which some of the ϕ_t, $t \in T$, are zero, the relation between λ, λ_μ, and ϕ is given by deleting the columns of P and P'_μ for which $\phi_t = 0$. Since P is nonsingular, it follows that there will be linear relations between λ_t, $t \in T_0$, and λ_μ as defined by the rows of the deleted P matrix and the deleted P'_μ vector.

■

For example, setting some $\phi_t = 0$ may result in sets of identical λ_t as in Example 9.3. In this case, we define a new P matrix—say, P^*—consisting of the linearly independent rows of the deleted P matrix, relating the distinct elements of λ—say, λ^*—to the nonzero elements of ϕ—say, ϕ^*—satisfying

$$\lambda^* = P^*\phi^*. \tag{9.54}$$

Thus, in Example 9.3, P^* is obtained from (9.23) by deleting the third column and the redundant third row of the resulting matrix. Thus,

$$P^* = \begin{bmatrix} 1 & 0 & 0 \\ 1 & na_2 & n \\ 1 & 0 & n \end{bmatrix}$$

and

$$\lambda^* = \begin{bmatrix} \lambda_0 \\ \lambda_1 \\ \lambda_{12} \end{bmatrix}.$$

The expression for V is now written as

$$V = \sum_{t \in T_0^*} \lambda_t^* A_t^* + \lambda_\mu A_\mu \tag{9.55}$$

where A_t^* denotes the sum of the A_t with $\lambda_t = \lambda_t^*$, and T_0^* represents the appropriate subset of T_0. We shall write A_μ separately as in (9.55), rather than combine it as in (9.52), even though we may have $\lambda_\mu = \lambda_t^*$ for some $t \in T_0^*$. The inverse of (9.55) is then given by

$$V^{-1} = \sum_{t \in T_0^*} (1/\lambda_t^*)A_t^* + (1/\lambda_\mu)A_\mu \tag{9.56}$$

since the matrices A_t^* have the same idempotency and orthogonality properties as the A_t.

In other situations, the linear relations on the λ_t may be more general, and, while we can write V in the form of (9.55), the resulting matrices A_t^* need not be idempotent or orthogonal. In this case, although V^{-1} is still given by Theorem 9.1, it does not have the form of (9.56). The consequences of this will be noted in Section 9.2.4.

9.2.2. Parameter Estimation

In Section 8.2.3, it was suggested that we consider the matrices in the expression for V^{-1} as candidates for use in developing moment estimators. The matrices A_t described in Theorem 9.1 are particularly interesting. Thus, we shall consider the quadratic forms

$$q_t = Y'A_tY \qquad t \in T_0. \tag{9.57}$$

The relevant properties are described in the next theorem. For now, we restrict the mean structure to the special case $E(Y) = \mu J$. Thus, in common terminology, we consider the random model.

THEOREM 9.4
Let $Y \sim N(\mu J, V)$ with V given by (9.38) and $\phi_t \neq 0$, $t \in T_0$. Then

(i) $q_t/\lambda_t \sim \chi^2(r_t) \qquad t \in T_0$

(ii) $Y'A_\mu Y/\lambda_\mu \sim \chi^2[1, N\mu^2/(2\lambda_\mu)]$

(iii) the statistics $\bar{y}$ and q_t, $t \in T_0$, are mutually independent and complete, sufficient statistics for μ and ϕ.

> ***Proof:*** From Theorem 9.1, we have $A_tV = \lambda_t A_t$ for $t \in T_0$. Using Theorem 2.3, the idempotency of A_t and the fact that $A_tJ = 0$ establishes (i), with r_t given by Theorem 9.2. Similarly, (ii) is established, and we note that $Y'A_\mu Y = N\bar{y}^2$. The independence is established by Theorems 2.4 and 2.5, and the complete sufficiency is noted by using the form of V^{-1} in Theorem 7.1 and writing the likelihood as
>
> $$L(Y, \mu, \phi) = (2\pi)^{-N/2}|V|^{-1/2} \exp\{(-1/2)[N(\bar{y} - \mu)^2/\lambda_\mu + \textstyle\sum q_t/\lambda_t]\}. \tag{9.58}$$
>
> The nonsingularity of P in (9.34) establishes the uniqueness of λ_t, $t \in T_0$. Application of the results on complete, sufficient statistics in Appendix B.I.5–7 finishes the proof.
>
> ■

From Theorem 9.4 we may now establish the following important results.

THEOREM 9.5
If $Y \sim N(\mu J, V)$ with V given by (9.38) and $\phi_t \neq 0$, $t \in T_0$, then

(i) $\hat{\lambda}_t = q_t/r_t \qquad t \in T_0$

are unbiased, minimum variance (UMV) estimates of λ_t, $t \in T_0$, with variances

(ii) $\text{Var}(\hat{\lambda}_t) = 2\lambda_t^2/r_t \qquad t \in T_0$.

Further, in view of the relation between λ and ϕ noted in Theorem 9.1, we may write

(iii) $\hat{\phi} = P^{-1}\hat{\lambda}$.

It follows that $\hat{\phi}$ is UMV for ϕ with covariance matrix

(iv) $\text{Var}(\hat{\phi}) = P^{-1}\text{Diag}(2\lambda_t^2/r_t)P'^{-1}$.

Proof: Application of the Lehmann-Scheffé Theorem, Appendix B.I.7, establishes (i) and (ii). The remaining results follow immediately from Theorem 9.4.

■

Note that the estimates of λ_t are positive; hence, the estimated covariance matrix

$$\hat{V} = \sum_{t \in T_0} \hat{\lambda}_t A_t + \hat{\lambda}_\mu A_\mu \tag{9.59}$$

is positive definite.

Theorems 9.4 and 9.5 extend immediately to cover certain special cases in which some of the ϕ_t, $t \in T$, are zero. The results are summarized as follows.

THEOREM 9.6

Let $Y \sim N(\mu J, V)$, and assume that some of the ϕ_t, $t \in T$, are zero. Suppose further that this results in sets of identical λ_t so that we may write V in the form (9.55) such that the A_t^* satisfy (9.45). Then

(i) $q_t^*/\lambda_t^* \sim \chi^2(r_t^*) \qquad t \in T_0^*$

where

$$q_t^* = Y'A_t^*Y$$

and r_t^* is the sum of the ranks of the A_t matrices whose sum is A_t^*.

(ii) The statistics $\bar{y}$ and q_t^*, $t \in T_0^*$, are mutually independent and complete, sufficient statistics for μ and ϕ^*.

Further,

(iii) $\hat{\lambda}_t^* = q_t^*/r_t^* \qquad t \in T_0$

are UMV for λ_t^* and

(iv) $\hat{\phi}^* = P^{*-1}\hat{\lambda}^*$.

is UMV for ϕ^*.

The variances and covariances of these estimates follow as in Theorem 9.5.

■

In the general situation with some $\phi_t = 0$ in which more complex linear relations exist among the λ_t, the results of Theorem 9.4 hold with the exception of completeness. As a result, the UMV property in Theorem 9.6 does not hold.

To illustrate Theorems 9.5 and 9.6, we provide two examples.

EXAMPLE 9.4
For the two-way classification structure discussed in Example 9.1 with A_t given in (9.20), P given in (9.23), and r_t given in Example 9.2, the UMV estimates for λ_t and ϕ_t are defined by the system of equations

$$\begin{bmatrix} \lambda_0 \\ \lambda_1 \\ \lambda_2 \\ \lambda_{12} \end{bmatrix} = \begin{bmatrix} 1 & 0 & 0 & 0 \\ 1 & na_2 & 0 & n \\ 1 & 0 & na_1 & n \\ 1 & 0 & 0 & n \end{bmatrix} \begin{bmatrix} \phi_0 \\ \phi_1 \\ \phi_2 \\ \phi_{12} \end{bmatrix} = \begin{bmatrix} q_0/[a_1a_2(n-1)] \\ q_1/(a_1-1) \\ q_2/(a_2-1) \\ q_{12}/[(a_1-1)(a_2-1)] \end{bmatrix}. \tag{9.60}$$

The UMV estimate of μ is $\bar{y}$.

■

EXAMPLE 9.5
For the two-fold nested structure discussed in Example 9.3 the UMV estimates of λ^* and ϕ^* are given by

$$\begin{bmatrix} \lambda_0 \\ \lambda_1 \\ \lambda_{12} \end{bmatrix} = \begin{bmatrix} 1 & 0 & 0 \\ 1 & na_2 & n \\ 1 & 0 & n \end{bmatrix} \begin{bmatrix} \phi_0 \\ \phi_1 \\ \phi_{12} \end{bmatrix} = \begin{bmatrix} Y'A_0Y/[a_1a_2(n-1)] \\ Y'A_1Y/(a_1-1) \\ Y'A_{12}^*Y/[a_1(a_2-1)] \end{bmatrix}. \tag{9.61}$$

As noted earlier, the matrix P^* in (9.61) is determined from P in (9.60) by deleting the column corresponding to ϕ_2 and then deleting the redundant row corresponding to λ_2, since now $\lambda_2 = \lambda_{12}$. Further, the estimate of λ_{12} is based on the matrix

$$A_{12}^* = A_2 + A_{12} \tag{9.62}$$

with divisor

$$a_1(a_2 - 1) = (a_2 - 1) + (a_1 - 1)(a_2 - 1). \tag{9.63}$$

Again $\bar{y}$ is UMV for μ.

■

A situation involving incomplete, sufficient statistics is given in Section 9.2.4.

9.2.3. A Comparison of ML, REML, and Moment Estimators

In Chapter 8, we described three procedures for estimating the vector ϕ for the mixed model given in Definition 8.1 and made some general comparisons of the methods. It is of interest to contrast these three procedures for the covariance structure defined in Theorem 9.1, assuming $E(Y) = \mu J$.

The moment estimators are given by Theorem 9.5. We shall discuss only the case for which $\phi_t \neq 0$; that is, the UMV estimators are obtained from the quadratic forms $q_t = Y'A_tY$, $t \in T_0$. Analogous results follow for the case in which some of the $\phi_t = 0$, as discussed in Theorem 9.6.

Consider first the REML equations defined by

$$E(\rho_t) \overset{S}{=} \rho_t \qquad t \in T_0 \tag{9.64}$$

where

$$\rho_t = Y'Q_tY \tag{9.65}$$

$$Q_t = V^{-1}MV_tM'V^{-1} \tag{9.66}$$

and

$$M = I - X(X'V^{-1}X)^{-1}X'V^{-1}. \tag{9.67}$$

In this case, since $X = J$, we obtain

$$M = I - A_\mu \tag{9.68}$$

and hence

$$M'V^{-1} = \sum_{t \in T_0} A_t/\lambda_t. \tag{9.69}$$

Using (9.43) to describe V_t in terms of A_t and using (9.45) and (9.46), we obtain from (9.66)

$$Q_t = \sum_{m \in T_0} p_{mt}A_m/\lambda_m^2. \tag{9.70}$$

It follows that

$$\rho_t = \sum_{m \in T_0} p_{mt}q_m/\lambda_m^2 \tag{9.71}$$

and

$$E(\rho_t) = \sum_{m \in T_0} p_{mt}r_m/\lambda_m. \tag{9.72}$$

Expressing (9.64) in matrix form, we may now write the REML equations as

$$P'\text{Diag}(r_t/\lambda_t^2)\lambda = P'\text{Diag}(\lambda_t^{-2})q \tag{9.73}$$

where q is the vector of quadratic forms q_t, $t \in T_0$. Since the matrix P defined in (9.35) is nonsingular, the REML equations are given by

$$\text{Diag}(r_t)\lambda = q \tag{9.74}$$

with $\hat{\lambda} = P\hat{\phi}$ providing the estimates of ϕ. Reference to Theorem 9.5 establishes that these estimators are equivalent to the moment estimators and hence are UMV.

From Theorems 9.3 and 9.6, we obtain the analogous results for certain special cases in which some ϕ_t are zero, and hence the REML estimators are given by Theorem 9.6.

The ML equations were related to the REML equations in Section 8.2.2 and are written as

$$E(\rho_t) + tr[V^{-1}(I - M)V_t] \overset{S}{=} \rho_t. \tag{9.75}$$

In this case, with M given by (9.68), we obtain, using (9.43),

$$tr[V^{-1}(I - M)V_t] = p_t/\lambda_\mu \qquad t \in T_0 \tag{9.76}$$

where p_t, $t \in T_0$, denotes the components of P_μ defined in (9.37). Using (9.73), we may write the ML equations, after simplification, as

$$\text{Diag}(r_t)\lambda + \text{Diag}(\lambda_t^2/\lambda_\mu)P'^{-1}P_\mu = q. \tag{9.77}$$

The contrast with the REML equations is evident. The equations, (9.77), will generally be nonlinear in λ, and no simple solution exists. The exceptions to this may occur in special cases in which some ϕ_t are zero. We shall see in Example 9.7 for the two-fold nested model that the equations determined by (9.77) are linear in λ. This characteristic is seen to hold for higher order nested models as well.

To illustrate the ML and REML equations, we again consider the two-factor and the two-fold nested structures discussed in Examples 9.1, 9.3, 9.4, and 9.5.

EXAMPLE 9.6

For the two-factor structure, with P given by (9.23), we have

$$P'^{-1} = \begin{bmatrix} 1 & 0 & 0 & -1/n \\ 0 & 1/na_2 & 0 & 0 \\ 0 & 0 & 1/na_1 & 0 \\ 0 & -1/na_2 & -1/na_1 & 1/n \end{bmatrix} \quad P_\mu = \begin{bmatrix} 1 \\ na_2 \\ na_1 \\ n \end{bmatrix}. \tag{9.78}$$

The ML equations are thus given by

$$\begin{bmatrix} a_1a_2(n-1)\,\lambda_0 \\ (a_1-1)\,\lambda_1 \\ (a_2-1)\,\lambda_2 \\ (a_1-1)(a_2-1)\,\lambda_{12} \end{bmatrix} + \begin{bmatrix} 0 \\ \lambda_1^2/\lambda_\mu \\ \lambda_2^2/\lambda_\mu \\ -\lambda_{12}^2/\lambda_\mu \end{bmatrix} = \begin{bmatrix} q_0 \\ q_1 \\ q_2 \\ q_{12} \end{bmatrix} \tag{9.79}$$

with $\lambda_\mu = \lambda_1 + \lambda_2 - \lambda_{12}$. The REML equations are obtained by deleting the nonlinear term in (9.79), and, as noted, they agree with the UMV estimators in (9.60). There is no closed form solution to (9.79), but the equations may be solved numerically.

■

EXAMPLE 9.7

For the two-fold nested structure, the REML equations are given in (9.61) since they agree with the UMV moment estimators. With

$$P^{*\prime} = \begin{bmatrix} 1 & 1 & 1 \\ 0 & na_2 & 0 \\ 0 & n & n \end{bmatrix} \quad P_\mu^* = \begin{bmatrix} 1 \\ na_2 \\ n \end{bmatrix} \tag{9.80}$$

we obtain, by analogy with (9.77), the ML equations

$$\begin{bmatrix} a_1a_2(n-1)\,\lambda_0 \\ (a_1-1)\,\lambda_1 \\ a_1(a_2-1)\,\lambda_{12} \end{bmatrix} + \begin{bmatrix} 0 \\ \lambda_1 \\ 0 \end{bmatrix} = \begin{bmatrix} q_0 \\ q_1 \\ q_2 + q_{12} \end{bmatrix} \tag{9.81}$$

since $\lambda_1 = \lambda_\mu$ when $\phi_2 = 0$. It follows that the ML estimates of λ_0 and λ_{12} agree with REML, but the estimate of λ_1 is given by

$$\hat{\lambda}_1 = q_1/a_1 \tag{9.82}$$

which is biased for λ_1.

■

These two examples suffice to illustrate the relative merits of ML and REML estimators for special cases of our covariance structure in the case $E(Y) = \mu J$. The ideas generalize in the sense that the REML estimators are always UMV (assuming Theorem 9.6 if some $\phi_t = 0$), while the ML equations are nonlinear for factorial structures and are linear for completely nested structures. In the latter case, *one* of the ML estimators will differ from REML, since the divisor differs. (The completely nested structure requires that the indices $t \in T_0^*$ be such that they may be ordered so that the ith index is always contained in the $(i + 1)$st.)

In Section 9.3, we shall consider more general structures, and in later sections we shall consider covariance structures for which V^{-1} does not have a simple expression.

9.2.4. Tests of Hypotheses on ϕ

The general theory for testing hypotheses on the parameter vector ϕ of the covariance matrix was developed in Section 8.3. There we noted that the likelihood ratio statistic was the ratio of the determinants of the estimated covariance matrix for the unconstrained and constrained cases. Equivalently, the statistic may be expressed as a ratio of products of the eigenvalues of the two matrices.

For the covariance structure discussed in Section 9.2.1, we shall see that it is possible to provide exact tests for certain hypotheses. The test statistics will be developed in terms of the sufficient statistics identified in Theorem 9.4 and will be related to the likelihood ratio tests. In other cases, exact tests are not available, but approximate tests will be developed. In all cases we continue to assume $E(Y) = \mu J$.

As a word of caution, we emphasize that, while it is possible to develop tests of hypotheses on ϕ or functions of ϕ, this does not imply that such tests are appropriate or meaningful. The researcher should give some thought to the consequences of rejecting or accepting the hypothesis. Since we have related ϕ to variances and covariances, it is possible to give an interpretation of the conclusion, but it may be of no practical importance. We shall comment on this in the examples below.

The ideas can best be illustrated by example, and we provide two in this section. Other examples for specific models will be noted in later sections.

Before we discuss the examples, however, we note a general situation in which exact tests on certain linear functions of the parameters are readily available. From Theorem 9.4, we recall that the statistics q_t are independent and distributed as

$$q_t/\lambda_t \sim \chi^2(r_t) \tag{9.83}$$

where λ_t and ϕ_t are related as in Theorem 9.1. It follows that, for any two of these quadratic forms, the distribution of the ratio is given by

$$(q_t/r_t)/(q_m/r_m) \sim (\lambda_t/\lambda_m)F(r_t, r_m). \tag{9.84}$$

For testing the null hypotheses

$$H_0\colon \lambda_t/\lambda_m = 1 \tag{9.85}$$

against an appropriate alternative, the ratio in (9.84) is compared to critical values for the central F-distribution. For two-sided alternatives, we would reject H_0 if the ratio were too large or too small, whereas one-sided alternatives would suggest one of these critical regions. Since the λ_t are linear in ϕ, it follows that the hypothesis (9.85) may be equivalent to a linear hypothesis on ϕ. This will be the case for some but not all of the possible linear hypotheses on ϕ, as the following examples illustrate.

EXAMPLE 9.8
For the two-factor structure illustrated in Examples 9.1, 9.4, and 9.6, we might consider the following hypotheses:

$$\begin{aligned} H_{12}&: \lambda_{12}/\lambda_0 = 1 \\ H_1&: \lambda_1/\lambda_{12} = 1 \\ H_2&: \lambda_2/\lambda_{12} = 1 \end{aligned} \tag{9.86}$$

with test statistics as described in (9.84). Reference to the P matrix in (9.23) shows that these three hypotheses are equivalent to three simple hypotheses on ϕ. The hypotheses and their test statistics are given by

Hypothesis	*Test Statistic*
$H_{12}: \phi_{12} = 0$	$F_{12} = \{q_{12}/[(a_1 - 1)(a_2 - 1)]\}/[q_0/a_1a_2(n - 1)]$
$H_1: \phi_1 = 0$	$F_1 = [q_1/(a_1 - 1)]/\{q_{12}/[(a_1 - 1)(a_2 - 1)]\}$ (9.87)
$H_2: \phi_2 = 0$	$F_2 = [q_2/(a_2 - 1)]/\{q_{12}/[(a_1 - 1)(a_2 - 1)]\}.$

The critical region depends on the alternative hypothesis. For example, since these parameters are described in terms of covariances in Example 8.4, it may be reasonable to consider the two-sided alternative. Thus, for testing

$$H_0: \phi_{12} = 0 \quad \text{versus} \quad H_A: \phi_{12} \neq 0 \tag{9.88}$$

we would reject H_0 if

$$F_{12} > F(\alpha_1) \quad \text{or} \quad F_{12} < F(1 - \alpha_2) \tag{9.89}$$

where $\alpha_1 + \alpha_2 = \alpha$.

It is of interest to ask how these test statistics compare with those that arise from the likelihood ratio test. Actually, we shall consider a modified likelihood ratio test in which we use the REML estimators, since closed form expressions for the ML estimators do not exist for this situation.

For illustration, we shall consider the hypothesis $H_{12}: \phi_{12} = 0$. The results for the other two hypotheses follow similarly. The unconstrained REML estimators of the λ_t are given in Example 9.6. Under the constraint $\phi_{12} = 0$ we see from the P matrix in (9.23) that $\lambda_0 = \lambda_{12}$; hence, we may write the expression for the covariance matrix under this hypothesis, using Theorem 9.3, as

$$V = \lambda_0(A_0 + A_{12}) + \lambda_1 A_1 + \lambda_2 A_2 + \lambda_\mu A_\mu. \tag{9.90}$$

Theorem 9.6 and the results of Section 9.2.3 immediately give the constrained REML estimators as solutions to the equations

$$
\begin{aligned}
[a_1a_2(n-1)+(a_1-1)(a_2-1)]\lambda_0 &= q_0+q_{12}\\
(a_1-1)\lambda_1 &= q_1\\
(a_2-1)\lambda_2 &= q_2.
\end{aligned}
\tag{9.91}
$$

From Section 8.3.2, we now obtain the squared, modified, likelihood ratio statistic given by (8.136). In this case, it is easily shown that the eigenvalues of $A'VA$ and of $A'V^*A$ are identical to those of V and V^* for $t \in T_0$. We emphasize that λ_μ is not an eigenvalue of $A'VA$ or $A'V^*A$ since $A'A_\mu = 0$. Thus,

$$\gamma^2 = (q_0/r_0)^{r_0}(q_{12}/r_{12})^{r_{12}}/[(q_0+q_{12})/(r_0+r_{12})]^{(r_0+r_{12})} \tag{9.92}$$

where $r_0 = a_1a_2(n-1)$ and $r_{12} = (a_1-1)(a_2-1)$. Simplifying, we find that (9.92) reduces to

$$\gamma^2 = [(r_0+r_{12})/r_0]^{r_0}[(r_0+r_{12})/r_{12}]^{r_{12}}(q_{12}/q_0)^{r_{12}}/(1+q_{12}/q_0)^{(r_0+r_{12})}. \tag{9.93}$$

Examination of (9.93) shows that the modified likelihood ratio criterion suggests rejection of the H_{12} if q_{12}/q_0 is either too small or too large. This conclusion agrees with (9.89) for the two-sided alternative. For one-sided alternatives we again reject if q_{12}/q_0 is extreme in one direction.

This development of the modified likelihood ratio test is primarily of academic interest for this example, since we already have a good test based on the expected values of the sufficient statistics. However, it does suggest a general approach that might be applied when the hypothesis of interest is not equivalent to a ratio of the form of (9.85). Such a situation is illustrated in the next example. Later, when we consider more complex covariance structures in the unbalanced data problem, we shall see that exact tests are rarely available.

Before leaving this example, we shall make a few comments on the hypotheses in (9.87) with reference to the machine-operators' experiment discussed in Example 8.4. Referring to the covariance structure (8.14), we see that $\phi_1 \neq 0$ says that responses on the same machine (row) are correlated, and $\phi_2 \neq 0$ says that responses for the same operator (column) are correlated. Further, if $\phi_{12} = 0$, we conclude that the covariances are additive for responses in the same cell (that is, same row and column). Thus, the hypothesis H_{12} has a logical and possibly meaningful interpretation. On the other hand, the hypothesis H_2: $\phi_2 = 0$ is less clear. Accepting this hypothesis while concluding $\phi_1 \neq 0$ and $\phi_{12} \neq 0$ suggests that observations within a column are not correlated unless they are in the same cell. Observations in the same row are correlated, but the correlation is different if they are in the same column. The analyst should evaluate whether this is meaningful, or even relevant, in the experiment.

As a final comment, we note that, if we conclude $\phi_2 = 0$, the covariance structure for this experiment is identical to the two-fold nested structure. We emphasize that this does not suggest that operators were nested in machines when the experiment was conducted, but it does suggest that the experiment could have been conducted in this way. The proper interpretation is in terms of covariances, as described above.

■

EXAMPLE 9.9

For this example, we consider the covariance structure defined by (9.17) with $k = 3$. We shall refer to this as the three-way classification structure. Following the notation developed for Theorem 9.1, we have

$$\phi' = (\phi_0, \phi_1, \phi_2, \phi_{12}, \phi_3, \phi_{13}, \phi_{23}, \phi_{123})$$

and the analogous expression for λ. From (9.35) we have $\lambda = P\phi$ with P given by

$$P = \begin{bmatrix} 1 & 0 & 0 & 0 & 0 & 0 & 0 & 0 \\ 1 & na_2a_3 & 0 & na_3 & 0 & na_2 & 0 & n \\ 1 & 0 & na_1a_3 & na_3 & 0 & 0 & na_1 & n \\ 1 & 0 & 0 & na_3 & 0 & 0 & 0 & n \\ 1 & 0 & 0 & 0 & na_1a_2 & na_2 & na_1 & n \\ 1 & 0 & 0 & 0 & 0 & na_2 & 0 & n \\ 1 & 0 & 0 & 0 & 0 & 0 & na_1 & n \\ 1 & 0 & 0 & 0 & 0 & 0 & 0 & n \end{bmatrix}. \tag{9.94}$$

Examination of the rows of P shows that certain simple hypotheses on ϕ are equivalent to hypotheses on ratios of the elements of λ of the form of (9.85), with test statistics given by (9.84). In particular, we have the following equivalences:

$$\begin{aligned} H_{123}\colon \phi_{123} &= 0 \leftrightarrow \lambda_{123}/\lambda_0 = 1 \\ H_{ij}\colon \phi_{ij} &= 0 \leftrightarrow \lambda_{ij}/\lambda_{123} = 1 \qquad i, j = (1, 2), (1, 3), (2, 3). \end{aligned} \tag{9.95}$$

On the other hand, hypotheses such as

$$H_i\colon \phi_i = 0 \qquad i = 1, 2, 3 \tag{9.96}$$

are not equivalent to simple ratios of elements of λ, and hence exact tests are not immediately available.

To illustrate an approximate solution to such problems, consider the hypothesis $H_3\colon \phi_3 = 0$. Note that this hypothesis is equivalent to the hypothesis

$$H_3\colon (\lambda_3 + \lambda_{123})/(\lambda_{13} + \lambda_{23}) = 1. \tag{9.97}$$

Following the approximation suggested in Section 8.4, we consider the statistics

$$\begin{aligned} s_1 &= q_3/r_3 + q_{123}/r_{123} \mathrel{\dot{\sim}} c_1\chi^2(d_1) \\ s_2 &= q_{13}/r_{13} + q_{23}/r_{23} \mathrel{\dot{\sim}} c_2\chi^2(d_2). \end{aligned} \tag{9.98}$$

It then follows that

$$(s_1/c_1d_1)/(s_2/c_2d_2) \mathrel{\dot{\sim}} F(d_1, d_2). \tag{9.99}$$

Here, d_1 and d_2 are given, approximately, from Section 8.4 as

$$\begin{aligned} d_1 &= (\hat{\lambda}_3 + \hat{\lambda}_{123})^2/(\hat{\lambda}_3^2/r_3 + \hat{\lambda}_{123}^2/r_{123}) \\ d_2 &= (\hat{\lambda}_{13} + \hat{\lambda}_{23})^2/(\hat{\lambda}_{13}^2/r_{13} + \hat{\lambda}_{23}^2/r_{23}) \end{aligned} \tag{9.100}$$

with $\hat{\lambda}_t = q_t/r_t$. Further, since $E(s_i) = c_id_i$, we have

$$s_1/s_2 \mathrel{\dot{\sim}} (\lambda_3 + \lambda_{123})/(\lambda_{13} + \lambda_{23})F(d_1, d_2). \tag{9.101}$$

Thus, the ratio s_1/s_2 follows, approximately, a central F distribution if $\phi_3 = 0$. The rejection region will depend upon the alternative hypothesis.

It is natural to ask whether the likelihood ratio test has anything better to offer. The unconstrained estimates of ϕ follow directly from Theorem 9.5. Under the hypothesis H_3: $\phi_3 = 0$, the REML estimates do not have a simple, closed form solution, since Theorem 9.6 does not apply in this case; that is, the constraint $\phi_3 = 0$ leads to the relation

$$\lambda_3 + \lambda_{123} = \lambda_{13} + \lambda_{23}$$

but not to sets of identical λ_t. At issue here is the fact that, while the statistics q_t, $t \in T_0$, are sufficient for ϕ, they are not complete. This is seen by noting that s_1 and s_2, as defined by (9.98), have the same expectation when $\phi_3 = 0$; hence,

$$E(s_1 - s_2) = 0. \tag{9.102}$$

The search for a complete set centers around the expression for V^{-1}. With $\phi_3 = 0$, $\lambda_3 = \lambda_{13} + \lambda_{23} - \lambda_{123}$; hence, we can write

$$\begin{aligned} V = {} & \lambda_0A_0 + \lambda_1A_1 + \lambda_2A_2 + \lambda_{12}A_{12} + \lambda_{13}(A_{13} + A_3) \\ & + \lambda_{23}(A_{23} + A_3) + \lambda_{123}(A_{123} - A_3) + \lambda_\mu A_\mu . \end{aligned} \tag{9.103}$$

Unfortunately, the matrices in (9.103) do not have the necessary idempotency and orthogonality properties that would enable us to identify the corresponding quadratic forms as complete sufficient statistics.

We are thus unable to obtain closed form estimates of ϕ and hence must resort to a numerical solution of the REML equations. The squared, modified, likelihood ratio statistic will have the form

$$\gamma^2 = \prod_{t \in T_0} (\hat{\lambda}_t/\tilde{\lambda}_t)^{r_t} \tag{9.104}$$

where $\hat{\lambda}_t$ denotes the unconstrained estimate and $\tilde{\lambda}_t$ the estimate constrained by $\phi_3 = 0$. Note that $\tilde{\lambda}_3 = \tilde{\lambda}_{13} + \tilde{\lambda}_{23} - \tilde{\lambda}_{123}$.

While closed form solutions of the REML equations under the hypothesis H_3: $\phi_3 = 0$ are not available, it seems reasonable to expect that the estimators of λ_t have the form

$$\tilde{\lambda}_t = \hat{\lambda}_t + a_t(s_1 - s_2) \qquad t = (13), (23), (123) \tag{9.105}$$

where the a_t are chosen to minimize the variance of $\tilde{\lambda}_t$. In particular, we obtain

$$\begin{aligned} a_{13} &= (\lambda_{13}^2/r_{13})/g(\lambda) \\ a_{23} &= (\lambda_{23}^2/r_{23})/g(\lambda) \\ a_{123} &= -(\lambda_{123}^2/r_{123})/g(\lambda) \\ g(\lambda) &= (\lambda_3 + \lambda_{123})^2/d_1 + (\lambda_{13} + \lambda_{23})^2/d_2 \\ &= \sum \lambda_t^2/r_t \end{aligned} \tag{9.106}$$

where this sum ranges over the sets of indices (3), (13), (23), and (123). Substitution of these estimates into (9.104) will give us an approximate likelihood ratio statistic. It is not clear that this result is equivalent to the test based on (9.101), but the algebra is tedious and we shall not pursue it further.

■

9.2.5. Confidence Intervals on ϕ

In Section 8.4.2 we indicated that, for certain linear functions of ϕ, we could obtain exact confidence intervals, and, based on these same functions, we could obtain approximate confidence intervals on general linear functions of ϕ. From Theorem 9.4 we now see that the λ_t represent the linear functions of ϕ for which exact confidence intervals are readily available; that is, we have the $(1 - \alpha)$ confidence interval

$$q_t/\chi_U^2 \leq \lambda_t \leq q_t/\chi_L^2 \tag{9.107}$$

where

$$\text{Prob}[\chi_L^2 \leq \chi^2(r_t) \leq \chi_U^2] = 1 - \alpha. \tag{9.108}$$

We also note, from (9.84), that we may write exact confidence intervals for certain ratios of linear functions of ϕ. In particular, we may write the $(1 - \alpha)$ confidence interval

$$F/F_U \leq \lambda_t/\lambda_m \leq F/F_L \tag{9.109}$$

where

$$F = (q_t/r_t)/(q_m/r_m) \tag{9.110}$$

and

$$\text{Prob}(F_L \leq F(r_t, r_m) \leq F_U) = 1 - \alpha. \tag{9.111}$$

These special cases, as well as confidence intervals on general linear functions of ϕ, will be illustrated in Example 9.10.

When writing confidence intervals for several functions of ϕ, we naturally raise the question of simultaneous confidence intervals; that is, we seek a scheme for writing sets of confidence intervals with the property that the intersection of the intervals in parameter space will have confidence coefficient of at least $1 - \alpha$. A procedure for developing simultaneous confidence intervals for continuous functions of ϕ was described by Khuri (1981). The procedure is closely related to the Scheffé method for simultaneous confidence intervals on linear functions of θ, described in Section 3.4.2, so we shall describe it in that context. Recall that the confidence intervals were determined by pairs of tangent planes to the ellipsoidal confidence region for θ (or for the set of linear functions, $H\theta$). The simultaneous confidence property of the resulting intervals rested on the fact that this region was convex and had confidence coefficient $1 - \alpha$. The region defined by the tangent planes thus contained the ellipsoidal region and had confidence coefficient of at least $1 - \alpha$.

In this case, the convex region we use is simply the intersection of the intervals defined by (9.107); that is, in the parameter space of the λ_t, $t \in T_0$, we consider the rectangular region, R, defined by the intersection of the regions

$$R_t\colon L_t = q_t/\chi_U^2 \leq \lambda_t \leq q_t/\chi_L^2 = U_t \qquad t \in T_0. \tag{9.112}$$

This region has confidence coefficient

$$1 - \alpha^* = (1 - \alpha)^m \tag{9.113}$$

where m is the number of intervals in (9.112).

We now consider the specification of a confidence interval for a linear function—say, $a'\lambda$ or, equivalently, $a'P\phi$. The concept of the tangent plane in the Scheffé intervals is replaced by the solution of the two problems,

$$\begin{aligned} &\min(\max)\ a'\lambda \\ &\text{subject to} \quad \lambda \in R \end{aligned} \tag{9.114}$$

where R is the rectangular region defined by (9.112). The solution of this problem is simply described. In (9.112), let (L_t, U_t), $t \in T_0$, denote the lower and upper limits. Then the solution is given by

$$\begin{aligned} \max a'\lambda &= \sum_{a_t>0} a_t U_t + \sum_{a_t<0} a_t L_t \\ \min a'\lambda &= \sum_{a_t>0} a_t L_t + \sum_{a_t<0} a_t U_t. \end{aligned} \tag{9.115}$$

The interval is then given by

$$\min a'\lambda \leq a'\lambda \leq \max a'\lambda. \tag{9.116}$$

This solution is easily visualized as the extreme vertices of the rectangular region, R, as determined by the plane $a'\lambda$.

Khuri (1981) notes that, if the components of λ are related by inequality constraints, such as $\lambda_{12} > \lambda_0$, then the region R in (9.114) should be the intersection of (9.112) and the region defined by these constraints. In this case, the solution is more complex, requiring numerical methods. We shall note such constraints in Chapter 10. The objective of using these constraints is to give shorter intervals than those defined by (9.116).

The concept extends directly to nonlinear functions of λ; that is, for the function $f(\lambda)$, the interval is defined by

$$\min f(\lambda) \leq f(\lambda) \leq \max f(\lambda) \tag{9.117}$$

where the limits are the solutions to

$$\begin{aligned} &\min(\max) f(\lambda) \\ &\text{subject to} \quad \lambda \in R. \end{aligned} \tag{9.118}$$

For example, we may be interested in a confidence interval for ϕ_t/ϕ_0. From the relation $\lambda = P\phi$, we may describe ϕ_t as a linear function of λ—say, $a'\lambda$. It follows that

$$\phi_t/\phi_0 = a'\lambda/\phi_0. \tag{9.119}$$

Khuri (1981) notes that the desired interval is given by

$$\min a'\lambda/U_0 \leq \phi_t/\phi_0 \leq \max a'\lambda/L_0 \tag{9.120}$$

where $\min a'\lambda$ and $\max a'\lambda$ are defined in (9.115).

As noted, this scheme provides for simultaneous confidence intervals for continuous functions of λ; hence, we would anticipate that they would be conservative. The following example will demonstrate various exact, approximate, and simultaneous confidence intervals for a simple situation.

EXAMPLE 9.10

We consider the two-parameter covariance matrix for the one-way classification model from Example 8.1; that is,

$$V = \phi_0 I_{an} + \phi_1(I_a \otimes J_n J_n'). \tag{9.121}$$

From Theorem 9.1 we have

$$\begin{bmatrix} \lambda_0 \\ \lambda_1 \end{bmatrix} = \begin{bmatrix} 1 & 0 \\ 1 & n \end{bmatrix} \begin{bmatrix} \phi_0 \\ \phi_1 \end{bmatrix} \tag{9.122}$$

and

$$
\begin{aligned}
A_0 &= V_0 - (1/n)V_1 \\
A_1 &= (1/n)V_1 - (1/an)JJ'.
\end{aligned}
\tag{9.123}
$$

Confidence intervals on λ_0 and λ_1 are immediately given by (9.107), and a confidence interval for λ_1/λ_0 is given by (9.109) as

$$F/F_U \leq (\phi_0 + n\phi_1)/\phi_0 \leq F/F_L \tag{9.124}$$

where

$$F = [q_1/(a - 1)]/[q_0/a(n - 1)] \tag{9.125}$$

with degrees of freedom determined from Theorem 9.2.

It follows from (9.124) that a confidence interval for ϕ_1/ϕ_0 is given by

$$(F/F_U - 1)/n \leq \phi_1/\phi_0 \leq (F/F_L - 1)/n. \tag{9.126}$$

Confidence intervals on functions of ϕ_1/ϕ_0 such as

$$\phi_1/(\phi_0 + \phi_1) \quad \text{and} \quad \phi_0/(\phi_0 + \phi_1) \tag{9.127}$$

are obtained immediately from (9.126).

An approximate confidence interval on ϕ_1 may be obtained from (8.150). Here,

$$
\begin{aligned}
h(\phi) &= \phi_1 = (\lambda_1 - \lambda_0)/n \\
q &= (q_1/r_1 - q_0/r_0)/n \\
c &= (\lambda_1^2/r_1 + \lambda_0^2/r_0)/n^2 \\
d &= (\lambda_1 - \lambda_0)^2/(\lambda_1^2/r_1 + \lambda_0^2/r_0)
\end{aligned}
\tag{9.128}
$$

hence, the interval is given by

$$(q_1/r_1 - q_0/r_0)d/(n\chi_U^2) \leq \phi_1 \leq (q_1/r_1 - q_0/r_0)d/(n\chi_L^2). \tag{9.129}$$

In this expression, the chi-square values are computed for degrees of freedom, d, where d is estimated from (9.128) with λ_t replaced by q_t/r_t.

For contrast, we describe the simultaneous intervals, (9.116), for ϕ_0 and ϕ_1 and (9.119) for ϕ_1/ϕ_0. We obtain, with simultaneous confidence coefficient $1 - \alpha^*$ given by (9.113), the intervals

$$
\begin{aligned}
L_0 &\leq \phi_0 \leq U_0 \\
(L_1 - U_0)/n &\leq \phi_1 \leq (U_1 - L_0)/n \\
(L_1 - U_0)/nU_0 &\leq \phi_1/\phi_0 \leq (U_1 - L_0)/nL_0.
\end{aligned}
\tag{9.130}
$$

These results are easily visualized by referring to Figure 9.1 with

$$\begin{aligned} \phi_1 &= f_1(\lambda) = (\lambda_1 - \lambda_0)/n \\ \phi_1/\phi_0 &= f_2(\lambda) = (\lambda_1 - \lambda_0)/n\lambda_0. \end{aligned} \tag{9.131}$$

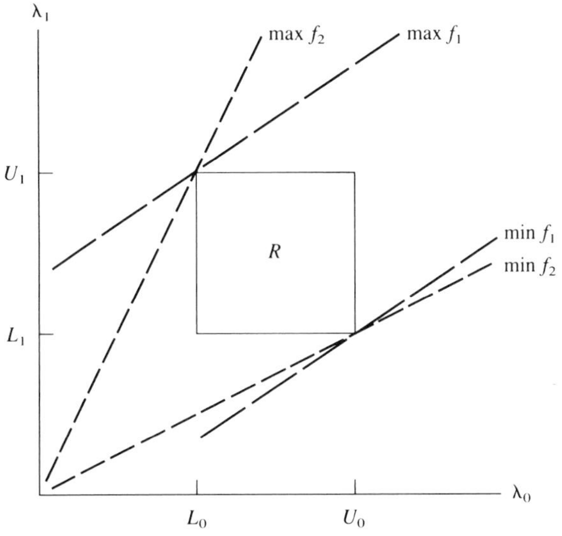

FIGURE 9.1
Simultaneous confidence intervals for $f(\lambda)$

We note that the approximate confidence interval for ϕ_1, proposed by Williams (1962; see also Searle, 1971a), arises by a process analogous to that used by Khuri (1981). The region R used by Williams had approximate coefficient $(1 - 2\alpha)$ and is unique to this one-way classification structure, while Khuri's method may be applied to any structure as defined by Theorem 9.1.

To illustrate the numerical differences, let $q_0 = 30.6$, $r_0 = 30$, $q_1 = 81.27$, $r_1 = 9$, and confidence coefficient $1 - \alpha = 0.90$ with equal tail probabilities. The exact individual intervals for ϕ_0 (9.107) and ϕ_1/ϕ_0 (9.126) are given by

$$\begin{aligned} 0.70 &\leq \phi_0 \leq 1.65 \\ 0.75 &\leq \phi_1/\phi_0 \leq 6.07. \end{aligned} \tag{9.132}$$

The approximate interval for ϕ_1 (9.129), with $d = 7$, is given by

$$0.995 \leq \phi_1 \leq 6.46. \tag{9.133}$$

The approximate interval given by Williams (1962) for this example is

$$0.53 \leq \phi_1 \leq 6.98. \tag{9.134}$$

To compute the simultaneous intervals (9.130), we determine U_t and L_t with $\alpha = 0.05$ to obtain $1 - \alpha^* = (1 - \alpha)^2 = 0.9025$. We obtain

$$\begin{aligned} 0.65 &\leq \phi_0 \leq 1.82 \\ 0.41 &\leq \phi_1/\phi_0 \leq 9.13 \\ 0.75 &\leq \phi_1 \leq 5.94. \end{aligned} \tag{9.135}$$

Comparing the more conservative simultaneous intervals with the exact or approximate intervals, we find that the simultaneous intervals are not substantially wider for this example. Given the guaranteed protection and ease of computation, they appear to deserve favorable consideration.

■

9.3. MIXED MODELS WITH SPECIAL STRUCTURE

9.3.1. A General Mean Structure

In the previous sections we have focused on the special case in which $E(Y) = \mu J$. We now consider more complex structures on the mean vector. In particular, we shall assume that the mean vector is expressed in the second canonical form described in Chapters 6 and 7.

To develop the general mean structure, we extend the description of the second canonical form. The design matrix is described as

$$X_0 = (J|X_t^0,\ t \in T_1) \tag{9.136}$$

where the subscripts are as defined in Definition 9.1 and T_1 is a subset of T in that definition. For example, with $k = 3$ and T given by (9.11), we might have $T_1 = (1, 2, 12)$. We define

$$X_t^0 = (Z_1 \otimes Z_2 \otimes \ldots \otimes Z_k) \tag{9.137}$$

where

$$Z_i = \begin{bmatrix} I_{a_i} & i \in t \\ J_{a_i} & i \notin t. \end{bmatrix} \tag{9.138}$$

Similarly, we define the parameter matrix as

$$P = \text{Diag}(1, P_t',\ t \in T_1) \tag{9.139}$$

where

$$P_t' = \nabla_1' \otimes \nabla_2' \otimes \ldots \otimes \nabla_k' \tag{9.140}$$

and

$$\nabla_i' = \begin{bmatrix} \Delta_{a_i}' & i \in t \\ 1 & i \notin t. \end{bmatrix} \tag{9.141}$$

For now, we consider the simple counting matrix

$$W = I_a \otimes J_n \tag{9.142}$$

where

$$a = \prod_{i=1}^{k} a_i = N/n. \tag{9.143}$$

In this case, we may write

$$X = WX_0P = (J_N | X_t, \ t \in T_1) \tag{9.144}$$

where

$$\begin{aligned} X_t &= (X_t^0 \otimes J_n)(P_t' \otimes 1) \\ &= (Z_1\nabla_1' \otimes Z_2\nabla_2' \otimes \ldots \otimes Z_k\nabla_k' \otimes J_n). \end{aligned} \tag{9.145}$$

Note that X_t has dimension $N \times r_t$ where $r_t = r(A_t)$ from Theorem 9.2. With this notation, we write the mean vector as

$$\begin{aligned} E(Y) &= X\beta \\ &= \beta_0 J + \sum_{t \in T_1} X_t\beta_t \end{aligned} \tag{9.146}$$

where

$$\beta' = (\beta_0, \ \beta_t, \ t \in T_1). \tag{9.147}$$

To further specify T_1, we shall require that, if $t \in T_1$, then any index set m contained in t is also in T_1. Thus, if the subset (12) $\in T_1$, then the subsets 1 and 2 must both be in T_1. Of course, we allow that some β_t may be zero.

Associated with this mean structure, we shall assume a covariance structure as in Theorem 9.1 of the form

$$V = \sum_{t \in T_2} \phi_t V_t \tag{9.148}$$

where the index set $T_2 = T_0 - T_1$. More particularly, we shall assume that V may be written as

$$V = \sum_{t \in T_2} \lambda_t A_t^* + \lambda_\mu A_\mu. \tag{9.149}$$

Here, the λ_t are as given in Definition 9.3, and the A_t^* may represent a sum of the A_t matrices given in that definition. Note that in (9.149) we could sum over $t \in T_0$, but some of the λ_t are identical.

9.3.2. Parameter Estimation

The general ML or REML equations from Chapter 8 may be applied, but we shall observe that simple UMV estimators are obtained. First, with regard to estimating β, note that V and X satisfy the conditions of Theorem 8.1, and hence β can be estimated independently of V; that is, we have the following theorem.

THEOREM 9.7
With $E(Y)$ and V as defined in (9.146) and (9.148), respectively, the ML and REML estimator for β is given by

$$\hat{\beta} = (X'X)^{-1}X'Y. \tag{9.150}$$

Proof: In view of Theorem 8.1, it suffices to establish the existence of a nonsingular matrix B satisfying

$$VX = XB. \tag{9.151}$$

From (9.28) and (9.145), we may write for $m \in T_0$ and $t \in T_1$

$$A_m X_t = (1/a_m^*)(G_1 Z_1 \nabla_1' \otimes G_2 Z_2 \nabla_2' \otimes \ldots \otimes G_k Z_k \nabla_k' \otimes J_n). \tag{9.152}$$

Now,

$$\begin{aligned} G_i Z_i \nabla_i' &= \Delta_{a_i}' && i \in t,\ i \in m \\ &= a_i J_{a_i} && i \notin t,\ i \notin m \\ &= 0 && \text{otherwise.} \end{aligned} \tag{9.153}$$

Therefore,

$$\begin{aligned} A_m X_t &= X_t && m = t \\ &= 0 && \text{otherwise.} \end{aligned} \tag{9.154}$$

Using V in the form (9.149) and noting that $A_\mu X_t = 0$, we obtain

$$\begin{aligned} VX_t &= \lambda_t X_t \\ VJ &= \lambda_\mu J. \end{aligned} \tag{9.155}$$

Thus,

$$VX = (\lambda_\mu J \mid \lambda_t X_t,\ t \in T_1) \\ = X\ \mathrm{Diag}(\lambda_\mu,\ \lambda_t I_t,\ t \in T_1) \tag{9.156}$$

where I_t has dimension r_t. Comparing (9.156) and (9.151) and noting that λ_μ and λ_t, $t \in T_1$, are nonzero establishes the result.

■

We may now establish the important theorem that identifies UMV estimates of β and ϕ.

THEOREM 9.8
Let $Y \sim N(X\beta, V)$ with mean and covariance given by (9.146) and (9.149), with $\phi_t \neq 0$, $t \in T_2$. Then

(i) $q_t = Y'A_tY \sim \lambda_t\chi^2(r_t)$, $t \in T_2$
(ii) $\hat{\beta} = (X'X)^{-1}X'Y \sim N[\beta, B(X'X)^{-1}]$

where

$$B = \mathrm{Diag}(\lambda_\mu,\ \lambda_t I_t,\ t \in T_1)$$

(iii) $q_m = Y'A_mY \sim \lambda_m\chi^2[r_m, (\beta_m'X_m'X_m\beta_m)/(2\lambda_m)]$, $m \in T_1$

(iv) the statistics $X_t'Y$, $t \in T_1$, and q_t, $t \in T_2$, are mutually independent and are complete, sufficient statistics for β and ϕ
(v) the statistics, q_t/r_t, $t \in T_2$, and $\hat{\beta}$ are UMV estimators for λ_t, $t \in T_2$, and β.

Proof: Results (i) and (ii) follow directly from Theorem 2.3 and Section 2.2, respectively. Theorem 2.3 also establishes (iii), where the noncentrality parameter follows from (9.154). The independence in (iv) follows from Theorems 2.4 and 2.5 using (9.154), (9.155), and the fact that $X_t'X_m = 0$, $t \neq m$.

The complete sufficiency is established by examining the likelihood function and applying the results in Appendix B.I.5. To do so, we first write

$$V^{-1} = \sum_{t \in T_2} (1/\lambda_t)A_t^* + (1/\lambda_\mu)A_\mu \tag{9.157}$$

where

$$A_t^* = A_t + \sum_m A_m \tag{9.158}$$

with the sum in (9.158) being over all $m \in T_0$ such that $\lambda_m = \lambda_t$. We then write

$$V^{-1} = \sum_{t \in T_2} A_t/\lambda_t + \sum_{m \in T_1} A_m/\lambda_m + A_\mu/\lambda_\mu \tag{9.159}$$

where the parameters λ_m in the second sum are identical to one of the parameters in the first sum. We may then write the exponent in the likelihood function as

$$\begin{aligned} Q(\beta, \phi) &= (Y - X\beta)'V^{-1}(Y - X\beta) \\ &= \sum_{t \in T_2} q_t/\lambda_t + \sum_{m \in T_1} (Y - X_m\beta_m)'A_m(Y - X_m\beta_m)/\lambda_m \qquad (9.160) \\ &\quad + N(\bar{y} - \beta_0)^2/\lambda_\mu . \end{aligned}$$

The simplifications in (9.160) follow by applying (9.154). The equivalence of the statistics A_mY and $X_m'Y$, $m \in T_1$, follows from (9.43) and the triangular structure of the P matrix in (9.35).

Finally, (v) follows by application of the Lehmann-Scheffé Theorem, Appendix B.I.7.

■

As in Theorem 9.6, it may happen that some of the ϕ_t, $t \in T_2$ are zero in (9.148) and, hence, in (9.149) A_t^* may contain more than one matrix with subscript in T_2. In this case, the statistics q_t are based on the sum of these matrices; that is, we have the following theorem.

THEOREM 9.9
If $Y \sim N(X\beta, V)$ with some of the $\phi_t = 0$, $t \in T_2$, so that V still has the form (9.149), then

$$\text{(i)}\ q_t = Y'A_t^*Y \sim \lambda_t\chi^2(r_t^*),\ t \in T_2$$

where

$$A_t^* = \sum_m A_m$$

$$r_t^* = \sum_m r_m$$

and the sum is on those $m \in T_2$ for which $\lambda_m = \lambda_t$. Also,

$$\text{(ii)}\ \hat{\lambda}_t = q_t/r_t^*,\ t \in T_2, \text{ is UMV for } \lambda_t.$$

■

To illustrate these results, we consider the mixed, two-way classification model proposed in Example 8.5.

EXAMPLE 9.11
From Example 8.5, we recall that

$$E(Y) = (I_{a_1} \otimes J_{na_2})\mu \qquad (9.161)$$

and, using the notation of Theorem 9.1,

$$V = \phi_0 I_N + \phi_2 V_2 + \phi_{12} V_{12}. \tag{9.162}$$

We note that, with regard to the means, we have a one-way classification structure and that the covariance matrix is that of a two-fold nested structure. Although not necessary in this case, we can express the mean vector in the canonical form (9.146) as

$$E(Y) = (J_N | X_1) \begin{bmatrix} \beta_0 \\ \beta_1 \end{bmatrix} \tag{9.163}$$

where

$$\begin{aligned} X_1 &= \Delta'_{a_1} \otimes J_{a_2} \otimes J_n \\ \beta_0 &= \bar{\mu}. \\ \beta_{1i} &= \mu_i - \bar{\mu}. \qquad i = 1, \ldots, a_1 - 1. \end{aligned} \tag{9.164}$$

Further, we may write

$$V = \lambda_0 A_0 + \lambda_2 A_2 + \lambda_{12}(A_1 + A_{12}) + \lambda_\mu A_\mu \tag{9.165}$$

using the results of Example 9.1 with $\phi_1 = 0$. Note that the index sets in this example are $T_1 = \{(1)\}$ and $T_2 = \{(0), (2), (12)\}$.

From Theorem 9.8, we immediately have the UMV estimates

$$\begin{aligned} \hat{\mu}_i &= \bar{y}_{i\cdot\cdot} \\ \hat{\lambda}_0 &= Y'A_0Y/[a_1a_2(n-1)] \\ \hat{\lambda}_2 &= Y'A_2Y/(a_2 - 1) \\ \hat{\lambda}_{12} &= Y'A_{12}Y/[(a_1 - 1)(a_2 - 1)]. \end{aligned} \tag{9.166}$$

From the relation (9.23) between λ and ϕ with $\phi_1 = 0$, we have the UMV estimates of ϕ as the solution to the equations

$$\begin{bmatrix} \hat{\lambda}_0 \\ \hat{\lambda}_2 \\ \hat{\lambda}_{12} \end{bmatrix} = \begin{bmatrix} 1 & 0 & 0 \\ 1 & na_1 & n \\ 1 & 0 & n \end{bmatrix} \begin{bmatrix} \hat{\phi}_0 \\ \hat{\phi}_2 \\ \hat{\phi}_{12} \end{bmatrix}. \tag{9.167}$$

Further, we note that

$$q_1 = Y'A_1Y \sim \lambda_{12}\chi^2[r_1, \beta_1'X_1'X_1\beta_1/(2\lambda_{12})] \tag{9.168}$$

since $\lambda_1 = \lambda_{12}$. The noncentrality parameter may be written, using (9.164), as

$$\beta_1' X_1' X_1 \beta_1 = a_2 n \sum_{i=1}^{a_1} (\mu_i - \bar{\mu}.)^2. \tag{9.169}$$

We shall return to this result in Section 9.3.3 when we consider tests of hypotheses on μ.

■

Since the results of Theorem 9.8 identify optimal estimators for this mixed model, it may seem redundant to examine other estimators. However, as in Section 9.2.3, we shall find interesting relations between ML and REML estimators and the UMV estimators of Theorem 9.8.

Considering first the REML estimators, we shall follow the development in Section 9.2.3. The essential step is to show that

$$\begin{aligned} M &= I - X(X'V^{-1}X)^{-1}X'V^{-1} \\ &= I - X(X'X)^{-1}X' \\ &= I - \sum_{t \epsilon T_1} A_t - A_\mu . \end{aligned} \tag{9.170}$$

The first step in (9.170) follows directly from Theorem 9.7 and, in particular, (9.151). To establish the final result, note from (9.144) and (9.145) that

$$X'X = \text{Diag}(N, X_t'X_t, t \epsilon T_1). \tag{9.171}$$

This follows since $J'X_t = 0$ and $X_m'X_t = 0$, $m \neq t$. Further, using (9.145), we may write

$$(X_t'X_t) = na_t^*(\nabla_1\nabla_1' \otimes \nabla_2\nabla_2' \otimes \ldots \otimes \nabla_k\nabla_k'). \tag{9.172}$$

For $i \epsilon t$, we have

$$\begin{aligned} (\nabla_i\nabla_i')^{-1} &= (\Delta_i\Delta_i')^{-1} = (I_{a_i-1} + JJ')^{-1} \\ &= I_{a_i-1} - (1/a_i)JJ' \end{aligned} \tag{9.173}$$

and

$$\nabla_i'(\nabla_i\nabla_i')^{-1}\nabla_i = I_{a_i} - (1/a_i)JJ'. \tag{9.174}$$

Recalling the expression for A_t from (9.28), we may now write

$$X_t(X_t'X_t)^{-1}X_t' = A_t. \tag{9.175}$$

It thus follows that

$$\begin{aligned} X(X'X)^{-1}X' &= (1/N)JJ' + \sum_{t \in T_1} X_t(X_t'X_t)^{-1}X_t \\ &= A_\mu + \sum_{t \in T_1} A_t. \end{aligned} \tag{9.176}$$

Continuing, with V given by (9.149), we obtain

$$\begin{aligned} V^{-1}M &= V^{-1} - \sum_{t \in T_1} (1/\lambda_t)A_t - (1/\lambda_\mu)A_\mu \\ &= \sum_{t \in T_2} (1/\lambda_t)A_t. \end{aligned} \tag{9.177}$$

From this result, we may establish that

$$\begin{aligned} Q_t &= V^{-1}MV_tM'V^{-1} \\ &= \sum_{m \in T_2} p_{mt}A_m/\lambda_m^2 \qquad t \in T_2 \end{aligned} \tag{9.178}$$

and hence that

$$\begin{aligned} \rho_t &= Y'Q_tY \\ &= \sum_{m \in T_2} p_{mt}q_m/\lambda_m^2 \qquad t \in T_2. \end{aligned} \tag{9.179}$$

Finally,

$$E(\rho_t) = \sum_{m \in T_2} p_{mt}r_m/\lambda_m \qquad t \in T_2 \tag{9.180}$$

hence, the REML equations are written in matrix form as

$$P^{*\prime}\text{Diag}(r_t/\lambda_t^2)\lambda^* = P^{*\prime}\text{Diag}(\lambda_t^{-2})q^* \tag{9.181}$$

where $\lambda^* = (\lambda_t,\ t \in T_2)$ is related to $\phi^* = (\phi_t,\ t \in T_2)$, the components of ϕ in (9.148), by

$$\lambda^* = P^*\phi^*. \tag{9.182}$$

Here P^* is defined as in (9.35) with columns deleted if $\phi_i \notin \phi^*$ and with redundant rows deleted. The vector q^* has components q_t, $t \in T_2$. It follows that (9.181) reduces to

$$\text{Diag}(r_t)\lambda^* = q^* \tag{9.183}$$

hence, the REML estimators for λ_t, $t \in T_2$, are identical to the UMV estimators in Theorem 9.8.

As noted in Section 8.2.2, the ML equations differ from the REML equations by the term

$$tr[V^{-1}(I - M)V_t] \qquad t \in T_2. \tag{9.184}$$

Utilizing (9.170), (9.159), and (9.43), we obtain

$$tr[V^{-1}(I - M)V_t] = \sum_{m \in T_1} p_{mt} r_m / \lambda_m + p_t / \lambda_\mu \qquad t \in T_2. \tag{9.185}$$

Noting that $p_{mt} = 0$, $m \in T_1$, $t \in T_2$, we have the ML equation in the form for comparison with (9.183); that is,

$$\text{Diag}(r_t)\lambda^* + \text{Diag}(\lambda_t^2/\lambda_\mu)P^{*\prime -1}P_\mu^* = q^*. \tag{9.186}$$

The second term on the left in (9.186) accounts for the difference in the two estimators and generally leads to nonlinear ML equations. In some cases, however, the equations may still be linear, as is the case with Example 9.11. This is illustrated in the next example.

EXAMPLE 9.12

Continuing with Example 9.11, P^* is given by (9.167) and P_μ^* by (9.25) with $\phi_1 = 0$. Also, we note that $\lambda_2 = \lambda_\mu$. It follows that the ML equations are given by

$$\begin{bmatrix} a_1 a_2(n-1)\,\lambda_0 \\ (a_2 - 1)\,\lambda_2 \\ (a_1 - 1)(a_2 - 1)\,\lambda_{12} \end{bmatrix} + \begin{bmatrix} 0 \\ \lambda_2 \\ 0 \end{bmatrix} = \begin{bmatrix} q_0 \\ q_2 \\ q_{12} \end{bmatrix}. \tag{9.187}$$

We note in this case that

$$\hat{\lambda}_2 = q_2 / a_2 \tag{9.188}$$

while $\hat{\lambda}_0$ and $\hat{\lambda}_{12}$ agree with REML.

■

9.3.3. Tests of Hypotheses

We shall discuss tests of hypotheses on both μ and ϕ. With regard to the latter, no new concepts occur. If the hypothesis on ϕ is equivalent to $\lambda_t/\lambda_m = 1$ for $t, m \in T_2$, then we may proceed as in Section 9.2.4 to obtain exact tests. In other cases, the approximate procedure in that same section may be applied. The relation to modified likelihood ratio tests is also unchanged.

For testing hypotheses on μ or, equivalently, β, we shall discuss three approaches. The first is based on the results in Theorem 9.8, the second is the modified likelihood ratio test described in Section 8.3.2, and the third is the approximate procedure described in Section 8.3.4.

Referring to Theorem 9.8, we see that for $m \in T_1$ and $t \in T_2$ such that $\lambda_m = \lambda_t$, the ratio

$$(q_m/r_m)/(q_t/r_t) \sim F[r_m, r_t, (\beta_m' X_m' X_m \beta_m)/(2\lambda_m)]. \tag{9.189}$$

It follows that for testing the hypothesis

$$H_0\colon \beta_m = 0 \tag{9.190}$$

the test statistic in (9.189) may be compared with critical values of the central F distribution in the usual way.

The modified likelihood ratio statistic has the form

$$\gamma^2 = \prod_{t \in T_2} (\hat{\lambda}_t/\tilde{\lambda}_t)^{r_t} \tag{9.191}$$

where $\hat{\lambda}_t$ and $\tilde{\lambda}_t$ denote the unconstrained estimates and the estimates constrained by the hypothesis, respectively. For hypotheses of the form (9.190) for which (9.189) is appropriate, we shall see that the modified likelihood ratio statistic leads to the same test statistic. This will be demonstrated in Example 9.13.

Finally, we consider the test statistic, given in Section 8.3.4, based on the case in which V is known apart from a constant; that is, for a hypothesis with rank v of the form

$$H_0\colon H\beta = 0 \tag{9.192}$$

we write

$$V = \phi_0 U \tag{9.193}$$

and consider the test statistic

$$F = \{(H\hat{\beta})'[H(X'\hat{U}^{-1}X)^{-1}H']^{-1}(H\hat{\beta})/(v\hat{\phi}_0)\}. \tag{9.194}$$

Here $\hat{U}$ denotes U in which the ratios ϕ_t/ϕ_0 are replaced by their REML or ML estimates. For illustration, we shall consider the hypothesis (9.190). Examining the numerator in (9.194) and utilizing (9.156) in the form

$$\phi_0 X B^{-1} = U^{-1} X \tag{9.195}$$

we obtain

$$\begin{aligned} H(X'U^{-1}X)^{-1}H' &= HB(X'X)^{-1}H'/\phi_0 \\ &= (X_m' X_m)^{-1}\lambda_m/\phi_0. \end{aligned} \tag{9.196}$$

Here, we have used the diagonal structure of $X'X$ given in (9.171) and $H = (0, \ldots, 0, I_{r_m}, 0, \ldots, 0)$. It follows that the numerator in (9.194) is given by

$$N_m = Y'X_m(X_m'X_m)^{-1}X_m'Y\hat{\phi}_0/\hat{\lambda}_m \tag{9.197}$$

and the rank of the hypothesis is r_m. Using (9.175), we write the test statistic, (9.194), as

$$F = (Y'A_mY)/(r_m\hat{\lambda}_m). \tag{9.198}$$

Recalling that $\lambda_m = \lambda_t$, $t \in T_2$, we see that (9.198) is identical to (9.189). Thus, in this special case, our approximate procedure leads to the same test statistic as the first two procedures. The only difference between the exact test (9.189) and the approximate test (9.194) is in the denominator degrees of freedom. The correct value is r_t, while the approximate procedure would use $N - r(X)$ where the rank of X is

$$r(X) = 1 + \sum_{m \in T_1} r_m. \tag{9.199}$$

To illustrate these results, we consider the two-way classification model described in Example 9.11.

EXAMPLE 9.13

Continuing with Example 9.11—the mixed, two-way classification model—we consider testing the hypothesis

$$H_0: \beta_1 = 0 \leftrightarrow \mu_i - \bar{\mu}. = 0 \qquad i = 1, \ldots, a_1 - 1. \tag{9.200}$$

From (9.166) and (9.168), we see that

$$\begin{aligned} F &= [q_1/(a_1 - 1)]/[q_{12}/(a_1 - 1)(a_2 - 1)] \\ &\sim F[(a_1 - 1), (a_1 - 1)(a_2 - 1), a_2 n \sum (\mu_i - \bar{\mu}.)^2/(2\lambda_{12})] \end{aligned} \tag{9.201}$$

represents a reasonable test statistic. We note that the approximate procedure (9.198) would suggest denominator degrees of freedom $a_1(na_2 - 1)$, rather than the correct value of $(a_1 - 1)(a_2 - 1)$ in (9.201).

To develop the modified likelihood ratio test for this hypothesis, we use $\hat{\lambda}$ from Example 9.11. Note that $\tilde{\lambda}$ is given as in Example 9.5, since the model, reduced by the hypothesis (9.200), has mean $E(Y) = \mu J$ and the nested covariance structure. Therefore,

$$\begin{aligned} \tilde{\lambda}_0 &= \hat{\lambda}_0 \\ \tilde{\lambda}_2 &= \hat{\lambda}_2 \\ \tilde{\lambda}_{12} &= Y'(A_1 + A_{12})Y/a_2(a_1 - 1) \\ &= (q_1 + q_{12})/(r_1 + r_{12}). \end{aligned} \tag{9.202}$$

It follows that

$$\begin{aligned}\gamma^2 &= (\hat{\lambda}_{12}/\tilde{\lambda}_{12})^{r_{12}} \\ &= [(r_1 + r_{12})/r_{12}]^{r_{12}}[1/(1 + q_1/q_{12})]^{r_{12}}.\end{aligned} \tag{9.203}$$

Thus, the modified likelihood ratio test suggests the rejection of H_0: $\beta_1 = 0$ when the ratio q_1/q_{12} is too large, and this test is identical to (9.201). In this case, this is also the ordinary likelihood ratio test, since only the estimate of λ_2 is changed, as indicated in Example 9.12. The properties of likelihood ratio tests thus apply to the intuitive test based on the ratio of quadratic forms, (9.201).

As noted, tests of hypotheses on ϕ involve no new concepts, but we illustrate one such test for demonstration and later reference. Consider,

$$H_0\colon \phi_2 = 0. \tag{9.204}$$

From (9.167), we see that this hypothesis is equivalent to the hypothesis

$$H_0\colon \lambda_2/\lambda_{12} = 1 \tag{9.205}$$

hence, as in Section 9.2.4, we suggest the statistic

$$F_2 = (q_2/r_2)/(q_{12}/r_{12}) \sim (\lambda_2/\lambda_{12})F(r_2, r_{12}). \tag{9.206}$$

This test statistic also arises using the likelihood ratio test.

Referring to Graybill (1961, p. 398) or Searle (1971a, p. 403), we see that some researchers disagree with this test statistic and indeed disagree with the use of (9.167) to estimate ϕ_t from $\hat{\lambda}_t$. A simple explanation for this controversy is given in Section 10.4.

■

EXAMPLE 9.14

Turning to a more complex example, we consider the mixed, three-way classification model. We first consider the case described as "*A* and *B* fixed, and *C* random"; that is,

$$E(Y) = (J\, X_1\, X_2\, X_{12})\beta \tag{9.207}$$

and

$$V = \phi_0 V_0 + \phi_3 V_3 + \phi_{13} V_{13} + \phi_{23} V_{23} + \phi_{123} V_{123}. \tag{9.208}$$

Referring to Example 9.9 and, in particular, to the P matrix given by (9.94), we find the relations

$$\begin{aligned}\lambda_1 &= \lambda_{13} \\ \lambda_2 &= \lambda_{23} \\ \lambda_{12} &= \lambda_{123}.\end{aligned} \tag{9.209}$$

It follows that the covariance matrix has the structure (9.149) as required for Theorem 9.8. Thus, closed form estimates are available for ϕ, and exact tests are available for the usual hypotheses. In particular, we have the following:

Hypothesis	*Test Statistic*	
H_1: $\beta_1 = 0 \leftrightarrow \bar{\mu}_{i\cdot} = \bar{\mu}_{\cdot\cdot}$ $i = 1, \ldots, a_1 - 1$	$(q_1/r_1)/(q_{13}/r_{13})$	
H_2: $\beta_2 = 0 \leftrightarrow \bar{\mu}_{\cdot j} = \bar{\mu}_{\cdot\cdot}$ $j = 1, \ldots, a_2 - 1$	$(q_2/r_2)/(q_{23}/r_{23})$	(9.210)
H_{12}: $\beta_{12} = 0 \leftrightarrow \mu_{ij} - \bar{\mu}_{i\cdot} - \bar{\mu}_{\cdot j} + \bar{\mu}_{\cdot\cdot} = 0$ $i = 1, \ldots, a_1 - 1$ $j = 1, \ldots, a_2 - 1.$	$(q_{12}/r_{12})/(q_{123}/r_{123}).$	

Alternatively, we might consider the case, "*A* fixed, *B* and *C* random"; that is,

$$E(Y) = (J\ X_1)\beta \tag{9.211}$$

$$V = \sum_{t \in T_0} \phi_t V_t \qquad \phi_1 = 0. \tag{9.212}$$

Referring to Example 9.9, we see that the relation on λ is

$$\lambda_1 + \lambda_{123} = \lambda_{12} + \lambda_{13}. \tag{9.213}$$

In this case, V does not satisfy the conditions of Theorem 9.8. As noted in Example 9.9, we do not have closed form estimates for ϕ. In addition, there is no exact test for the hypotheses H_1: $\beta_1 = 0$. Khuri (1982) develops simultaneous confidence intervals for β in such situations, and these could be used for testing this hypothesis. Alternatively, we could use the likelihood ratio statistic (9.191) with approximate large sample distribution or use the intuitive approach (9.194). Clearly, more research is needed in this area.

Finally, we note that if, for example, we are willing to assume $\phi_{13} = 0$, then $\lambda_1 = \lambda_{12}$ and $\lambda_{13} = \lambda_{123}$. Thus, Theorem 9.8 applies, and an exact test for $\beta_1 = 0$ is available. We emphasize that this result should not be used to justify the assumption.

■

9.3.4. Confidence Intervals

Assuming that the covariance matrix satisfies the conditions of Theorem 9.8, then

$$\begin{aligned} \hat{\beta}_m &\sim N[\beta_m, \lambda_m(X_m'X_m)^{-1}] \qquad m \in T_1 \\ q_t &\sim \lambda_t \chi^2(r_t) \qquad t \in T_2. \end{aligned} \tag{9.214}$$

Since $\lambda_m = \lambda_t$ for some $t \in T_2$ and β_m and q_t are independent, it follows that individual confidence intervals on the components of β_m are given by

$$\beta_{m_i}\colon \hat{\beta}_{m_i} \pm t(\alpha/2, r_t)(\hat{\lambda}_t C_{m_{ii}})^{1/2}. \tag{9.215}$$

Here, $C_{m_{ii}}$ is the ith diagonal element of $C_m = (X_m' X_m)^{-1}$.

Simultaneous confidence intervals on linear functions of β_m follow directly from (9.194), using the Scheffé procedure described in Section 3.4.2. In particular, we obtain

$$a'\beta_m\colon a'\hat{\beta}_m \pm [r_m F(\alpha; r_m, r_t)\hat{\lambda}_t a'(X_m' X_m)^{-1} a]^{1/2}. \tag{9.216}$$

Individual and simultaneous confidence intervals on functions of ϕ follow as in Section 9.2.5 as applied to q_t, λ_t, and ϕ_t, $t \in T_2$.

9.4. TWO EXAMPLES

In this section, we present two examples that are more complex to illustrate the analysis of mixed models. Both are special cases of the general mean and covariance structure described in Sections 9.1 and 9.2. We also use this section to introduce the Analysis of Variance table to summarize some of the results of the analysis.

9.4.1. A Partially Nested Model

For our first example, we consider a balanced version of the detergent study, Example 7.4, described in Section 7.2.2.

Let us suppose that we are interested in four specific detergents ($a_1 = 4$) and three particular types of washing machines ($a_2 = 3$). We have, for the experiment, six laundromats each containing eight machines. Two of the laundromats use machine-type 1, two use type two, and two use type 3, ($a_3 = 2$). Within a laundromat, the eight machines are allocated at random, two to each detergent, ($n = 2$).

We assume that the mean response, if detergent i is used on machine j, is μ_{ij}. Thus, if we let y_{ijkr} denote the rth response for laundromat k within machine-type j receiving detergent i, we have

$$E(y_{ijkr}) = \mu_{ij} \qquad \begin{aligned} i &= 1, 2, 3, 4 \\ j &= 1, 2, 3 \\ k &= 1, 2 \\ r &= 1, 2. \end{aligned} \tag{9.217}$$

Further, we assume that responses within a given laundromat are correlated and that this correlation is different for machines using the same detergent than for machines using different detergents. We assume, too, that these correlations are the same for all laundromats and that responses from different laundromats are uncorrelated. This covariance structure is written as

$$\begin{aligned}
\operatorname{cov}(y_{ijkr}, y_{i^*j^*k^*r^*}) &= \phi_0 + \phi_{23} + \phi_{123} && i = i^*,\ j = j^*\\
&&& k = k^*,\ r = r^*\\
&= \phi_{23} + \phi_{123} && i = i^*,\ j = j^*\\
&&& k = k^*,\ r \neq r^*\\
&= \phi_{23} && i \neq i^*,\ j = j^*\\
&&& k = k^*\\
&= 0 && \text{otherwise.}
\end{aligned} \tag{9.218}$$

Writing the mean and covariance structure using the notation of Sections 9.2 and 9.3, we have

$$E(Y) = (J\ X_1\ X_2\ X_{12})\beta \tag{9.219}$$

with X_t defined by (9.145) and β defined by (9.147). Further,

$$V = \phi_0 I + \phi_{23} V_{23} + \phi_{123} V_{123} \tag{9.220}$$

where V_{23} and V_{123} are defined by (9.15). We note that this covariance structure is a special case of the three-way classification structure in which $\phi_1 = \phi_2 = \phi_{12} = \phi_3 = \phi_{13} = 0$. Referring to Example 9.9 and, in particular, to (9.94), we see that

$$\begin{aligned}
\lambda_1 &= \lambda_{12} = \lambda_{13} = \lambda_{123}\\
\lambda_2 &= \lambda_3 = \lambda_{23}
\end{aligned} \tag{9.221}$$

hence, V has the form (9.149); that is,

$$\begin{aligned}
V = {} & \lambda_0 A_0 + \lambda_{23}(A_2 + A_3 + A_{23})\\
& + \lambda_{123}(A_1 + A_{12} + A_{13} + A_{123}) + \lambda_\mu A_\mu .
\end{aligned} \tag{9.222}$$

The index sets are $T_1 = (1, 2, 12)$ and $T_2 = (0, 3, 13, 23, 123)$. Noting that $\phi_3 = \phi_{13} = 0$, we see that Theorem 9.9 is appropriate. We thus have the following results:

$$\begin{aligned}
q_0 &= Y'A_0Y \sim \lambda_0\chi^2(r_0)\\
q_{23} &= Y'(A_3 + A_{23})Y \sim \lambda_{23}\chi^2(r_3 + r_{23})\\
q_{123} &= Y'(A_{13} + A_{123})Y \sim \lambda_{123}\chi^2(r_{13} + r_{123}).
\end{aligned} \tag{9.223}$$

Estimates of ϕ_0, ϕ_{23}, and ϕ_{123} are given by solving

$$\begin{bmatrix} q_0/r_0 \\ q_{23}/(r_3 + r_{23}) \\ q_{123}/(r_{13} + r_{123}) \end{bmatrix} = \begin{bmatrix} 1 & 0 & 0 \\ 1 & na_1 & n \\ 1 & 0 & n \end{bmatrix} \begin{bmatrix} \phi_0 \\ \phi_{23} \\ \phi_{123} \end{bmatrix}. \tag{9.224}$$

From Theorem 9.8 we have

$$\begin{aligned} \hat{\beta}_m &= (X_m'X_m)^{-1}X_m'Y \sim N[\beta_m, \lambda_m(X_m'X_m)^{-1}] \\ q_m &= Y'A_mY \sim \lambda_m\chi^2(r_m, \beta_m'X_m'X_m\beta_m/2\lambda_m) \\ m &= 1, 2, 12. \end{aligned} \tag{9.225}$$

Tests of hypotheses on ϕ and β are determined from (9.224) and (9.225) as described in Sections 9.2.4 and 9.3.3. This information is conveniently summarized in Table 9.1.

TABLE 9.1 Analysis of variance table for the detergent study.

Description	*Hypothesis*	*df*	*SS*	*EMS*
Detergents	H_1	$a_1 - 1$	q_1	$\lambda_{123} + \beta_1'X_1'X_1\beta_1/r_1$
Machines	H_2	$a_2 - 1$	q_2	$\lambda_{23} + \beta_2'X_2'X_2\beta_2/r_2$
Detergent × machine type	H_{12}	$(a_1 - 1)(a_2 - 1)$	q_{12}	$\lambda_{123} + \beta_{12}'X_{12}'X_{12}\beta_{12}/r_{12}$
Estimate of λ_{23}	H_{23}	$a_2(a_3 - 1)$	q_{23}	$\lambda_{23} = \phi_0 + na_1\phi_{23} + n\phi_{123}$
Estimate of λ_{123}	H_{123}	$a_2(a_1 - 1)(a_3 - 1)$	q_{123}	$\lambda_{123} = \phi_0 + n\phi_{123}$
Estimate of λ_0	—	$a_1a_2a_3(n - 1)$	q_0	$\lambda_0 = \phi_0$

Referring to Table 9.1, we see that it is just a convenient summary of the information for estimating ϕ, testing hypotheses about ϕ, and testing hypotheses about β. The last three rows contain the information on ϕ as described in (9.223) and (9.224). The first three rows contain the information for testing hypotheses about β in the sense described in Section 9.3.3. Thus, we have the following summary of hypotheses and test statistics:

Hypothesis	*Test Statistic*	
H_1: $\bar{\mu}_{i\cdot} = \bar{\mu}_{\cdot\cdot}$	$(q_1/r_1)/(q_{123}/r_{123}^*)$	
H_2: $\bar{\mu}_{\cdot j} = \bar{\mu}_{\cdot\cdot}$	$(q_2/r_2)/(q_{23}/r_{23}^*)$	
H_{12}: $\mu_{ij} - \bar{\mu}_{i\cdot} - \bar{\mu}_{\cdot j} + \bar{\mu}_{\cdot\cdot} = 0$	$(q_{12}/r_{12})/(q_{123}/r_{123}^*)$	(9.226)
H_{23}: $\phi_{23} = 0$	$(q_{23}/r_{23}^*)/(q_{123}/r_{123}^*)$	
H_{123}: $\phi_{123} = 0$	$(q_{123}/r_{123}^*)/(q_0/r_0)$.	

We note that, for the two main effect hypotheses, the denominators are different. This may seem curious, but it is a consequence of the model that reflects the way the experiment was conducted.

Confidence intervals on β and ϕ follow directly from Sections 9.2.5 and 9.3.4 and so will not be spelled out in detail.

To conclude this example, we note that, if we assume no detergent-by-machine interaction, then

$$q_{123} = Y'(A_{12} + A_{13} + A_{123})Y \sim \lambda_{123}\chi^2(r_{12} + r_{13} + r_{123}) \tag{9.227}$$

as a direct consequence of (9.221). The appropriate changes should be made in the Analysis of Variance table.

9.4.2. The Split-Plot Model

Consider now the experiment described as Example 8.7. Let y_{ijk} denote the response from the kth subplot of the jth whole plot in the ith field. In our current notation, the ranges on i, j, and k are a_1, a_2, and a_3, respectively. Letting u_{jk} denote the mean response from the jth variety and kth fertilizer, we have

$$E(y_{ijk}) = \mu_{jk} \tag{9.228}$$

or, in vector form

$$E(Y) = (J_{a_1} \otimes I_{a_2} \otimes I_{a_3})\mu. \tag{9.229}$$

The covariance structure, rewritten in the current notation, is

$$\begin{aligned} \operatorname{cov}(y_{ijk}, y_{i^*j^*k^*}) &= \phi_0 + \phi_1 + \phi_{12} && i = i^*, j = j^*, k = k^* \\ &= \phi_1 + \phi_{12} && i = i^*, j = j^*, k \neq k^* \\ &= \phi_1 && i = i^*, j \neq j^* \\ &= 0 && \text{otherwise.} \end{aligned} \tag{9.230}$$

We may write this covariance structure in matrix form as

$$V = \phi_0 I + \phi_1 V_1 + \phi_{12} V_{12} \tag{9.231}$$

with V_1 and V_{12} defined by (9.15). The mean structure (9.229) may be written in canonical form as a two-way classification with interaction as

$$E(Y) = (J\ X_2\ X_3\ X_{23})\beta \tag{9.232}$$

with $\beta' = (\beta_0, \beta_2', \beta_3', \beta_{23}')$ using (9.174) with $n = 1$. Although the subscripting seems somewhat curious, we note that this structure satisfies the requirements of Theorem 9.9 with index sets $T_1 = (2, 3, 23)$ and $T_2 = (0, 1, 12, 13, 123)$ and with

$\phi_{13} = \phi_{123} = 0$. Further, the basic relations between parameters are again defined by Example 9.9, with $\phi_2 = \phi_3 = \phi_{23} = \phi_{13} = \phi_{123} = 0$. Thus, we have

$$\begin{aligned} \lambda_2 &= \lambda_{12} \\ \lambda_0 &= \lambda_3 = \lambda_{13} = \lambda_{23} = \lambda_{123}. \end{aligned} \tag{9.233}$$

We may now write V in the form of (9.149) as

$$\begin{aligned} V = \lambda_0(A_0 + A_3 + A_{13} + A_{23} + A_{123}) + \lambda_1 A_1 \\ + \lambda_{12}(A_2 + A_{12}) + \lambda_\mu A_\mu. \end{aligned} \tag{9.234}$$

Application of Theorems 9.8 and 9.9 then gives the following results:

$$\begin{aligned} q_0 &= Y'(A_0 + A_{13} + A_{123})Y \sim \lambda_0\chi^2(r_0 + r_{13} + r_{123}) \\ q_1 &= Y'A_1Y \sim \lambda_1\chi^2(r_1) \\ q_{12} &= Y'A_{12}Y \sim \lambda_{12}\chi^2(r_{12}) \end{aligned} \tag{9.235}$$

and

$$\begin{aligned} \hat{\beta}_m &= (X_m'X_m)^{-1}X_m'Y \sim N[\beta_m, \lambda_m(X_m'X_m)^{-1}] \\ q_m &= Y'A_mY \sim \lambda_m\chi^2[r_m, \beta_m'X_m'X_m\beta_m/(2\lambda_m)] \\ m &= 2, 3, 23. \end{aligned} \tag{9.236}$$

The estimates of ϕ_0, ϕ_1, and ϕ_{12} are given by solving

$$\begin{bmatrix} q_0/(r_0 + r_{13} + r_{123}) \\ q_1/r_1 \\ q_{12}/(r_{12}) \end{bmatrix} = \begin{bmatrix} 1 & 0 & 0 \\ 1 & a_2a_3 & a_3 \\ 1 & 0 & a_3 \end{bmatrix} \begin{bmatrix} \phi_0 \\ \phi_1 \\ \phi_{12} \end{bmatrix}. \tag{9.237}$$

(Note: $n = 1$; hence, $r_0 = 0$.) The information for estimating ϕ and testing hypotheses on ϕ and β is summarized in Table 9.2.

TABLE 9.2 Analysis of variance table for the split-plot model.

Description	*Hypothesis*	*df*	*SS*	*EMS*
Fields	H_1	$a_1 - 1$	q_1	$\lambda_1 = \phi_0 + a_2a_3\phi_1 + a_3\phi_{12}$
Varieties	H_2	$a_2 - 1$	q_2	$\lambda_{12} + \beta_2'X_2'X_2\beta_2/r_2$
Estimate of λ_{12}	H_{12}	$(a_1 - 1)(a_2 - 1)$	q_{12}	$\lambda_{12} = \phi_0 + a_3\phi_{12}$
Fertilizers	H_3	$a_3 - 1$	q_3	$\lambda_0 + \beta_3'X_3'X_3\beta_3/r_3$
Variety × fertilizer	H_{23}	$(a_2 - 1)(a_3 - 1)$	q_{23}	$\lambda_0 + \beta_{23}'X_{23}'X_{23}\beta_{23}/r_{23}$
Estimate of λ_0	—	$a_2(a_1 - 1)(a_3 - 1)$	q_0	$\lambda_0 = \phi_0$

Table 9.2 contains the information in (9.237) for estimating ϕ and for testing hypotheses about ϕ and β. These hypotheses and their test statistics are summarized as follows:

Hypothesis	*Test Statistic*
H_2: $\bar{\mu}_{j\cdot} = \bar{\mu}_{\cdot\cdot}$ $j = 1, \ldots, a_2 - 1$	$(q_2/r_2)/(q_{12}/r_{12})$
H_3: $\bar{\mu}_{\cdot k} = \bar{\mu}_{\cdot\cdot}$ $k = 1, \ldots, a_3 - 1$	$(q_3/r_3)/(q_0/r_0^*)$
H_{23}: $\mu_{jk} - \bar{\mu}_{j\cdot} - \bar{\mu}_{\cdot k} + \bar{\mu}_{\cdot\cdot} = 0$ $j = 1, \ldots, a_2 - 1,$ $k = 1, \ldots, a_3 - 1$	$(q_{23}/r_{23})/(q_0/r_0^*)$ (9.238)
H_1: $\phi_1 = 0$	$(q_1/r_1)/(q_{12}/r_{12})$
H_{12}: $\phi_{12} = 0$	$(q_{12}/r_{12})/(q_0/r_0^*)$.

9.5. UNBALANCED MIXED MODELS

9.5.1. Definition of the Unbalanced Model

The mean and covariance structure, developed in Sections 9.2 and 9.3, corresponds to what are classically called *balanced models*. The mathematical development yielding closed form expressions for V^{-1}, complete sufficient statistics for the parameters, UMV estimates, and exact tests for certain hypotheses was a consequence of this balance.

To introduce the concept of unbalanced models, we recall the general mean and covariance structure using slightly different notation. First, from Section 9.3.1, we note that we can write (9.145) as

$$X_t = WX_t^0P_t' \qquad t \in T_1 \tag{9.239}$$

where

$$W = I_a \otimes J_n = D_a(J_n) \tag{9.240}$$

with $a = \Pi_{i=1}^k a_i$. Similarly, from (9.15) we may write

$$V_t = W(L_1 \otimes L_2 \otimes \ldots \otimes L_k)W' \tag{9.241}$$

where again W is given by (9.240).

The balance in the model refers to the fact that W is written as (9.240). When we consider unbalanced models, the only notational change is that W is replaced by

$$W = D_a(J_{n_{i_1 i_2 \ldots i_k}}). \tag{9.242}$$

EXAMPLE 9.15
To illustrate, recall Example 9.11, where we considered the two-way classification, mixed model. For the balanced case, we may write X_1 and V_2 as

$$\begin{aligned} X_1 &= (I_{a_1} \otimes J_{a_2} \otimes J_n)(\Delta'_{a_1} \otimes 1 \otimes 1) \\ &= W(I_{a_1} \otimes J_{a_2})(\Delta'_{a_1} \otimes 1) \end{aligned} \tag{9.243}$$

and

$$\begin{aligned} V_2 &= U_{a_1} \otimes I_{a_2} \otimes U_n \\ &= W(U_{a_1} \otimes I_{a_2})W'. \end{aligned} \tag{9.244}$$

The balanced model is written, using either form, with

$$W = I_{a_1} \otimes I_{a_2} \otimes J_n \tag{9.245}$$

in the second form for X_1 and V_2. The unbalanced model is written, using the second form only, with

$$W = D_{a_1 a_2}(J_{n_{ij}}, i = 1, \ldots, a_1, j = 1, \ldots, a_2.). \tag{9.246}$$

Allowing for a variable range on the subscripts would appear to be a minor change, but it results in greatly increased difficulty in the analysis. The result is that, except in special cases, we cannot obtain a closed form expression for the inverse of V. As a consequence, we cannot identify sufficient statistics and must resort to numerical methods for solving the likelihood or modified likelihood equations. In general, we do not have exact distributional results, exact tests of hypotheses on either ϕ or β, nor exact confidence regions.

This notation will be further illustrated in the next section in which we examine a model for which we have some success with the analytical approach.

9.5.2. The Mixed, Balanced Incomplete Block Model

The balanced incomplete block model was discussed in Section 7.3 as a special case of the two-way classification model without interaction and with missing cells. As such, it is seen to violate the definition of balance given in the previous section. The name for this model arises, however, because the design has some features of balance. In particular, we noted that differences in cell means within a column are all estimated with the same precision. As a consequence of the design, we were also able to obtain closed form expressions for the parameter estimates and exact tests for certain hypotheses.

In this section, we consider the analysis of the mixed, BIB model. Note that this is a special case of the two-way classification mixed model in which $\phi_{12} = 0$ and in which we have at most one observation per cell.

To motivate the model, consider the suntan-lotion experiment from Example 7.5, in which we had $a_1 = 3$ lotions (treatments), $a_2 = 3$ subjects (blocks), and the experimental units were the subjects' arms. Let y_{ij} denote the response of the jth individual to the ith lotion if the individual receives that lotion. The vector of responses is

$$Y' = (y_{11}\, y_{12}\, y_{21}\, y_{23}\, y_{32}\, y_{33}) \tag{9.247}$$

as indicated by the incidence matrix in Table 7.17. Let μ_i denote the mean response to the ith lotion. Then the mean structure is given by

$$E(Y) = Z_t\mu \tag{9.248}$$

where

$$Z_t = I_3 \otimes J_2. \tag{9.249}$$

Note that this matrix arises from the mean structure (9.161) in Example 9.11 by premultiplying by the counting matrix W of the form (9.246). Here, $a_1 = a_2 = 3$ and $J_{n_{ij}}$ has length one if the cell is observed and zero otherwise. Thus,

$$Z_t = W(I_{a_1} \otimes J_{a_2}). \tag{9.250}$$

In canonical form, we may write

$$\begin{aligned} E(Y) &= W(J_a | \Delta'_{a_1} \otimes J_{a_2}) \begin{bmatrix} \beta_0 \\ \beta_1 \end{bmatrix} \\ &= \beta_0 J + X_1\beta_1. \end{aligned} \tag{9.251}$$

We use X_1 as defined here for notational convenience and recognize its relation to X_1 as defined in Section 9.3.

The covariance structure arises by noting that responses on an individual are assumed to be correlated; that is, we write

$$\begin{aligned} \operatorname{cov}(y_{ij}, y_{i^*j^*}) &= \phi_0 + \phi_2 & i = i^*, j = j^* \\ &= \phi_2 & i \neq i^*, j = j^* \\ &= 0 & \text{otherwise.} \end{aligned} \tag{9.252}$$

In this example, it is reasonable to assume that ϕ_2 is nonnegative, but it is easy to visualize situations in which negative covariance is reasonable. Thus, we shall not restrict the sign of ϕ_2. We may write (9.252) in matrix notation as

$$V = \phi_0 I + \phi_2 V_2 \tag{9.253}$$

where

$$\begin{aligned} V_2 &= W(U_{a_1} \otimes I_{a_2})W' \\ &= W(J_{a_1} \otimes I_{a_2})(J'_{a_1} \otimes I_{a_2})W' \\ &= Z_b Z'_b \end{aligned} \tag{9.254}$$

and

$$Z_b = W(J_{a_1} \otimes I_{a_2}). \tag{9.255}$$

We note that Z_t and Z_b defined here are identical to X_t and X_b as defined in Section 7.3; hence, we may use the properties of these matrices in our analysis. Also, to conform to convention, we shall let $t = a_1$ and $b = a_2$ denote the range on i and j in this discussion. We shall use the mean structure (9.248), since there is no advantage to the canonical form for this simple structure.

Our first step is to attempt to obtain the inverse of V. Guided by the results in Theorem 9.1, we seek an expression for V in terms of orthogonal, idempotent matrices. In this case, we are successful and write

$$\begin{aligned} V &= \phi_0 I + \phi_2 Z_b Z_b' \\ &= \phi_0[I - (1/k)Z_b Z_b'] + \lambda_2[(1/k)Z_b Z_b'] \\ &= \phi_0 A_0 + \lambda_2 A_2 \end{aligned} \tag{9.256}$$

where

$$\lambda_2 = \phi_0 + k\phi_2. \tag{9.257}$$

It is easily verified that A_0 and A_2 are idempotent and orthogonal; hence,

$$V^{-1} = (1/\phi_0)A_0 + (1/\lambda_2)A_2. \tag{9.258}$$

[Note that the notation used here is for convenience in relating to the balanced case but that the eigenvalues, ϕ_0 and λ_2, and the matrices, A_0 and A_2, are as defined in (9.256) and (9.257).]

It is easily verified that V and Z_t do not enjoy the relation described in Theorem 8.1; hence, we do not have the result described in Theorem 9.7. As a further consequence, we do not have the simple results of Theorem 9.8. We can, however, identify sufficient statistics for μ and ϕ as described in the next theorem.

THEOREM 9.10

A set of sufficient statistics for the parameters in the BIB is given by

(i) $q_0 = Y'A_0Y$ and $q_2 = Y'A_2Y$,

(ii) the components of the vector $A'Y$, where

$$\begin{aligned} A' &= [Z_t' - (1/k)NZ_b'] \\ N &= Z_t'Z_b \end{aligned}$$

(iii) the components of the vector, $NZ_b'Y$.

Proof: Using the properties of the matrices A and N as given in Section 7.3, we may verify that

$$Z_t'V^{-1} = A'/\lambda_0 + (1/k)NZ_b'/\lambda_2. \tag{9.259}$$

The exponent in the likelihood function is then written as

$$\begin{aligned} Q(\mu, \phi) &= (Y - Z_t\mu)'V^{-1}(Y - Z_t\mu) \\ &= q_0/\lambda_0 + q_2/\lambda_2 - 2\mu'Z_t'V^{-1}Y + \mu'Z_t'V^{-1}Z_t\mu. \end{aligned} \tag{9.260}$$

Substituting from (9.259) and applying the results of Appendix B.I establishes the sufficiency.

■

Examination of these statistics reveals that the number of statistics exceeds the number of parameters; hence, they are not complete. More satisfying expressions for these statistics are obtained by noting the components of the two vectors

$$Z_t'Y = (y_{i\cdot}, i = 1, \ldots, t) \tag{9.261}$$

$$Z_b'Y = (y_{\cdot j}, j = 1, \ldots, b). \tag{9.262}$$

Letting n_{ij} denote the elements of N, we identify the minimal set of sufficient statistics for the case $b > t$ as

$$\begin{aligned} &q_0, q_2 \\ &y_{i\cdot}, \qquad i = 1, \ldots, t-1 \\ &\sum_j n_{ij}y_{\cdot j}, \qquad i = 1, \ldots, t. \end{aligned} \tag{9.263}$$

The redundancy deleted in (9.263) is a consequence of the fact that

$$kJ'Z_t'Y = J'NZ_b'Y = kJ'Y. \tag{9.264}$$

In the special case $b = t$, $\Sigma_j n_{ij}y_{\cdot j}$, $i = 1, \ldots, t$ can be replaced by $y_{\cdot j}$, $j = 1, \ldots, b - 1$. (These statistics are equivalent to those identified by Hultquist and Graybill, 1965.)

The lack of completeness precludes the simple identification of UMV estimators; hence, we shall proceed with the likelihood analysis as described in Chapter 8. It will be useful, however, to first describe the distributional properties of certain functions of the sufficient statistics. This is done in the following theorem.

THEOREM 9.11
With $Y \sim N(Z_t\mu, V)$, we have

(i) $q_0 \sim \phi_0\chi^2[b(k-1)]$

$q_2 \sim \lambda_2\chi^2[b, \mu'NN'\mu/(2k\lambda_2)]$

where q_0 and q_2 are independent,

(ii) $\hat{\mu}_1 = (1/r)Z_t'Y \sim N\{\mu, [(\phi_0/r)I_t + (\phi_2/r^2)NN']\}$

$\hat{\mu}_2 = (NN')^{-1}NZ_b'Y \sim N[\mu, k\lambda_2(NN')^{-1}]$

$\text{cov}(\hat{\mu}_1, \hat{\mu}_2) = (\lambda_2/r)I_t$

(iii) $s_0 = q_0 - (k/\lambda t)Y'AA'Y \sim \phi_0\chi^2(f) \qquad f = bk - b - t + 1$

$s_2 = q_2 - Y'[Z_bN'NZ_b' - (\lambda t/b)JJ']Y/k(r - \lambda) \sim \lambda_2\chi^2(b - t)$

(iv) $p_0 = Y'[k(r - \lambda)AA' - \lambda tZ_bN'A']Y/\lambda t(r - \lambda)$

$E(p_0) = \phi_0(t - 1)$

$p_2 = Y'[rZ_bN'NZ_b' - (r - \lambda)Z_bN'Z_t' - \lambda_kJJ']Y/\lambda t(r - \lambda)$

$E(p_2) = \lambda_2(t - 1)$

(v) p_0 and p_2 are not correlated with s_0 and s_2.

Proof: These results follow directly from the theorems on linear and quadratic functions of normal variables, established in Chapter 2. ■

The statistics identified in Theorem 9.11 were motivated by our search for functions of the sufficient statistics to be used to estimate μ, ϕ_0, and ϕ_2. We now turn to the likelihood equations for further guidance. We shall find that, while the nonlinear equations do not yield closed form solutions, they do suggest reasonable estimates. As a byproduct of our examinations of the likelihood equations, we shall identify the interblock and intrablock estimates (see Yates, 1940, or Graybill, 1961), and recognize that the likelihood equations lead us directly to the estimates without the confusion of the classical analysis.

Looking first for an estimate of μ, it is natural to attempt a combination of $\hat{\mu}_1$ and $\hat{\mu}_2$ from Theorem 9.11. We consider the linear combination

$$\hat{\mu} = w\hat{\mu}_1 + (1 - w)\hat{\mu}_2 \tag{9.265}$$

where w is chosen to minimize the variance of $p'\hat{\mu}$, with $p'J = 0$; that is, we consider minimizing the variance of contrasts on $\hat{\mu}$. Using Theorem 9.11 (ii), it is easily shown that

$$w = r\lambda_2/(r\phi_0 + \lambda t\phi_2) \tag{9.266}$$

hence, our suggested estimator, as a function of the unknown ϕ_0 and ϕ_2, is

$$\hat{\mu} = (w/r)Z_t'Y + (1 - w)(NN')^{-1}NZ_b'Y. \tag{9.267}$$

It is easily verified that this function is the solution to the likelihood equations

$$(Z_t'V^{-1}Z_t)\mu = Z_t'V^{-1}Y \tag{9.268}$$

as a function of ϕ_0 and ϕ_2.

Using the properties of Z_t, Z_b and the expression for NN' from (7.102), we may rewrite (9.267) as

$$\hat{\mu} = v\tilde{\mu}_1 + (1 - v)\tilde{\mu}_2 \tag{9.269}$$

where

$$\begin{aligned} \tilde{\mu}_1 &= (k/\lambda t)A'Y + \bar{y}J \\ \tilde{\mu}_2 &= (NZ_b'Y - \lambda t\bar{y}J)/(r - \lambda) \end{aligned} \tag{9.270}$$

and

$$v = \lambda tw/rk. \tag{9.271}$$

The motivation for this form of $\hat{\mu}$ is found in the fixed effects analysis from Section 7.3. In particular, if we write the mean vector in canonical form as in (9.251) and use the estimates of $\beta_0(\mu)$ and $\beta_1(\tau)$ from (7.105) and (7.106), then $\hat{\mu} = \hat{\beta}_0 + \hat{\beta}_1$. Thus, $\tilde{\mu}_1$ is based on the estimates of the analogous quantities from the fixed effects model. In the literature, $\tilde{\mu}_1$ is known as the *intrablock estimator* of μ. The second term, $\tilde{\mu}_2$, is known as the *interblock estimator* of μ. We see that (9.269), as determined from the likelihood equations, is a weighted average of these two estimators. In the classical treatment of the BIB model, we are first led to the intrablock estimator from the fixed effects analysis and then shown how we can recover interblock information on μ by combining the two estimators. From the likelihood analysis, we are led directly to this combined form of the estimator, and there is no reason to prefer the form (9.269) over (9.265).

The weight function w in (9.266) requires estimates of ϕ_0 and ϕ_2. Theorem 9.11 suggests two potential sets of estimators. The first is

$$\begin{aligned} \tilde{\phi}_0 &= s_0/f \\ \tilde{\lambda}_2 &= s_2/(b - t). \end{aligned} \tag{9.272}$$

We note that $\tilde{\phi}_0$ is the residual mean square from the fixed model in Section 7.3, and it is commonly referred to as the *intrablock estimate* of ϕ_0. From Theorem 9.11, we see that another set of estimates is given by

$$\begin{aligned} \phi_0^* &= p_0/(t - 1) \\ \lambda_2^* &= p_2/(t - 1). \end{aligned} \tag{9.273}$$

If we let $\tilde{\zeta}$ denote the vector of estimates in (9.272) and ζ^* the vector of estimates in (9.273), then a natural estimator for $\zeta' = (\phi_0, \lambda_2)$ is

$$\hat{\zeta} = \begin{bmatrix} \hat{\phi}_0 \\ \hat{\lambda}_2 \end{bmatrix} = \tilde{\zeta} + B(\zeta^* - \tilde{\zeta}) \tag{9.274}$$

where B is chosen in an optimal way—say, to minimize the variance of the estimates of ϕ_0 and λ_2.

Rather than pursue this approach, we shall turn to the modified likelihood equations. (Based on our experience with the balanced case, we prefer these to the usual likelihood equations.) In the notation of Chapter 8, we find

$$\begin{aligned}\rho_0 &= s_0/\phi_0^2 + [(r-\lambda)/k^2\delta^2][(r-\lambda)p_0 + \lambda t p_2]\\ \rho_2 &= s_2/\lambda_2^2 + (\lambda t/k^2\delta^2)[(r-\lambda)p_0 + \lambda t p_2]\\ \delta &= r\phi_0 + \lambda t\phi_2.\end{aligned} \tag{9.275}$$

Recalling that the modified likelihood equations are given by $E(\rho_i) \overset{s}{=} \rho_i$, we obtain, after some simplification, the equation

$$\Omega\zeta = \Omega\tilde{\zeta} + [(t-1)/k^2\delta^2]dd'(\zeta^* - \tilde{\zeta}) \tag{9.276}$$

where

$$d = \begin{bmatrix} r-\lambda \\ \lambda t \end{bmatrix}$$

$$\Omega = \begin{bmatrix} f/\phi_0^2 & 0 \\ 0 & (b-t)/\lambda_2^2 \end{bmatrix} + [(t-1)/k^2\delta^2]dd'. \tag{9.277}$$

Multiplying by the inverse of Ω yields

$$\zeta = \tilde{\zeta} + \begin{bmatrix} (r-\lambda)\phi_0^2/f \\ \lambda t\lambda_2^2/(b-t) \end{bmatrix} d'(\zeta^* - \tilde{\zeta})/c \tag{9.278}$$

with

$$c = k^2\delta^2/(t-1) + (r-\lambda)^2\phi_0^2/f + (\lambda t)^2\lambda_2^2/(b-t). \tag{9.279}$$

The MINQUE estimator arises from (9.278) by letting $\phi_0 = \lambda_2 = 1$. The REML estimator requires an iterative solution of these equations. It is of interest to note that, if we take the MINQUE estimator and make certain degree of freedom adjustments such as assuming $\lambda t = rk$, rather than $\lambda(t-1) = r(k-1)$, and $k(b-1) = t(r-1)$ as opposed to $kb = tr$, we obtain the Yates (1940) estimators

$$\begin{aligned}\hat{\phi}_0 &= \tilde{\phi}_0\\ \hat{\lambda}_2 &= [R(\beta|\mu, \tau) - (t-k)\tilde{\phi}_0]/t(r-1).\end{aligned} \tag{9.280}$$

Here, $R(\beta|\mu, \tau)$ is the column effect sum of squares from Table 7.18. This estimator arises by equating $R(\beta|\mu, \tau)$ to its expected value and solving for λ_2. Kutner (1971) noted that there may be substantial differences between these estimators and the iterative solution of (9.278).

Looking next at the problem of testing hypotheses, we consider the hypothesis

$$H_0: \mu_i = \bar{\mu}. \qquad i = 1, \ldots, t-1. \tag{9.281}$$

Consider first the modified likelihood ratio test described in Section 8.3.1, recognizing that we do not have closed form estimates for ϕ_0 and λ_2, the eigenvalues of V. Under the hypothesis constraint, (9.281), the REML estimates of ϕ_0 and λ_2 are

$$\begin{aligned} \phi_0^{**} &= q_0/b(k-1) \\ \lambda_2^{**} &= (q_2 - n\bar{y}^2)/(b-1). \end{aligned} \tag{9.282}$$

For convenience, we let

$$\begin{aligned} v_0 &= q_0 - s_0 \\ v_2 &= q_2 - s_2 - n\bar{y}^2. \end{aligned} \tag{9.283}$$

Using (9.272) for the unconstrained estimates, we obtain the approximate, modified, likelihood ratio statistic

$$\begin{aligned} \gamma^2 &= (\tilde{\phi}_0/\phi_0^{**})^{b(k-1)}(\tilde{\lambda}_2/\lambda_2^{**})^b \\ &= \text{constant } x(s_0/q_0)^{b(k-1)}[s_2/(s_2 + v_2)]^b. \end{aligned} \tag{9.284}$$

It can be shown the ratios s_0/q_0 and $s_2/(s_2 + v_2)$ are independent beta variables. Approximations to the distribution of products of beta variables are common in multivariate analysis (see Kshirsagar, 1972).

We shall not pursue this idea but merely suggest it as an exercise. Instead, we recognize that (9.284) suggests rejection of H_0 if v_0/s_0 and/or v_2/s_2 are large. Note that, under H_0,

$$\begin{aligned} F_1 &= [v_0/(t-1)]/(s_0/f) \sim F(t-1, f) \\ F_2 &= [v_2/(t-1)]/[s_2/(b-t)] \sim F(t-1, b-t) \end{aligned} \tag{9.285}$$

and F_1 and F_2 are independent. Either of these ratios could be used to test the hypothesis (9.281). To combine them, let U_i denote the c.d.f. transform of independent central F statistics with degrees of freedom as in (9.285). Then

$$Z = -2 \log U_1 U_2 \sim \chi^2(4). \tag{9.286}$$

Graybill (1961) suggests the rejection of (9.281) at level α if

$$z = -2 \log(u_1 u_2) > \chi^2(\alpha; 4) \tag{9.287}$$

where u_1 and u_2 are the significance levels for F_1 and F_2.

As a third procedure, we recall the suggestion made in Section 8.3.4, which leads us to consider the test statistic

$$F = (\hat{w}/\hat{\phi}_0)\hat{\mu}'[I - (1/t)JJ']\hat{\mu}/r(t-1) \tag{9.288}$$

which is approximately $F(t-1, f)$ under H_0 if we use (9.282) to estimate ϕ_0 and w. The relative merits of the three test statistics, (9.284), (9.287), and (9.288) have not been examined.

CHAPTER 9 EXERCISES

1. For each of the models described in Examples 8.1 through 8.7, do the following:
   (a) Express the covariance matrices (9.1) through (9.7) in the two forms given in Theorem 9.1. Identify the index sets T and T_0 and the parameters ϕ_t and λ_t.
   (b) Determine the relation between ϕ and λ as given in (9.34) and (9.36) and relate it to Theorem 9.3.
   (c) Determine the inverse of V.
   (d) Express V_t and A_t in the form of Definitions 9.2 and 9.3.
   (e) Verify the orthogonality properties given by (9.45).
   (f) Verify the ranks of A_t and A_μ given in Theorem 9.2, and verify that λ_t and λ_μ are the eigenvalues.
2. Spell out the details of the proof of Theorem 9.4.
3. For Examples 8.2, 8.4, and 8.7, show that the quadratic forms suggested by V^{-1} may be used to generate an estimate of ϕ.
4. For the three-way classification, random model in Example 9.9, develop tests of the hypotheses H_{123}: $\phi_{123} = 0$ and H_{12}: $\phi_{12} = 0$. Use the modified likelihood ratio test for this, and observe that the results are based on ratios of quadratic forms.
5. For Example 9.9, show that the modified likelihood ratio procedure does not lead to a simple expression for the test statistic under the hypothesis H_3: $\phi_3 = 0$.
6. For the random, two-fold nested model, Example 8.3, describe confidence intervals for ϕ_1, $\phi_1 + \phi_3$, and ϕ_3/ϕ_0 based on quadratic forms, ratios of quadratic forms, or approximations. Contrast these intervals with the simultaneous intervals described in Section 9.2.5. Illustrate with a numerical example.
7. Show that the mean and covariance structure in Examples 8.2, 8.6, and 8.7 satisfy the requirements of the special mixed model described in Section 9.3.1.
8. Develop the parameter estimates for Examples 8.2 and 8.6.
9. Develop the test statistic for the hypothesis H: $\mu_i = \mu_{i*}$ for the two-fold nested model in Example 8.2 and for the randomized block model in Example 8.6.
10. Develop the modified likelihood equations for the one-way classification model given by

$$E(Y) = \mu J$$

$$\mathrm{Var}(Y) = \phi_0 I + \phi_1 D_a(J_{n_i} J'_{n_i}).$$

    Identify a set of sufficient statistics for μ, ϕ_0, and ϕ_1. Is this a complete set?
11. For the mixed BIB model described in Section 9.4.2, verify the sufficiency of the statistics identified in Theorem 9.10 and the distributional properties of the statistics identified in Theorem 9.11.

CHAPTER TEN
VARIANCE COMPONENT MODELS

10.1. DEFINITION AND EXAMPLES

In Chapters 8 and 9, we studied models in which the data vector, Y, was assumed to be normally distributed with mean vector

$$E(Y) = X\theta \tag{10.1}$$

and covariance matrix

$$\text{Var}(Y) = V = \sum_{t=0}^{T} \phi_t V_t. \tag{10.2}$$

Specific examples were given to illustrate how such covariance matrices arise, and the interpretation of the parameters, ϕ_t, was discussed. We now describe another way in which such models may arise. In this case, we shall see that the parameters, ϕ_t, are naturally restricted to be nonnegative by the nature of the model formulation. These parameters in such models are generally referred to as *variance components*. We have reserved this term for the model formulation to be discussed here, but it could well be applied to the less restrictive models of Chapters 8 and 9 since the parameters are components of the covariance matrix, V.

EXAMPLE 10.1
To motivate the model formulation, we recall the canonical model for the unconstrained two-way classification model from Section 6.2.3. Recall that in terms of the new parameters, we can write the model as

$$y_{ijk} = \mu + \alpha_i + \beta_j + (\alpha\beta)_{ij} + e_{ijk} \qquad \begin{aligned} i &= 1, \ldots, a_1 \\ j &= 1, \ldots, a_2 \\ k &= 0, 1, \ldots, n_{ij}. \end{aligned} \tag{10.3}$$

In the fixed effects model, the parameters, α_i, β_j, and $(\alpha\beta)_{ij}$ are linear functions of the cell means that satisfy certain linear relations depending on the choice of canonical form. These parameters provide a measure of the row, column, and interaction effects.

We now consider a model of the form (10.3) in which μ is an unknown parameter but α_i, β_j, and $(\alpha\beta)_{ij}$ are assumed to be random variables. In particular, we shall assume the following properties:

$$\begin{array}{ll} \alpha_i \sim N(0, \phi_1) & \beta_j \sim N(0, \phi_2) \\ (\alpha\beta)_{ij} \sim N(0, \phi_{12}) & e_{ijk} \sim N(0, \phi_0). \end{array} \tag{10.4}$$

It is assumed that the variables in (10.4) are mutually independent. We emphasize that these variables *do not* satisfy the linear relations of the fixed effects canonical models.

With these assumptions, the random model associated with (10.3) may be stated by specifying the mean and covariance structure as follows:

$$\begin{aligned} E(y_{ijk}) &= \mu \\ \text{Cov}(y_{ijk}, y_{i^*j^*k^*}) &= \phi_0 + \phi_1 + \phi_2 + \phi_{12} && i = i^*, j = j^*, k = k^* \\ &= \phi_1 + \phi_2 + \phi_{12} && i = i^*, j = j^*, k \neq k^* \\ &= \phi_1 && i = i^*, j \neq j^* \\ &= \phi_2 && i \neq i^*, j = j^* \\ &= 0 && i \neq i^*, j \neq j^*. \end{aligned} \tag{10.5}$$

Reference to Example 8.4 shows that this is precisely the mean and covariance structure associated with that model.

Alternatively, we may express the model (10.3) in matrix form as

$$Y = \mu J_N + U_1\alpha + U_2\beta + U_{12}(\alpha\beta) + e. \tag{10.6}$$

Here, $U_t = WX_t^0$ with W and X_t^0 as defined in (9.242) and (9.137). In this notation, the assumptions are

$$\begin{array}{ll} \alpha \sim N(0, \phi_1 I_{a_1}) & \beta \sim N(0, \phi_2 I_{a_2}) \\ (\alpha\beta) \sim N(0, \phi_{12} I_{a_1 a_2}) & e \sim N(0, \phi_0 I_N). \end{array} \tag{10.7}$$

All variables in (10.7) are assumed to be independent. The covariance structure is then specified by

$$\text{Var}(Y) = \phi_0 I + \phi_1 U_1 U_1' + \phi_2 U_2 U_2' + \phi_{12} U_{12} U_{12}'. \tag{10.8}$$

Note that this is the covariance structure associated with Example 8.4 as given in (9.4) if we are in the unbalanced case defined in Section 9.5.

■

The simple idea behind the classical variance component models is to write the fixed effects model either in algebraic form (10.3) or matrix form (10.6) and then designate some of the parameters to be random variables with appropriate distributional properties. We shall give a more precise definition later in this section. First, we spell out the seven examples discussed in Section 8.1 and again in Section 9.1 in this format. The source of the names associated with those models will now be clear. Note that the terms *fixed, random,* and *mixed models* are not really precise, since all models have at least one fixed parameter, μ, and one random factor, e. We have retained this terminology, however, for reference to the literature.

EXAMPLE 8.1 The Random, One-Way Classification Model

$$\begin{aligned} y_{ij} &= \mu + \alpha_i + e_{ij} \\ E(y_{ij}) &= \mu \\ \alpha_i &\sim N(0, \phi_1) \\ e_{ij} &\sim N(0, \phi_0) \end{aligned} \tag{10.9}$$

where $\alpha_i, i = 1, \ldots, a; e_{ij}, i = 1, \ldots, a, j = 1, \ldots, n$ are mutually independent.

EXAMPLE 8.2 The Mixed, Two-Fold Nested Model

$$\begin{aligned} y_{ijk} &= \mu + \alpha_i + \beta_{ij} + e_{ijk} \\ E(y_{ijk}) &= \mu_i = \mu + \alpha_i \\ \beta_{ij} &\sim N(0, \phi_{12}) \\ e_{ijk} &\sim N(0, \phi_0) \end{aligned} \tag{10.10}$$

where β_{ij} and e_{ijk}, $i = 1, \ldots, a, j = 1, \ldots, b, k = 1, \ldots, n$, are mutually independent. We may assume any canonical form for μ and α_i—for example, $\alpha. = 0$ or $\alpha_a = 0$.

EXAMPLE 8.3 The Random, Two-Fold Nested Model
Same as Example 8.2, except that

$$\begin{aligned} E(y_{ijk}) &= \mu \\ \alpha_i &\sim N(0, \phi_1) \end{aligned} \tag{10.11}$$

where all random variables are mutually independent.

EXAMPLE 8.4 The Random, Two-Way Classification Model
See Example 10.1.

EXAMPLE 8.5 The Mixed, Two-Way Classification Model

$$\begin{aligned} y_{ijk} &= \mu + \alpha_i + \beta_j + (\alpha\beta)_{ij} + e_{ijk} \\ E(y_{ijk}) &= \mu_i = \mu + \alpha_i \\ \beta_j &\sim N(0, \phi_2) \\ (\alpha\beta)_{ij} &\sim N(0, \phi_{12}) \\ e_{ijk} &\sim N(0, \phi_0) \end{aligned} \tag{10.12}$$

where all random variables are assumed independent and any canonical form can be assumed for μ and α_i.

EXAMPLE 8.6 The Randomized Block Model

$$\begin{aligned} y_{ij} &= \mu + \alpha_i + \beta_j + e_{ijk} \\ E(y_{ij}) &= \mu_i = \mu + \alpha_i \\ \beta_j &\sim N(0, \phi_2) \\ e_{ijk} &\sim N(0, \phi_0) \end{aligned} \tag{10.13}$$

where all random variables are independent.

EXAMPLE 8.7 The Split-Plot Model

$$\begin{aligned} y_{ijk} &= \mu + \alpha_i + \beta_j + (\alpha\beta)_{ij} + \gamma_k + (\beta\gamma)_{jk} + e_{ijk} \\ E(y_{ijk}) &= \mu_{jk} = \beta_j + \gamma_k + (\beta\gamma)_{jk} \\ \alpha_i &\sim N(0, \phi_1) \\ (\alpha\beta)_{ij} &\sim N(0, \phi_{12}) \\ e_{ijk} &\sim N(0, \phi_0). \end{aligned} \tag{10.14}$$

Again, any canonical form may be assumed for the fixed effects. For example, we might assume $\beta. = \gamma. = (\beta\gamma)_j. = (\beta\gamma)._k = 0$. Also, all random variables are assumed independent.

■

The relation between the variance component models and those discussed in Chapters 8 and 9 is now clear. It is easily verified that the covariance structure implied by the current model formulation is identical to that described earlier. The essential distinction is that the parameters, ϕ_t, in this section are variances of random variables and, hence, are inherently positive. The formulation in Chapter 8 made no such requirement. The basis for the model in Chapter 8 was simply that certain observations were assumed to be correlated without any restriction on the sign of the correlation. The reader is urged to reflect on the basic differences in these model formulations.

It appears that the less restrictive model may be applicable in some cases, and we shall return to a discussion of this topic in Section 10.3. For now, we shall accept the classical variance component model to relate to the existing literature and methods of analysis.

A general definition of the variance component model is now given, using the notation developed in Sections 9.3.1 and 9.5.

DEFINITION 10.1

The variance component model is written as

$$\begin{aligned} Y &= X\beta + \sum_{t \in T - T_1} U_t e_t + e_0 \\ &= X\beta + Ue + e_0. \end{aligned} \tag{10.15}$$

Here, $X = (J_N | X_t, t \in T_1)$ and $\beta' = (\beta_0, \beta_t', t \in T_1)$ are as defined in Sections 9.3.1 and 9.5.1. These fixed effects are generally expressed in canonical form but could be written in terms of cell means. Here we use β rather than θ to denote the vector of fixed effects parameters to conform to standard notation. The vectors e_0 and e_t, $t \in T - T_1$ are independently distributed as

$$e_t \sim N(0, \phi_t I_t) \qquad t = 0, t \in T - T_1 \tag{10.16}$$

and we let e denote the concatenated column vector

$$e = (e_t, t \in T - T_1). \tag{10.17}$$

Further, we let $U = (U_t, t \in T - T_1)$ denote the matrix of U_t written in partitioned form to conform with e where

$$U_t = WX_t^0 \qquad t \in T - T_1 \tag{10.18}$$

with W defined as in (9.242).

The index sets T and T_1 are as defined in Section 9.3.1. For $t \in T_1$, X_t has dimension $N \times r_t$, and we shall assume the connected case in which these matrices have full column rank. As before, we shall assume that, if $t \in T_1$, then any index contained in t is also in T_1. For $t \in T - T_1$, X_t^0 is defined in (9.137), and U_t has dimension $N \times \Pi_{i \epsilon t} a_i$. Of course, some of the β_t and/or ϕ_t may be zero.

■

We note that an alternative statement of Definition 10.1 is given by specifying that

$$Y \sim N(X\beta, V)$$

where

$$V = \sum_{t \in T - T_1} \phi_t U_t U_t' + \phi_0 I. \tag{10.19}$$

This covariance matrix is equivalent to (9.148) with $V_t = U_t U_t'$. (In this case, the restriction $\phi_t \geq 0$ is implicit in the definition.)

In the methods to be described in this chapter, we shall have occasion to discuss the fixed effects analog of this model. The idea is that we shall do certain computations as if the e_t were unknown parameters. In this case we shall have to view the model as if it were in canonical form and write

$$U_t e_t = U_t P_t' \beta_t = X_t \beta_t \qquad t \in T - T_1. \tag{10.20}$$

DEFINITION 10.2
The fixed effects analog of the variance component model is written as

$$\begin{aligned} Y &= X\beta + \sum_{t \in T - T_1} X_t \beta_t + e_0 \\ &= \beta_0 J_N + \sum_{t \in T} X_t \beta_t + e_0 . \end{aligned} \tag{10.21}$$

■

This model will be used to develop quadratic forms for use in estimating the variance components.

10.2. ANALYSIS OF VARIANCE ESTIMATORS OF VARIANCE COMPONENTS

Definition 10.2 suggests a natural source of quadratic forms to be used to develop moment estimators for the variance components; that is, we shall generate an AOM table for the fixed effects analog and then equate the mean squares from this table with their expected values computed using the variance component model of Definition 10.1. We shall use the term ANOVA *estimators* for any estimators computed in this way, recognizing that this definition is not unique. In the case of balanced data we shall see that the AOM for testing main effects and interaction is optimal for this purpose. With unbalanced data, we have seen in Chapters 6 and 7 that there is not a natural choice for the AOM table. At this time there does not appear to be an optimal choice. (Note that the term ANOVA *estimator* is used by Searle, 1971a, to refer to a particular estimator of this type. We prefer the general definition.) The following example will illustrate the ideas.

EXAMPLE 10.2
Consider the balanced, two-way classification random model from Example 10.1; that is, $k = 1, \ldots, n$ for all i, j. The AOM for the fixed effects model was given as Table 6.5. The essential facts from that table are repeated in Table 10.1 with an added column showing the expected values of the quadratic forms under the random model. Note that now we have expressed the sums of squares in terms of the general notation of Theorem 9.1. It is easily established that these are in agreement with those given for Table 6.5. We recognize that the sums of squares in Table 10.1 are precisely the sufficient statistics identified in Theorem 9.4.

TABLE 10.1 Analysis of variance table for the two-way classification, random model.

Description	*df*	*SS*	*MS*	*EMS*
Rows	$r_1 = a_1 - 1$	$Y'A_1Y$	q_1/r_1	$\lambda_1 = \phi_0 + n\phi_{12} + na_2\phi_1$
Columns	$r_2 = a_2 - 1$	$Y'A_2Y$	q_2/r_2	$\lambda_2 = \phi_0 + n\phi_{12} + na_1\phi_2$
Interaction	$r_{12} = (a_1 - 1)(a_2 - 1)$	$Y'A_{12}Y$	q_{12}/r_{12}	$\lambda_{12} = \phi_0 + n\phi_{12}$
Residual	$r_0 = a_1a_2(n - 1)$	$Y'A_0Y$	q_0/r_0	ϕ_0

The mean squares in Table 10.1 are the UMV estimates of the eigenvalues, λ_t, identified in Theorem 9.5. These eigenvalues, or expected mean squares, are related to the ϕ_t by the P-matrix identified in Theorem 9.1. Reference to Example 9.4 establishes the equivalence of the UMV, REML, and ANOVA estimators for this example.

We note that the term *interaction* is not really appropriate for the random model in that it merely reflects the additional correlation between observations in the same cell. If we have only one observation per cell, this correlation cannot be estimated, but we can still estimate ϕ_1 and ϕ_2. This should be contrasted with the fixed effects model in which the lack of multiple observations on the treatment combinations forced us to make a strong assumption on the relation between the cell means.

■

The procedure implied by Example 10.2 generalizes to any balanced model; that is, we consider the fixed effects analog from Definition 10.2 and adopt the second canonical form as defined in Section 9.3.1. The AOM table is developed for the hypotheses on the individual effects (main effects and interactions). The mean squares associated with the random effects from Definition 10.1 are seen to be the complete, sufficient statistics for ϕ and are the UMV estimators of the eigenvalues, λ_t. The estimates of ϕ_t are computed in the usual way. The mean squares associated with the fixed effects are the numerators for testing the hypotheses, $\beta_t = 0$. The appropriate denominator is suggested by the linear combination of the variance components in the expected mean square as noted in Section 9.3.3. These results are established in the following theorem.

THEOREM 10.1
For the special case of the balanced mixed model the ANOVA estimators of β_t and ϕ_t are identical to the UMV estimators identified in Theorem 9.8. The ANOVA tests of the hypothesis $\beta_t = 0$ are identical to the modified likelihood ratio test noted in Section 9.3.3.

Proof: From Definition 10.2 we write the fixed effects analog as

$$Y = \beta_0 J_N + \sum_{t \in T} X_t \beta_t + e_0 \tag{10.22}$$

where we are assuming the second canonical form as in Section 9.3.1. It follows from the notation developed in that section that

$$X_t'X_m = 0 \qquad t \neq m$$
$$X_t(X_t'X_t)^{-1}X_t' = A_t \qquad t \in T \tag{10.23}$$

where A_t is given in Definition 9.3. It follows directly from the general results of Chapter 3 that the numerator sum of squares for testing $\beta_t = 0$, $t \in T$, in the fixed effects model is

$$\begin{aligned} N_t &= Y'X_t(X_t'X_t)^{-1}X_t'Y \\ &= Y'A_tY. \end{aligned} \tag{10.24}$$

The quadratic forms in (10.24) are precisely those noted in Theorem 9.8 for the balanced mixed model. Application of Theorem 9.8 establishes the results of the present theorem.

■

To illustrate this general result, we consider the two-way classification, mixed model for which the estimates and test statistics were developed in Examples 9.11 and 9.13.

EXAMPLE 10.3
Consider the balanced mixed model defined by (10.12). Referring to Table 10.1 for the fixed effects AOM table, we need only express the expected mean squares that arise under the mixed model. The results are determined from Theorem 9.8 and are given in Table 10.2.

TABLE 10.2 Analysis of variance table for the two-way classification, mixed model.

Description	*MS*	*EMS*
Row	q_1/r_1	$\phi_0 + n\phi_{12} + na_2\Sigma\alpha_i^2/r_1$
Column	q_2/r_2	$\phi_0 + n\phi_{12} + na_1\phi_2$
Interaction	q_{12}/r_{12}	$\phi_0 + n\phi_{12}$
Residual	q_0/r_0	ϕ_0

The ANOVA estimates of ϕ_t and μ_i agree with those obtained in Example 9.11. The test statistic for the hypothesis H: $\alpha_i = \mu_i - \bar{\mu}. = 0$, $i = 1, \ldots, a_1 - 1$, as suggested by the ANOVA, is

$$F = (q_1/r_1)/(q_{12}/r_{12}). \tag{10.25}$$

This result agrees with our conclusion in Example 9.13. As in that example, we note that the suggested test statistic for the hypothesis H: $\phi_2 = 0$ is

$$F_2 = (q_2/r_2)/(q_{12}/r_{12}). \tag{10.26}$$

As noted earlier, there is some disagreement on this last result. This point is resolved in Section 10.4.

■

There is no assurance that all of the variance components will have nonnegative estimates, which violates the assumptions of this model. We address this problem in Section 10.3.

The fact that the ANOVA estimators are optimal (apart from the requirement $\phi_t \geq 0$) for the balanced case suggests that the AOM table may be a good place to look for quadratic forms for moment estimators in the unbalanced case. However, there are two major drawbacks. First, as noted in Chapters 6 and 7, there is not a uniformly accepted AOM table in the unbalanced case. Second, the problem of determining the expected values of the mean squares is not trivial. Some of the methods associated with this approach to the unbalanced data problem are discussed in Section 10.5.

10.3. THE PROBLEM OF NEGATIVE ESTIMATES OF VARIANCE COMPONENTS

We have noted on several occasions that the estimated covariance matrix will be positive definite in the balanced case but that there is no guarantee that the estimates of the variance components will be positive. In the formulation of Chapters 8 and 9, we noted that the variance components, with the exception of ϕ_0, could logically be negative.

In this chapter, the model, as described in Definition 10.1, specifies that $\phi_t \geq 0$. It appears that this may often be a consequence of this description of the model as opposed to a serious *a priori* consideration. Since negative estimates do arise in practice, it is natural to ask (a) whether they are a consequence of highly variable data, (b) whether they occurred because of spurious or incorrect observations, (c) whether it is possible that a negative correlation is actually feasible, or (d) whether the model is incorrect.

In this section, we make a simple observation about the estimator for certain variance components that allows us to develop a diagnostic procedure for evaluating the estimate, whether or not it is negative. We shall present this idea in terms of a simple example and numerical illustration (see Hocking, 1983a).

EXAMPLE 10.4

We consider the randomized block model, which, in the notation of this chapter, is written as

$$y_{ij} = \mu_i + \beta_j + e_{ij} \qquad \begin{array}{l} i = 1, \ldots, t \\ j = 1, \ldots, b \end{array} \tag{10.27}$$

where

$$\beta_j \sim N(0, \phi_2) \qquad e_{ij} \sim N(0, \phi_0). \tag{10.28}$$

Alternatively, the model might be specified as

$$\begin{aligned} E(y_{ij}) &= \mu_i \\ \text{cov}(y_{ij}, y_{i^*j^*}) &= \phi_0 + \phi_2 \qquad & i = i^*, j = j^* \\ &= \phi_2 & i \neq i^*, j = j^* \\ &= 0 & j \neq j^*. \end{aligned} \tag{10.29}$$

In this second formulation, it makes sense to allow ϕ_2 to be negative. Since ϕ_2 represents the covariance between observations within the same block, it could well be that there is a competition within a block for a certain resource not related to the treatments. In this case, the units might respond differently, quite apart from the treatment effect. In the case of positive covariance, we would expect all observations in a block to lie on the same side of their mean value.

Using the results of Chapter 9, we see that the estimates of the variance components are given by

$$\hat{\phi}_0 = q_0/(t - 1)(b - 1) \tag{10.30}$$

and

$$\hat{\phi}_2 = (\hat{\lambda}_2 - \hat{\phi}_0)/t \tag{10.31}$$

where

$$\hat{\lambda}_2 = q_2/(b - 1). \tag{10.32}$$

Here,

$$\begin{aligned} q_0 &= Y'A_0Y = Y'[I - (1/b)V_1 - (1/t)V_2 + (1/tb)JJ']Y \\ &= \sum_i \sum_j (y_{ij} - \bar{y}_{i\cdot} - \bar{y}_{\cdot j} + \bar{y}_{\cdot\cdot})^2 \end{aligned}$$

and (10.33)

$$\begin{aligned} q_2 &= Y'A_2Y = Y'[(1/t)V_2 - (1/tb)JJ']Y \\ &= \sum_i \sum_j (\bar{y}_{\cdot j} - \bar{y}_{\cdot\cdot})^2. \end{aligned}$$

After some simplifying algebra, we may write

$$\hat{\phi}_2 = \sum_i \sum_{i^*(i \neq i^*)} C_{ii^*}/t(t - 1) \tag{10.34}$$

where

$$C_{ii^*} = \sum_j (y_{ij} - \bar{y}_{i\cdot})(y_{i^*j} - \bar{y}_{i^*\cdot})/(b-1). \tag{10.35}$$

We note that C_{ii^*} is just the sample covariance computed using the data for treatments i and i^*. Since our model says that ϕ_2 is the covariance between elements in the same block, C_{ii^*} is an estimate of this covariance based on these two rows of the two-way table of data. Returning to the estimate (10.34) of ϕ_2, we see that it is just the average of the estimates of ϕ_2 based on all pairs of treatments. It follows that, if there is sufficient data and it is good in the sense that there are no spurious observations, then, if ϕ_2 is positive, we would expect primarily positive values of C_{ii^*}. On the other hand, if there is competition within a block, suggesting $\phi_2 < 0$, we would expect negative values of C_{ii^*} to dominate.

An obvious diagnostic procedure for evaluating both the model and the data is to examine the values of C_{ii^*} and to inspect the two-dimensional plots of the data from each pair of treatments.

To illustrate, we use an example from Snedecor and Cochran (1967, p. 300). The data are shown in Table 10.3.

TABLE 10.3 Data for Example 10.4.

		Blocks					
		1	2	3	4	5	$\bar{y}_{i\cdot}$
	1	8	10	12	13	11	10.8
	2	2	6	7	11	5	6.2
Treatments	3	4	10	9	8	10	8.2
	4	3	5	9	10	6	6.6
	5	9	7	5	5	3	5.8

To illustrate a situation leading to a negative estimate of ϕ_2, we change the observation y_{51} from 9 to 13. The resulting ANOVA is shown in Table 10.4.

TABLE 10.4 ANOVA for Example 10.4 with $y_{51} = 13$.

Description	*df*	*MS*	*EMS*
Treatments	$t - 1 = 4$	18.16	$\phi_0 + 5\Sigma(\mu_i - \bar{\mu}_\cdot)^2/4$
Blocks	$b - 1 = 4$	8.46	$\phi_0 + 5\phi_2$
Residual	$(t-1)(b-1) = 16$	8.81	ϕ_0

The estimates of ϕ_0 and ϕ_2, obtained from Table 10.4, are

$$\begin{aligned}\hat{\phi}_0 &= 8.81\\ \hat{\phi}_2 &= -0.07.\end{aligned} \tag{10.36}$$

Various suggestions have been made for altering the estimation procedure to remove the negative estimate. Thompson and Moore (1963) suggest a pooling procedure. In this case, their procedure sets $\phi_2 = 0$ and pools the block and residual sums of squares to yield the estimates

$$\begin{aligned}\hat{\phi}_0 &= 8.74\\ \hat{\phi}_2 &= 0.\end{aligned} \tag{10.37}$$

Others such as Hartley and Rao (1967) suggest a constrained maximum likelihood procedure, forcing ϕ_2 to be nonnegative. In this example, that procedure yields the estimates

$$\begin{aligned}\hat{\phi}_0 &= 6.99\\ \hat{\phi}_2 &= 0.\end{aligned} \tag{10.38}$$

Our approach is to attempt to locate the cause of the negative estimate, using the form of the estimator given in (10.34).

In Table 10.5 we show the covariance matrix—that is, the matrix of C_{ii^*}, $i \neq i^*$—for this modified data. Recalling that the average of the elements of this matrix gives the estimate of ϕ_2, we see that the apparent cause of the negative estimate is the data for the fifth treatment; that is, all covariances involving row five are negative. This suggests a critical examination of the data in that row or perhaps a

TABLE 10.5 Covariance matrix for Example 10.4 with $y_{51} = 13$.

		Row			
		2	3	4	5
Row	1	5.8	2.8	5.4	−6.1
	2		3.7	8.6	−7.9
	3			3.1	−8.4
	4				−7.7

more critical examination of the assumption that the observations in a block have covariance ϕ_2 for all blocks.

To see whether the problem lies with the data, we examine the two-dimensional plots of the data in row five against the data in each of the other rows. In Figure 10.1 (a–d), we show the data from rows one through four plotted against the data from row five. It is clear that the observation y_{51} is primarily responsible for these coefficients being negative.

Having located the problem, we next consider what action might be taken. Perhaps, as in this case, the observation was erroneously recorded and a simple check of the records would resolve the difficulty. In other cases, we may be led to question the model. For example, is it reasonable to assume that the correlation between elements in a block is constant for all blocks or for all pairs of treatments within a block? Whatever the cause, it seems clear that simply reporting an estimate, $\phi_2 = 0$, is not the proper action.

Returning to the original data set shown in Table 10.3, we obtain the ANOVA given by Snedecor and Cochran (1967), shown in Table 10.6.

TABLE 10.6 ANOVA for Example 10.4.

Description	*df*	*MS*
Treatments	4	20.96
Blocks	4	12.46
Residual	16	5.41

The estimates of ϕ_0 and ϕ_2 obtained from this table are

$$\begin{aligned} \hat{\phi}_0 &= 5.41 \\ \hat{\phi}_2 &= 1.41 \end{aligned} \qquad (10.39)$$

suggesting a positive correlation between observations within a block. This seems reasonable in the agricultural setting of the Snedecor-Cochran example. However, it is natural to ask whether our diagnostic procedure raises any questions about the reliability of this estimate.

In Table 10.7, we show the covariance matrix for this data. We again note that there is an apparent difficulty in row five. Reference to Figure 10.1 with $y_{51} = 9$, rather than 13, reveals that the same basic problem exists; that is, observation y_{51} is having a dramatic effect in reducing the estimate of ϕ_2.

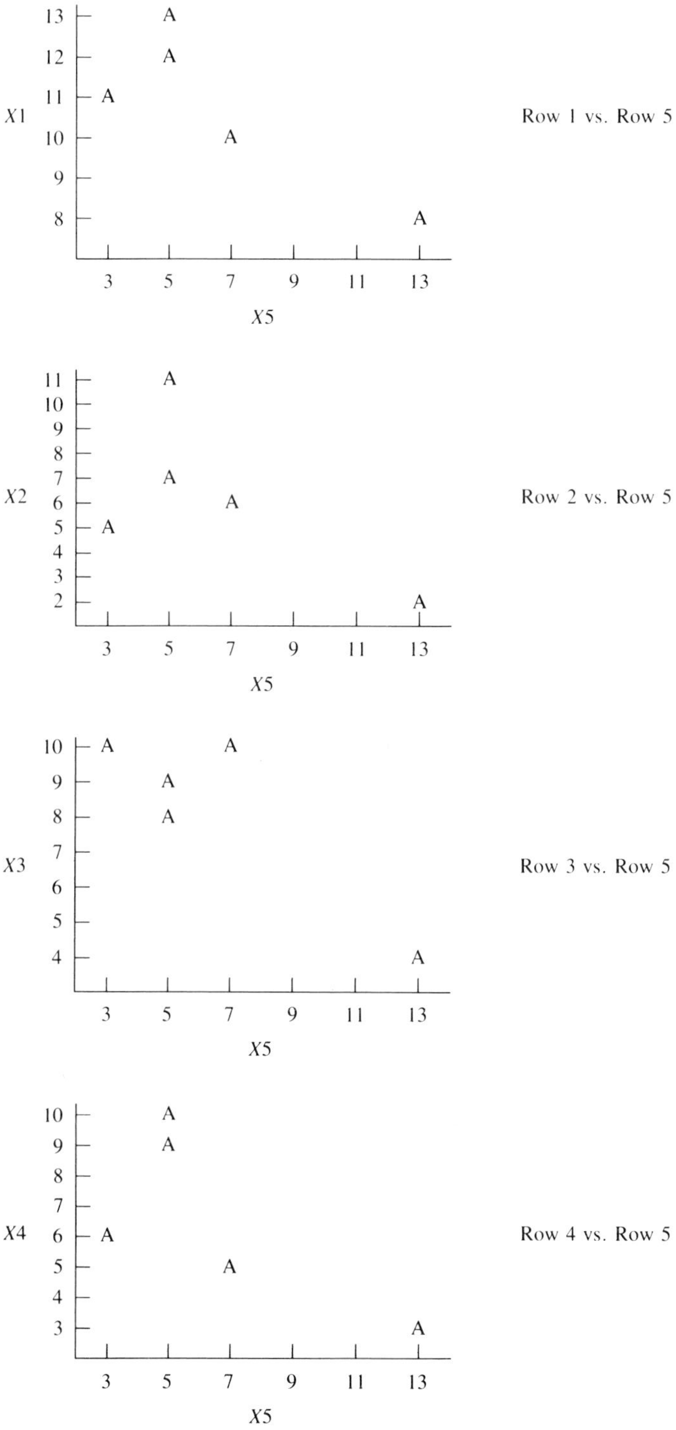

FIGURE 10.1
Diagnostic plots for variance component estimation

TABLE 10.7 Covariance matrix for Example 10.4.

		Row			
		2	3	4	5
Row	1	5.8	2.8	5.4	−3.3
	2		3.7	8.6	−3.7
	3			3.1	−4.2
	4				−4.1

Again, we can only speculate as to the cause of the problem. For example, it may be that there was a recording error and that the proper value should have been $y_{51} = 5$. If so, then the estimates would be

$$\begin{aligned}\hat{\phi}_0 &= 3.29\\ \hat{\phi}_2 &= 2.89.\end{aligned} \tag{10.40}$$

Several points should be clear from this example. First, the estimates of the variance components are extremely sensitive to the single observation y_{51}. (The same point could be made with certain other observations in this example, such as y_{31}.) Further, when negative estimates are noted, artificially constraining them to be nonnegative seems to be inappropriate, at least as an initial solution. Examination of the data and the model assumptions using the diagnostic procedure described here may well reveal the cause of the problem. In general, even if the estimates are positive, the diagnostic procedure may reveal information that casts doubt on the reliability of the estimate. Reviewing the data and the model assumptions may provide much better estimates.

■

We have provided an expression for the block variance component in the randomized block model, but the basic ideas described here may be extended (see Hocking, 1983a). For the main effect variance components in multifactor models, the expressions for the estimates are similar to (10.34) with covariances computed in terms of sample cell means. For interaction and nested components, the expressions are somewhat more complex, but the basic ideas remain the same and valuable information may be obtained with little effort by utilizing these procedures.

10.4. RESOLUTION OF A CONTROVERSY ON THE TWO-WAY CLASSIFICATION MIXED MODEL

In Example 9.13 and again in Example 10.3, we alluded to a difference in opinion on the analysis of the two-way classification mixed model. This difference is best exemplified by illustrating an ANOVA table that is often given for this model. Table

10.8 is found, for example, in Graybill (1961, p. 398), but is also given in many other references. The notation is simplified for comparison with our notation.

TABLE 10.8 An alternative ANOVA table for the two-way classification mixed model.

Description	*MS*	*EMS*
Row	q_1/r_1	$\zeta_0 + n\zeta_{12} + na_2\Sigma(\mu_i - \bar{\mu}.)^2/r_1$
Column	q_2/r_2	$\zeta_0 + na_1\zeta_2$
Interaction	q_{12}/r_{12}	$\zeta_0 + n\zeta_{12}$
Residual	q_0/r_0	ζ_0

In Table 10.8, we have used ζ_t to denote the variance components for reasons that will be clear later. For now, we recognize that the difference between Tables 10.2 and 10.8 lies in the expected mean square for columns. Since the expected mean square in Table 10.8 does not contain the term in ζ_{12}, it is clear that the estimate of ζ_2 will differ from the estimate of ϕ_2 from Table 10.2. For that same reason, the test of the hypothesis H: $\zeta_2 = 0$ based on Table 10.8 suggests the ratio

$$F = (q_2/r_2)(q_0/r_0) \tag{10.41}$$

which disagrees with the test statistic for H: $\phi_2 = 0$ given by (10.26).

A discussion of this difference is found in Hocking (1973) and Searle (1971a, p. 400). We shall provide a simple summary of the essential ideas and identify the source of the difference.

Since the mean squares in the two tables are identical, it is obvious that the difference must be a consequence of a difference in the models. We shall see that it is really just a difference in the definition of parameters. To identify this difference, we use the model as given in Graybill (1961), written, with slight notational changes, as

$$\begin{aligned}
y_{ijk} &= \mu_i + b_j + (ab)_{ij} + e_{ijk} \\
b_j &\sim N(0, \zeta_2) \\
(ab)_{ij} &\sim N\{0, [(a_1 - 1)/a_1]\zeta_{12}\} \\
e_{ijk} &\sim N(0, \zeta_0) \\
\mathrm{cov}[(ab)_{ij}, (ab)_{i*j*}] &= -\zeta_{12}/a_1 \qquad i \neq i^*, j = j^* \\
&= 0 \qquad i = i^*, j \neq j^*
\end{aligned} \tag{10.42}$$

with all other covariances being zero. Comparing this model with (10.12) reveals that the apparent difference is in the terms b_j, β_j, $(ab)_{ij}$, and $(\alpha\beta)_{ij}$. Searle (1971a) notes the relation between these terms.

A clearer relation between the two models is given by spelling out their covariance structures. From Example 8.5, or directly from (10.12), we obtain

$$\begin{aligned} \text{cov}(y_{ijk}, y_{i^*j^*k^*}) &= \phi_0 + \phi_2 + \phi_{12} && i = i^*, j = j^*, k = k^* \\ &= \phi_2 + \phi_{12} && i = i^*, j = j^*, k \neq k^* \\ &= \phi_2 && i \neq i^*, j = j^* \\ &= 0 && j \neq j^*. \end{aligned} \tag{10.43}$$

From (10.42) we obtain

$$\begin{aligned} \text{cov}(y_{ijk}, y_{i^*j^*k^*}) &= \zeta_0 + \zeta_2 + (a_1 - 1)\zeta_{12}/a_1 && i = i^*, j = j^*, k = k^* \\ &= \zeta_2 + (a_1 - 1)\zeta_{12}/a_1 && i = i^*, j = j^*, k \neq k^* \\ &= \zeta_2 - \zeta_{12}/a_1 && i \neq i^*, j = j^* \\ &= 0 && j \neq j^*. \end{aligned} \tag{10.44}$$

A comparison of (10.43) and (10.44) shows that the two models are identical in the sense that they both allow for observations in the same column to be correlated and for observations in the same cell to have a different correlation. The only difference is in the choice of parameters to describe the covariances. The simple relations are

$$\begin{aligned} \phi_0 &= \zeta_0 \\ \phi_2 &= \zeta_2 - \zeta_{12}/a_1 \\ \phi_{12} &= \zeta_{12}. \end{aligned} \tag{10.45}$$

Returning to Table 10.8, we see that, with this relation, the expected mean square for the column effect is

$$\begin{aligned} \zeta_0 + na_1\zeta_2 &= \phi_0 + na_1(\phi_2 + \phi_{12}/a_1) \\ &= \phi_0 + n\phi_{12} + na_1\phi_2 \end{aligned} \tag{10.46}$$

which is identical to that found in Table 10.2, and the controversy is resolved.

Apart from explaining the relation between these two, apparently different, model formulations, this exercise emphasizes the advantage of formulating the model by specifying the mean vector and covariance structure as opposed to basing the definition on an interpretation of the fixed effects model as in Definition 10.1.

The covariance structure (10.42), as described by Graybill (1961), seems quite unusual, and it is natural to ask how it arose. The source is apparently found in Scheffé (1959). He describes a very general covariance structure in terms of the symmetric matrix

$$\Sigma = (\sigma_{ii^*}, i, i^* = 1, \ldots, a_1) \tag{10.47}$$

obtaining

$$\begin{aligned}
\operatorname{cov}(y_{ijk}, y_{i^*j^*k^*}) &= \sigma_{ii} + \phi_0 && i = i^*,\ j = j^*,\ k = k^* \\
&= \sigma_{ii^*} && j = j^*,\ k \neq k^* \\
&= 0 && j \neq j^*.
\end{aligned} \tag{10.48}$$

Scheffé focuses attention on two functions of the matrix, Σ—namely,

$$\begin{aligned}
\delta_2 &= \bar{\sigma}.. \\
\delta_{12} &= \sum_i (\sigma_{ii} - \bar{\sigma}..)/(a_1 - 1)
\end{aligned} \tag{10.49}$$

which he calls variance components. He then suggests that a possible structure for Σ is

$$\sum = pI + qJJ' \tag{10.50}$$

where $p > 0$ and $p + a_1 q > 0$ to assure positive definiteness. In this case, we see that

$$\begin{aligned}
\sigma_{ii} &= p + q & \sigma_{ii^*} &= q \\
\delta_2 &= q + p/a_1 & \delta_{12} &= p.
\end{aligned} \tag{10.51}$$

In this special case, the Scheffé covariance structure (10.48) is given by

$$\begin{aligned}
\operatorname{cov}(y_{ijk}, y_{i^*j^*k^*}) &= \phi_0 + p + q && i = i^*, j = j^*, k = k^* \\
&= p + q && i = i^*, j = j^*, k \neq k^* \\
&= q && i \neq i^*, j = j^* \\
&= 0 && j \neq j^*
\end{aligned} \tag{10.52}$$

or as

$$\begin{aligned}
\operatorname{cov}(y_{ijk}, y_{i^*j^*k^*}) &= \phi_0 + \delta_2 + (a_1 - 1)\delta_{12}/a_1 && i = i^*, j = j^*, k = k^* \\
&= \delta_2 + (a_1 - 1)\delta_{12}/a_1 && i = i^*, j = j^*, k \neq k^* \\
&= \delta_2 - \delta_{12}/a_1 && i \neq i^*, j = j^* \\
&= 0 && j \neq j^*.
\end{aligned} \tag{10.53}$$

The relations are now evident. The two structures (10.43) and (10.52) are identical if we let

$$p = \phi_2 \qquad q = \phi_{12} \tag{10.54}$$

while the two structures (10.44) and (10.53) are identical if we let

$$\delta_2 = \zeta_2 \qquad \delta_{12} = \zeta_{12}. \tag{10.55}$$

The formulation (10.42) also follows from this special case of the Scheffé model with appropriate definition of terms as given in Scheffé (1959). However, we shall not spell this out since our point has been made. The models, as defined by (10.12) and (10.42), are identical apart from the definition, (10.45), of the parameters. Since (10.12) is simpler and more intuitive, especially when expressed in terms of the covariance structure, we prefer to use it and also allow the possibility that ϕ_2 and ϕ_{12} may be negative in (10.43).

10.5. ANALYSIS OF VARIANCE ESTIMATORS WITH UNBALANCED DATA

10.5.1. Discussion and Example

The problem of estimating variance components in mixed models was considered in Chapter 8, where the ML, REML, and moment estimators were discussed. We have seen that, with balanced data, the REML estimators were optimal in the sense of Theorem 9.8 and that they agree with the ANOVA estimators described in Section 10.2. With unbalanced data we may still compute REML estimators, but it is natural to ask whether the ANOVA approach might be worth pursuing. In this section, we give a brief introduction to the topic and refer you to Searle (1971a) for details.

There are two essential problems. The first, which is computational, is that of determining the coefficients for the expected mean squares from a given ANOVA table. A simple procedure for doing this, called *synthesis,* was introduced by Hartley (1967) and is described in Section 10.5.2. A procedure appropriate for a special class of ANOVA tables was proposed by Henderson (1953), and this method is discussed in Section 10.5.3. A third procedure, from Speed and Hocking (1974), is described in Section 10.5.4. The question of choosing the "correct" ANOVA table has not been resolved, and this is probably not a tractable mathematical problem. Simulation studies, such as the one by Speed (1979), are suggestive but inconclusive.

To focus on the issue of evaluating expected mean squares, we give an example.

EXAMPLE 10.5

Consider the random, one-way classification model from Example 8.1, described in (10.9); that is,

$$y_{ij} = \mu + \alpha_i + e_{ij} \qquad \begin{array}{l} i = 1, \ldots, a \\ j = 1, \ldots, n_i \end{array} \tag{10.56}$$

where $\alpha_i \sim N(0, \phi_1)$ and $e_{ij} \sim N(0, \phi_0)$. In matrix form, we write the model as

$$Y = \mu J + U\alpha + e \tag{10.57}$$

where

$$U = D_a(J_{n_i}). \tag{10.58}$$

The AOM table for the fixed effects analog of this model is given in Table 5.1. The ANOVA for estimating ϕ_0 and ϕ_1 is given in Table 10.9. The mean squares are identical to those in Table 5.1, but the expected mean squares are determined using

$$\mathrm{Var}(Y) = V = \phi_0 I + \phi_1 UU'. \tag{10.59}$$

TABLE 10.9 ANOVA for the one-way classification, random model.

Description	*df*	*MS*	*EMS*
Between groups	$a - 1$	$Y'B_1Y$	$\phi_0 + n^*\phi_1$
Residual	$N - a$	$Y'B_0Y$	ϕ_0

The mean squares in Table 10.9 are defined by the matrices

$$\begin{aligned} B_0 &= [I - UD_a^{-1}(n_i)U']/(N - a) \\ B_1 &= [UD_a^{-1}(n_i)U' - JJ'/N]/(a - 1). \end{aligned} \tag{10.60}$$

We note the relation of B_0 and B_1 to A_0 and A_1 as given in Definition 9.3 for the special case $n_i = n$.

The expected mean squares are developed directly from the general results of Chapter 2. We obtain

$$\begin{aligned} E(Y'B_1Y) &= tr(B_1V) + \mu^2 J'B_1J \\ &= \phi_0 tr(B_1) + \phi_1 tr(U'B_1U) \\ &= \phi_0 + \phi_1(N^2 - \textstyle\sum n_i^2)/N(a - 1) \\ &= \phi_0 + n^*\phi_1 \end{aligned} \tag{10.61}$$

and

$$E(Y'B_0Y) = \phi_0. \tag{10.62}$$

The ANOVA estimators are computed, as usual, by equating the mean squares to their expected values.

Two observations are in order. First, the expected mean squares were easily evaluated in this case, since simple matrix expressions were available for the mean squares and the algebra was quite easy. Second, we note from (10.61) that we could have written

$$\begin{aligned} E(Y'B_1Y) &= \phi_0 + \phi_1 tr(U'B_1U) \\ &= \phi_0 + \phi_1 \sum_{i=1}^{a} U_i'B_1U_i \end{aligned} \tag{10.63}$$

where U_i denotes the ith column of U. This form suggests a very simple numerical procedure for evaluating the coefficient of ϕ_1; that is, if we fit the model (10.57) using the ith column of U as the response vector, then the between-group mean square in Table 10.9 is

$$m_i = U_i'B_1U_i. \tag{10.64}$$

Thus, if we repeat the analysis a times using the columns of U as the response vectors, the coefficient of ϕ_1 is given by the sum of these mean squares; that is,

$$n^* = \sum_{i=1}^{a} m_i. \tag{10.65}$$

This observation forms the basis for the method of *synthesis* discussed in Section 10.5.2.

■

Attempts to compare ANOVA estimators with ML or REML estimators are generally unsuccessful because of the algebraic complexity of the latter two procedures with unbalanced data. Even for the simple model in Example 10.5, the ML and REML equations are highly nonlinear. Searle (1971a) gives expressions for variances of the ANOVA estimators and the large sample ML estimators for the model in Example 10.5, but, apart from confirming that the latter provide lower bounds on the variance of unbiased estimators, this information does not allow a good comparison of the methods. Searle (1971a) also provides estimators and their variances for some other special models, but, again, the results are not conclusive. As a general principle, ANOVA estimators are justified only by the fact that they are unbiased and agree with the optimal REML estimators for balanced data. As we depart from the balanced case, it seems more natural to appeal to the general good properties of ML estimation, and perhaps the REML estimators may be preferable by having less bias. We also recognize that the optimal properties of ML estimation are large sample properties, and we rarely enjoy that luxury in experimental situations (see Hartley and Rao, 1967).

The ANOVA methods given in the next three sections are presented for completeness and as potential alternatives to ML and REML, but no claim is made as to their optimality.

10.5.2. The Method of Synthesis for Computing Expected Mean Squares

Hartley (1967) proposed a clever procedure for evaluating the expected mean squares for random models, and Rao (1968) clarified and extended the method to general mixed models. The distinct advantage of the method is that any computer program for analyzing fixed effects models may be easily adapted to evaluate the expected mean squares for the analogous mixed model. As we shall see, the expressions for the expected mean squares depend on the nature of the computing algorithm used for the fixed effects program; hence, different programs may yield different results.

The method is described in terms of the variance component model and its fixed effects analog from Definitions 10.1 and 10.2. For reference, these models are, respectively,

$$Y = X\beta + \sum_{t \epsilon T - T_1} U_t e_t + e_0 \tag{10.66}$$

and

$$Y = \beta_0 J + \sum_{t \epsilon T} X_t \beta_t + e_0. \tag{10.67}$$

The AOM table for the fixed effects model consists of rows describing the numerator sums of squares for testing certain hypotheses. For example, these hypotheses may be

$$H_t\colon \beta_t = 0 \qquad t \epsilon T. \tag{10.68}$$

If we have used the second canonical form in (10.67), these hypotheses correspond to the usual main effects and interaction hypotheses, which were described in Chapters 6 and 7 and which extend to more general models. As noted in Chapter 6, there are many AOM tables available corresponding to different sets of hypotheses. The method of synthesis may be applied with any algorithm used to generate the mean squares in the AOM table. Of course, the expected mean squares will differ depending on the mean squares generated.

Let $M(Y) = Y'BY$ denote the mean square for any row of the AOM table. Then, with V given by (10.19), the expected value of this mean square under the mixed model, (10.66), is

$$\begin{aligned} E[M(Y)] &= E(Y'BY) = tr(BV) + \beta'X'BX\beta \\ &= \sum_{t \epsilon T - T_1} \phi_t tr(U_t'BU_t) + \phi_0 tr(B) + \beta'X'BX\beta. \end{aligned} \tag{10.69}$$

We shall be interested in mean squares with the property $BX = 0$. This property always holds for the random model—that is, $X = J$ in (10.66)—but otherwise may not hold.

It remains to evaluate

$$tr(U_t'BU_t) = \sum_j U_{tj}'BU_{tj} \tag{10.70}$$

where U_{tj} denotes the jth column of U_t. Note that

$$U_{tj}'BU_{tj} = M(U_{tj}). \tag{10.71}$$

Thus, we can compute $M(U_{tj})$ by using the column U_{tj} as the response vector and computing the AOM table as if we had the model

$$U_{tj} = \beta_0 J + \sum_{t \epsilon T} X_t \beta_t + e_0. \tag{10.72}$$

We may thus evaluate (10.70) by performing an AOM for each column of U_t. Recall that the number of columns of U_t is

$$c_t = \prod_{i \epsilon t} a_i. \tag{10.73}$$

Thus, the total number of AOM calculations that must be performed is

$$c = \sum_{t \epsilon T - T_1} c_t \qquad t \neq 0. \tag{10.74}$$

Since we are assuming $Y'BY$ is a mean square, $tr(B) = 1$.

To summarize, let

$$k_t = tr(U_t' B U_t) \qquad t \epsilon T - T_1. \tag{10.75}$$

Then,

$$E(Y'BY) = \phi_0 + \sum_{t \epsilon T - T_1} k_t \phi_t. \tag{10.76}$$

For mean squares such that $BX \neq 0$, the expected mean square will contain terms in the fixed effects. Such mean squares are of no value in estimating the variance components but may be of interest for testing hypotheses about the fixed effects. For example, we see from (10.69) that the mean square $Y'BY$ is an appropriate numerator mean square for testing the hypothesis

$$H\colon TX\beta = 0 \tag{10.77}$$

where

$$T'T = B.$$

Since it is not generally true that this hypothesis is the same as the hypothesis in the fixed effects model that led to $Y'BY$, it may be of interest to evaluate the second term of (10.69) for such mean squares.

Note that

$$\beta' X' B X \beta = \sum_i \beta_i^2 X_i' B X_i + \sum_{i \neq j} \beta_i \beta_j X_i' B X_j \tag{10.78}$$

where β_i and X_i are the components of β and the columns of X excluding β_0 and J. (Note further that we could write $E(Y) = W\mu$ and express the mean squares directly in terms of the cell means, rather than β.)

Now

$$\begin{aligned} X_i'BX_i &= M(X_i) \\ 2X_i'BX_j &= M(X_i + X_j) - M(X_i) - M(X_j). \end{aligned} \tag{10.79}$$

If we let

$$d_i = M(X_i) \qquad d_{ij} = X_i'BX_j \tag{10.80}$$

then

$$E(Y'BY) = \phi_0 + \sum_{t \epsilon T - T_1} k_t \phi_t + \sum_i d_i \beta_i^2 + \sum_{i \neq j} d_{ij} \beta_i \beta_j. \tag{10.81}$$

Thus, with X having h columns in addition to the column of ones, it follows that the total number of AOM computations required to evaluate (10.81) is

$$c + h + h(h - 1)/2. \tag{10.82}$$

While this appears to be an exorbitant amount of arithmetic, the computations are efficiently performed, and the simplicity of the method may well outweigh the cost of computing.

Rao (1968) also shows how to evaluate the variances of the estimates, but the required computations for this may now be unreasonable.

10.5.3. Henderson's Method for Computing Expected Mean Squares

Henderson (1953) proposed three methods for estimating variance components from ANOVA tables. These three methods are described in detail in Searle (1968) and (1971a,b). The most popular and the most generally applicable of these is Method 3, sometimes called the *method of fitting constants*. We shall describe this method for comparison with the other methods of this section. The term *fitting constants* refers to the fact that the estimating equations are developed by fitting the fixed effects analog, Definition 10.2, and various submodels and then using the resulting reductions in the regression sum of squares as the quadratic forms in the method of moments. The R-notation described in Section 6.3.1.3 will be convenient for describing these quadratics.

The key to the method is an interesting observation on the expected value of certain quadratic forms. To describe this, assume that the fixed effects analog, Definition 10.2, is written in matrix form as

$$Y = Z\gamma + e_0 \tag{10.83}$$

where

$$Z = (J_N, X_t, t \epsilon T). \tag{10.84}$$

The matrix Z will generally be partitioned as

$$Z = (Z_1 | Z_2) \tag{10.85}$$

where

$$\begin{aligned} Z_1 &= (J_N, X_t, t \in T_1, X_t, t \in T_{11}) \\ Z_2 &= (X_t, t \in T_{12}) \\ T &= T_1 + T_{11} + T_{12}. \end{aligned} \tag{10.86}$$

The notation implies that Z_1 contains all the matrices corresponding to the fixed effects and some of those associated with the random effects. The vector γ is similarly partitioned as $\gamma' = (\gamma_1' \gamma_2')$. We shall be interested in the sum of squares

$$R(\gamma_2 | \gamma_1) = R(\gamma_1, \gamma_2) - R(\gamma_1) \tag{10.87}$$

where we recall that

$$\begin{aligned} R(\gamma_1, \gamma_2) &= Y'Z(Z'Z)^{-1}Z'Y \\ R(\gamma_1) &= Y'Z_1(Z_1'Z_1)^{-1}Z_1'Y. \end{aligned} \tag{10.88}$$

To simplify the notation, we shall write the mixed model from Definition 10.1 as

$$Y = L\theta + e_0 \tag{10.89}$$

where

$$L = (X | U) \qquad \theta' = (\beta', e'). \tag{10.90}$$

As above, we shall partition L as

$$L = (L_1 | L_2) \tag{10.91}$$

with

$$\begin{aligned} L_1 &= (X | U_1) = (X | U_t, t \in T_{11}) \\ L_2 &= (U_2) = (U_t, t \in T_{12}). \end{aligned} \tag{10.92}$$

The vector $\theta' = (\theta_1', \theta_2')$ is partitioned conformably. To relate the two models, (10.83) and (10.89), we use the parameter matrix, P, defined in Section 9.3.1 and write

$$Z = LP = (L_1 P_1 | L_2 P_2). \tag{10.93}$$

(Since X was already assumed in canonical form, the corresponding block of P is the identity. The remaining blocks corresponding to U_t may correspond to any canonical form.)

The fact noted by Henderson (1953) is given in Theorem 10.2.

THEOREM 10.2
Assume that the partition of L is such that, if the index set $t \in T_{11}$ and the index set t^* is contained in t, then $t^* \in T_{11}$. It then follows that

$$E[R(\gamma_2|\gamma_1)] = \sum_{t \in T_{12}} \phi_t tr\{U_t'[I - Z_1(Z_1'Z_1)^{-1}Z_1']U_t\} + \phi_0[r(Z) - r(Z_1)]. \quad (10.94)$$

The result is invariant to the choice of parameter matrix, P.

Proof: The proof rests on the following important fact:

$$L'Z(Z'Z)^{-1}Z' = L'. \quad (10.95)$$

To establish this, we write

$$P_0 = (P|P^*) \quad (10.96)$$

where P^* denotes a set of columns adjoined to P so that P_0 is nonsingular. (For example, in the second canonical form we adjoin the column $(0, \ldots, 0, 1)'$ to Δ_a'.) Now, let

$$Z_0 = (Z|Z^*) = LP_0. \quad (10.97)$$

Then

$$\begin{aligned} L'Z(Z'Z)^{-1}Z' &= P_0'^{-1}Z_0'Z(Z'Z)^{-1}Z' \\ &= P_0'^{-1}\begin{bmatrix} Z'Z \\ Z^{*\prime}Z \end{bmatrix}(Z'Z)^{-1}Z' \\ &= P_0'^{-1}\begin{bmatrix} Z' \\ Z^{*\prime}Z(Z'Z)^{-1}Z' \end{bmatrix} \\ &= P_0'^{-1}\begin{bmatrix} Z' \\ Z^{*\prime} \end{bmatrix} \\ &= L'. \end{aligned} \quad (10.98)$$

The next-to-the-last step in (10.98) is important. Since

$$r(Z_0) = r(L) = r(Z) \quad (10.99)$$

it follows that the columns of Z^* are linear combinations of the columns of Z—say, $Z^* = ZK$. The result then follows.

By the same argument, we may show that

$$L_1'Z_1(Z_1'Z_1)^{-1}Z_1' = L_1' \quad (10.100)$$

assuming the condition on the indices is satisfied. This is necessary to ensure that $r(Z_1) = r(L_1)$.

Now, using (10.89), the expected value of $Y'BY$ is written as

$$E(Y'BY) = tr[L'BLE(\theta\theta')] + \phi_0 tr(B). \tag{10.101}$$

Now, with

$$B = Z(Z'Z)^{-1}Z' - Z_1(Z_1'Z_1)^{-1}Z_1' \tag{10.102}$$

we have, in view of (10.98) and (10.100),

$$\begin{aligned} L'BL &= L'L - \begin{bmatrix} L_1'L_1 & L_1'L_2 \\ L_2'L_1 & L_2'Z_1(Z_1'Z_1)^{-1}Z_1'L_2 \end{bmatrix} \\ &= \begin{bmatrix} 0 & 0 \\ 0 & L_2'[I - Z_1(Z_1'Z_1)^{-1}Z_1']L_2 \end{bmatrix}. \end{aligned} \tag{10.103}$$

From (10.101) we have

$$E[R(\gamma_2 | \gamma_1)] = tr\{L_2'[I - Z_1(Z_1'Z_1)^{-1}Z_1']L_2E(\theta_2\theta_2')\} + \phi_0[r(Z) - r(Z_1)]. \tag{10.104}$$

The desired result, (10.94), follows directly by noting the form of L_2 in (10.92).

The invariance to the choice of P follows as in Chapter 6, since the different canonical forms are just full rank transformations of the full or reduced cell means model.

■

The essential point of this theorem is that the expected value of the sum of squares, $R(\gamma_2 | \gamma_1)$, does not depend on the fixed effects, β, and depends only on the variance components ϕ_t, $t \in T_{12}$.

Since this sum of squares satisfies the condition $BX = 0$ noted in the discussions of synthesis, a computer program generating this sum of squares could be used in the synthesis method to compute the coefficients of ϕ_t in (10.94). However, we can take advantage of our knowledge of the structure of the matrix in the quadratic form to develop a more efficient computing scheme.

Reference to (10.94) shows that the coefficient of ϕ_t—say, k_t—is given by

$$k_t = \sum_j U_{tj}'[I - Z_1(Z_1'Z_1)^{-1}Z_1']U_{tj}. \tag{10.105}$$

The terms in this sum are just the residual sums of squares for fitting the columns of U_t to the model

$$U_{tj} = Z_1\gamma_1 + e. \tag{10.106}$$

This suggests a synthesis approach that is more efficient than the one described in Section 10.5.2 in that the model is of lower dimension [$r(Z_1)$, rather than $r(Z)$], and only the residual sum of squares from each AOM must be computed.

The Sweep Operator (Appendix A.I.10 and A.I.11) allows us to efficiently perform these computations in such a way that the total amount of computation is essentially that required for fitting one fixed effects model with dimension $r(Z)$.

To see this, we write the augmented sum of squares and cross products matrix for the mixed model [Definition 10.1 or (10.89)] as

$$A = \begin{bmatrix} L'L & L'Y \\ Y'L & Y'Y \end{bmatrix} = \begin{bmatrix} X'X & X'U & X'Y \\ U'X & U'U & U'Y \\ Y'X & Y'U & Y'Y \end{bmatrix}. \tag{10.107}$$

After sweeping on the fixed effects part of the model, we obtain the matrix

$$A_x = \begin{bmatrix} (X'X)^{-1} & — & — \\ — & U'[I - X(X'X)^{-1}X']U & — \\ — & — & RSS(\beta) \end{bmatrix}. \tag{10.108}$$

Here we show only the relevant parts of the swept matrix. In particular, we note that the diagonal elements of the matrix

$$U'[I - X(X'X)^{-1}X']U \tag{10.109}$$

are just the residual sums of squares for fitting the model (10.106) for the special case $Z_1 = X$. The expected value of $R(\gamma_2 | \gamma_1)$, with $\gamma_1 = \beta$, follows immediately from (10.94); that is, the coefficients of ϕ_t are just the traces of the appropriate diagonal blocks in (10.109).

Continuing with the sweep operations, we effectively partition the A matrix (10.107) as

$$A = \begin{bmatrix} L_1'L_1 & L_1'U_2 & L_1'Y \\ U_2'L_1 & U_2'U_2 & U_2'Y \\ Y'L_1 & Y'U_2 & Y'Y \end{bmatrix}. \tag{10.110}$$

We note that $L_1'L_1$ is singular, but the sweep procedure will invert as much of it as possible. This will yield the inverse of $Z_1'Z_1$, in which we have assumed the first canonical form to transform U_t to X_t, $t \in T_{11}$. The matrix, after sweeping on the upper left corner, has the form

$$A_{Z_1} = \left[\begin{array}{cc|cc} (Z_1'Z_1)^{-1} & 0 & — & — \\ 0 & 0 & — & — \\ \hline — & & U_2'[I - Z_1(Z_1'Z_1)^{-1}Z_1']U_2 & — \\ — & & — & RSS(\gamma_1) \end{array}\right]. \tag{10.111}$$

Again, the central block in (10.111) yields the coefficients of ϕ_t in the expected value of $R(\gamma_2|\gamma_1)$ by taking the trace of the appropriate diagonal blocks. Note that $RSS(\gamma_1) = Y'Y - R(\gamma_1)$ so that, as we proceed, we are obtaining the information for evaluating the sums of squares as well as their expected values. After sweeping the entire matrix, we will obtain $RSS(\gamma)$ and, hence, have a system of linear equations in ϕ_t and ϕ_0 of the form required for the method of moments estimators—that is, a system of equations of the form

$$\Omega\phi = \rho \tag{10.112}$$

where the components of ρ are the reductions $R(\gamma_2|\gamma_1)$ and the elements of Ω are as defined in Theorem 10.2. If we proceed sequentially with the sweep as implied above, Ω will be a triangular matrix.

To illustrate, consider the two-way classification, mixed model in which, for simplicity, we assume $n_{ij} = n$.

EXAMPLE 10.6
Let

$$y_{ijk} = \mu_i + \beta_j + (\alpha\beta)_{ij} + e_{ijk} \tag{10.113}$$

or, in matrix form, let

$$Y = W(I_a \otimes J_b)\mu + W(J_a \otimes I_b)e_2 + W(I_{ab})e_{12} + e_0 \tag{10.114}$$

where $W = I_{ab} \otimes J_n$. We obtain

$$A = \begin{bmatrix} bnI_a & nJ_aJ_b' & nI_a \otimes J_b' \\ — & anI_b & nJ_a' \otimes I_b \\ (\text{sym}) & & nI_{ab} \end{bmatrix}. \tag{10.115}$$

(For simplicity, we have deleted the last row and column of A.) After sweeping on the fixed effects, we obtain

$$A_x = \begin{bmatrix} (1/bn)I_a & — & — \\ — & an[I_b - (1/b)J_bJ_b'] & — \\ — & — & n[I_{ab} - (1/b)I_a \otimes J_bJ_b'] \end{bmatrix}. \tag{10.116}$$

Evaluating the traces of the two matrices corresponding to the two random factors and noting that $r(Z) = ab$ and $r(X) = a$, we obtain

$$E[R(\gamma_2, \gamma_{12}|\mu)] = a(b - 1)(\phi_0 + n\phi_{12} + n\phi_2). \tag{10.117}$$

Sweeping on the next block, we obtain

$$A_z = \left[\begin{array}{ccc|c} (I_a + (b-1)J_aJ_a'/a)/b & -J_aJ_{b-1}'/a & 0 & — \\ — & (I_{b-1} + J_{b-1}J_{b-1}')/a & 0 & — \\ — & — & 0 & — \\ \hline — & — & — & n(I_a - J_aJ_a'/a) \otimes (I_b - J_bJ_b'/b) \end{array}\right]. \tag{10.118}$$

The trace of the nonswept block yields the coefficient of ϕ_{12} and $r(Z_1) = a + b - 1$; hence,

$$E[R(\gamma_{12} | \mu, \gamma_2)] = (a - 1)(b - 1)(\phi_0 + n\phi_{12}). \tag{10.119}$$

Relating these results to those of Table 10.2 and noting that

$$\begin{aligned} R(\gamma_{12} | \mu, \gamma_2) &= q_{12} \\ R(\gamma_2, \gamma_{12} | \mu) &= q_2 + q_{12} \end{aligned} \tag{10.120}$$

we see that the estimating equations are given by

$$\begin{aligned} q_2 + q_{12} &= a(b - 1)(\phi_0 + n\phi_{12} + n\phi_2) \\ q_{12} &= (a - 1)(b - 1)(\phi_0 + n\phi_{12}) \\ RSS &= ab(n - 1)\phi_0. \end{aligned} \tag{10.121}$$

It is easily verified that these equations are equivalent to those obtained from Table 10.2.

■

A valid criticism of this method is that it does not specify a unique set of quadratic forms. Any partition of γ into $(\gamma_1 | \gamma_2)$, satisfying the conditions of Theorem 10.2, such that γ_1 contains the fixed effects will suffice. To illustrate, consider the two-way classification random model given in Example 10.1.

EXAMPLE 10.7
For the random model

$$y_{ijk} = \mu + \alpha_i + \beta_j + (\alpha\beta)_{ij} + e_{ijk} \tag{10.122}$$

the sweep procedure described above will yield the quadratics

$$\begin{aligned} &R(\gamma_1, \gamma_2, \gamma_{12} | \mu) \\ &R(\gamma_2, \gamma_{12} | \mu, \gamma_1) \\ &R(\gamma_{12} | \mu, \gamma_1, \gamma_2) \end{aligned} \tag{10.123}$$

which, along with the residual sum of squares, will yield a valid set of estimating equations. By reversing the order in which we enter the main effect components, the second equation in (10.123) would be replaced by

$$R(\gamma_1, \gamma_{12} | \mu, \gamma_2) \tag{10.124}$$

yielding an equally valid set of equations. Most proponents of this method would use a combination of these two choices given by

$$\begin{aligned} &R(\gamma_1 | \mu, \gamma_2) = R(\gamma_1, \gamma_{12} | \mu, \gamma_2) - R(\gamma_{12} | \mu, \gamma_1, \gamma_2) \\ &R(\gamma_2 | \mu, \gamma_1) = R(\gamma_2, \gamma_{12} | \mu, \gamma_1) - R(\gamma_{12} | \mu_1, \gamma_1, \gamma_2) \\ &R(\gamma_{12} | \mu, \gamma_1, \gamma_2). \end{aligned} \tag{10.125}$$

Reference to Table 6.3 shows that the quadratics in (10.125), along with the residual sum of squares, correspond to using the AOM based on the partially sequential method. There appears to be no theoretical justification for this choice. ■

Since we generally preferred the marginal means AOM in Table 6.1 for the fixed effects model, it is natural to ask why we do not use it for variance component estimation. There is no good reason why we should not, but we note that the Henderson method does not include these quadratic forms. The reason for this observation is that Theorem 10.2 rules out the use of quadratics such as $R(\gamma_1 | \mu, \gamma_2, \gamma_{12})$, since the index set (1, 2) is in T_{11} and the index set (1) is contained in (1, 2).

It seems intuitively reasonable that AOM tables based on acceptable hypotheses might be preferred as ANOVA tables for variance component estimation. The Henderson method has a distinct computational advantage over synthesis but restricts the choice of quadratic forms. In the next section, we present a method that removes this restriction while at the same time requiring about the same effort as the Henderson method.

10.5.4. The Fixed Effects Hypothesis Method for Computing Expected Mean Squares

The computational efficiency of Henderson's method relative to synthesis was primarily due to the fact that the method is based on a specific set of quadratic forms. In our discussion of fixed effects models, we emphasized the importance of specifying the hypotheses that led to the quadratic forms in the AOM table. Speed and Hocking (1974) described a method for computing expected mean squares based on an expression for the quadratic form in terms of the hypothesis tested in the fixed model. The advantage of this method is that it is as general as synthesis in that any AOM table may be used. Goodnight and Speed (1980) described an improvement on the original method such that the total computational effort is, as in the Henderson method, essentially that of one AOM computation. The important ideas are contained in the following theorem.

THEOREM 10.3
Let N_H denote the numerator sum of squares for testing the hypothesis H_0: $H\gamma = 0$ in the fixed effects analog model, (10.83). Then there exists a matrix,

$$C = (C_0, C_t, t \in T - T_1) \tag{10.126}$$

such that under the mixed model, (10.89), the expected value of N_H is given by

$$E(N_H) = \beta' C_0' C_0 \beta + \sum_{t \in T - T_1} \phi_t tr(C_t' C_t) + \phi_0 r(H). \tag{10.127}$$

Proof: From Chapter 3, we write

$$\begin{aligned} N_H &= (H\hat{\gamma})'[H(Z'Z)^{-1}H']^{-1}H\hat{\gamma} \\ &= Y'BY \end{aligned} \tag{10.128}$$

with

$$B = Z(Z'Z)^{-1}H'[H(Z'Z)^{-1}H']^{-1}H(Z'Z)^{-1}Z'. \tag{10.129}$$

In view of the expression (10.69) for $E(Y'BY)$, it follows that we need only determine the diagonal blocks of the matrix $L'BL$, where L is defined in (10.90). To do so, note that we may use (10.97) and the fact that $Z^* = ZK$ to write

$$\begin{aligned} L &= Z_0 P_0^{-1} \\ &= Z(I|K)P_0^{-1}. \end{aligned} \tag{10.130}$$

Thus, we define the matrix Λ by writing

$$\begin{aligned} H(Z'Z)^{-1}Z'L &= H(I|K)P_0^{-1} \\ &= \Lambda. \end{aligned} \tag{10.131}$$

Note that Λ has the column dimension of L; hence, we may write Λ in partitioned form as

$$\Lambda = (\Lambda_0, \Lambda_t, t \in T - T_1). \tag{10.132}$$

Using (10.131) with (10.129), we have

$$\begin{aligned} L'BL &= \Lambda'[H(Z'Z)^{-1}H']^{-1}\Lambda \\ &= C'C \end{aligned} \tag{10.133}$$

where $C = (S')^{-1}\Lambda$. The matrix S is obtained by the factorization

$$H(Z'Z)^{-1}H' = S'S \tag{10.134}$$

using, for example, the Cholesky decomposition from Appendix A.I.9.

In view of (10.132), we may write C in that same partitioning yielding (10.126). The diagonal blocks of $L'BL = C'C$ are then used as in (10.127) to write the expected value of N_H.

■

To clarify the results of this theorem and to note its contribution to computing expected mean squares, we make some simple observations.

First, the matrix Λ, defined in (10.131), is just the hypothesis matrix corresponding to H in the overparameterized version of the fixed effects model; that is, we have the equivalent statements

$$H\gamma = 0 \leftrightarrow \Lambda P\gamma = 0. \tag{10.135}$$

The vector $P\gamma$ includes such redundant parameters as those noted in Section 6.2.3.

Second, the amount of computation required to evaluate the coefficients in (10.127) is essentially that required for the Cholesky decomposition in (10.134). The sum of squares, N_H, may be generated at the same time since

$$N_H = [(S')^{-1}H\hat{\gamma}]'[(S')^{-1}H\hat{\gamma}]. \tag{10.136}$$

It thus follows that the amount of computation required for developing a set of estimating equations is roughly equal to that of performing an AOM on the fixed effects model (10.83).

Finally, to ensure that the expected value of N_H, given by (10.127), does not depend on the fixed effects, we need only ensure that $C_0 = 0$ or, equivalently, $\Lambda_0 = 0$. In view of (10.131), it is sufficient to require that the hypothesis $H\gamma = 0$ does not involve the fixed effects terms from the mixed model.

The computational advantages of this method might appear to be offset by the need to specify the hypotheses. Indeed, that is a factor if we consider complex hypotheses such as those in the sequential or partially sequential AOM tables in Chapters 6 and 7. As we noted in Section 10.5.3, the Henderson method is particularly good for those sums of squares and does not require specification of the hypotheses. The advantage of the method of this section is that it allows us to use the sums of squares, such as those in the AOM shown in Table 6.1. The hypotheses involved in that table are particularly simple to describe for the model written in the second canonical form.

To illustrate the method of this section, we turn again to the two-way classification mixed model discussed in Examples 10.3 and 10.6.

EXAMPLE 10.8

We write the model in the fixed effects analog as

$$Y = J_N\mu + Z_1\gamma_1 + Z_2\gamma_2 + Z_{12}\gamma_{12} + e_0. \tag{10.137}$$

Here,

$$\begin{aligned} Z_1 &= (I_a \otimes J_{bn})\Delta_a' \qquad Z_2 = (J_a \otimes I_b \otimes J_n)\Delta_b' \\ Z_{12} &= (I_{ab} \otimes J_n)(\Delta_a' \otimes \Delta_b') \end{aligned} \tag{10.138}$$

where Δ_a' is defined in (5.7).

As is evident from (10.138), we are assuming the balanced model $n_{ij} = n$ for simplicity. The sums of squares in Table 10.2 arise by testing the hypotheses

$$H_1\colon \gamma_1 = 0 \qquad H_2\colon \gamma_2 = 0 \qquad H_{12}\colon \gamma_{12} = 0. \tag{10.139}$$

For illustration, we consider H_2. In this case, the hypothesis matrix is

$$H = (0\ 0\ I_{b-1}\ 0). \tag{10.140}$$

Reference to (6.53) shows that

$$\begin{aligned} H(Z'Z)^{-1}H' &= (1/nab)(bI - JJ') \\ &= S'S. \end{aligned} \tag{10.141}$$

Utilizing (10.137), we may write

$$\Lambda = (0|0|\Lambda_2|(1/a)J_a' \otimes \Lambda_2) \tag{10.142}$$

where

$$\Lambda_2 = (1/b)[(bI - JJ')|-J]. \tag{10.143}$$

It follows that

$$C = [0|0|S'^{-1}\Lambda_2|(1/a)J_a' \otimes S'^{-1}\Lambda_2]. \tag{10.144}$$

The relation between Λ_2 and $S'S$ in this simple example allows us to evaluate the necessary matrices without explicitly determining S. Thus,

$$\begin{aligned} C_2'C_2 &= \Lambda_2'(S'S)^{-1}\Lambda_2 \\ &= \begin{bmatrix} (na)^2S'S & -naJ/b \\ -naJ'/b & (1/b^2)J'(S'S)^{-1}J \end{bmatrix} \end{aligned} \tag{10.145}$$

and

$$C_{12}'C_{12} = (1/a^2)J_aJ_a' \otimes \Lambda_2'(S'S)^{-1}\Lambda_2. \tag{10.146}$$

The expected value of q_2, the numerator sum of squares corresponding to H_2, is given by noting that

$$\begin{aligned} tr(C_2'C_2) &= na(b - 1) \\ tr(C_{12}'C_{12}) &= n(b - 1) \\ r(H_2) &= b - 1. \end{aligned} \tag{10.147}$$

Thus,

$$E(q_2/r_2) = \phi_0 + n\phi_{12} + an\phi_2 \tag{10.148}$$

in agreement with Table 10.2.

■

The simplicity of this result is a consequence of the balance in the model. However, the matrix Λ given by (10.142) is not affected by the cell frequencies; hence, for this hypothesis, the matrix Λ can be a permanent part of a computer program as a function of the dimension b. The hypothesis matrices for H_1 and H_{12} are also easily described.

As a final comment, we note that it is not necessary to determine the matrix S or its inverse explicitly. As noted in Appendix A.I.9, performing row reductions on $S'S$ to obtain S is equivalent to multiplying by $(S')^{-1}$. Thus, these same row reductions, when applied to Λ, will yield C and, when applied to $H\hat{\gamma}$, will yield the vector needed to compute N_H.

To summarize, it appears that the method of this section, combined with the Henderson method, will allow for efficient generation of the most common sums of squares and their expected values for the purpose of computing ANOVA estimates of variance components. At the present time, there does not appear to be a valid criterion for choosing between the methods. In a limited simulation study, Speed (1979) suggests that, for the two-way classification random model, the hypotheses leading to Table 6.1 are superior to those generated by the Henderson method.

It is natural to ask why we even consider the ANOVA method as an alternative to ML and REML. The primary reason appears to be ease of computing and the optimality properties for the balanced case. Harville (1977) appears to favor ML, or perhaps REML, but the final word on this matter appears to be in the future.

CHAPTER 10 EXERCISES

1. For Examples 6.1 through 6.7, verify that the covariance structures that arise from the models as stated in (10.9) through (10.13) are the same as those given in Chapters 8 and 9, apart from the assumption of $\phi_t > 0$.
2. Determine the estimates of the variance components for the mixed, two-fold nested model, Example 8.2, using the Analysis of Variance method described in Section 10.2.
3. Determine the estimates of the variance components for the split-plot model, Example 8.7, using the Analysis of Variance method.
4. The following data are assumed to arise from the model

$$y_{ij} = \mu + \alpha_i + \beta_j + e_{ij} \qquad i = 1, 2, 3, 4$$
$$j = 1, 2, 3, 4$$

where $\alpha_i \sim N(0, \phi_1)$, $\beta_j \sim N(0, \phi_2)$, $e_{ij} \sim N(0, \phi_0)$ and all random variables are independent.

		B			
		1	2	3	4
A	1	70.5	57.0	66.0	78.5
	2	70.5	62.6	66.0	71.0
	3	63.5	56.5	70.0	67.5
	4	67.0	65.5	67.0	76.0

(a) Use the ANOVA method to estimate the variance components.
(b) Apply the results of Section 10.3 to diagnose any problems with the data or inconsistencies with the model.

5. Consider the random two-way classification model with unequal cell frequencies given by

$$y_{ijk} = \mu + \alpha_i + \beta_j + (\alpha\beta)_{ij} + e_{ijk} \qquad \begin{aligned} i &= 1, \ldots, a \\ j &= 1, \ldots, b \\ k &= 1, \ldots, n_{ij}. \end{aligned}$$

(a) Use the results of Chapter 2 and equations (6.120) and (6.124) to determine the expected values of the main effect sums of squares.
(b) For the special case $a = 2$, use equation (6.65) to find the expected value of the interaction mean square.
(c) Apply the method of synthesis to answer parts (a) and (b).

6. Apply Henderson's method to determine the expected mean squares for the unbalanced, two-fold nested model, Example 8.3, given by

$$y_{ijk} = \mu + \alpha_i + \beta_{ij} + e_{ijk} \qquad \begin{aligned} i &= 1, \ldots, a \\ j &= 1, \ldots, b \\ k &= 1, \ldots, n_{ij}. \end{aligned}$$

Verify the results using either synthesis or the general results of Chapter 2.

7. Relate the mean squares in Table 10.9 to the sufficient statistics identified in Exercise 10 of Chapter 9. Suggest alternative functions of these statistics that might be used to estimate ϕ_0 and ϕ_1, which are motivated by choosing linear, minimum variance functions of the sufficient statistics.

8. Reference to Table 6.5 for the fixed effects, two-factor model and to Table 10.2 for the mixed, two-factor model raises an interesting question. If we treat all effects as fixed, the test for the row main effect is given by (6.119), while in the mixed model the test for the analogous effect is (10.25). The point to notice is that different denominators, and hence different degrees of freedom, are used for the two tests. The different test statistics could easily lead to different conclusions with regard to the significance of this effect. How would you respond to a client who notes that the treatment differences are significant in one model and not in the other?

APPENDIX A
MATHEMATICAL FACTS

This appendix contains a summary of results from matrix algebra and multivariate optimization. The facts are presented without proof, and it is assumed that the reader is familiar with the basic results from matrix algebra and multivariable calculus. Most of the results on matrices are in Graybill (1969). The optimization results are found in most texts on advanced calculus.

I. MATRIX ALGEBRA

I.1. Notation

I.1.1. General Notation

The $m \times n$ matrix A with elements a_{ij}, $i = 1, \ldots, m$, $j = 1, \ldots, n$, will be written as

$$A = A_{mn} = (a_{ij})$$

as needed to emphasize the dimensions or the elements of A. All matrices are assumed to have real elements. The transpose and inverse (if it exists) of A will be denoted by A' and A^{-1}, respectively. The column vector a with elements a_i, $i = 1, \ldots, n$, may be written as (a_i) to emphasize its elements. Row vectors will be written as $a' = (a_i)'$.

I.1.2. Row and Column Vectors and Diagonal Matrices

The column vector $x = (x_i)$ of length a will sometimes be denoted by

$$C_a(x_i)$$

and the corresponding row vector by

$$R_a(x_i) = C'_a(x_i).$$

Diagonal matrices with elements x_i, $i = 1, \ldots, a$, will be denoted by

$$D_a(x_i) = \begin{bmatrix} x_1 & & & \\ & x_2 & & 0 \\ & & \cdot & \\ 0 & & & \cdot \\ & & & & x_a \end{bmatrix}.$$

This notation may be extended to include multiple subscripts with combinations of these matrices. For example,

$$R_a[D_b(x_{ij})] = [D_b(x_{1j})D_b(x_{2j}), \ldots, D_b(x_{aj})]$$

and

$$C_a[R_b(x_{ij})] = \begin{bmatrix} R_b(x_{1j}) \\ R_b(x_{2j}) \\ \vdots \\ R_b(x_{aj}) \end{bmatrix}.$$

The a-vector of ones is denoted by

$$J_a = C_a(1).$$

I.1.3. Vector Norm

The Euclidean norm of a vector is written as

$$\|a\|^2 = a'a = \sum_i a_i^2.$$

I.1.4. Partitioned Matrices

A rectangular partition of A will be written

$$A = \begin{bmatrix} A_{11} & A_{12} \\ A_{21} & A_{22} \end{bmatrix}.$$

I.2. Rank

I.2.1. Definition

The rank of A is the number of linearly independent rows or columns of A and is denoted by $r(A)$.

I.2.2. Useful Facts

(i) If A and C are nonsingular, then

$$r(ABC) = r(B)$$

(ii) $r(A) = r(A'A) = r(AA')$
(iii) If $r(A) = r(A_1)$ and

$$A = \begin{bmatrix} A_1 \\ A_2 \end{bmatrix}$$

then there exists a matrix B such that $A_2 = BA_1$.
(iv) $r(AB) \leq \min[r(A), r(B)]$
(v) If $AB = 0$, then either A and B are both singular or one of them is the zero matrix.

I.3. Trace

I.3.1. Definition

The trace of a square matrix, A, is the sum of the diagonal elements of A denoted by $tr(A)$.

I.3.2. Cyclic Property

Assuming both multiplications are defined,

$$tr(AB) = tr(BA).$$

I.4. Eigenvalues and Eigenvectors

I.4.1. Definition

If A is a square matrix, then the scalar, λ, is called an eigenvalue of A if λ is a root of the polynomial defined by

$$|A - \lambda I| = 0.$$

Associated with λ is an eigenvector, v, determined by

$$Av = \lambda v.$$

If A is $n \times n$, then there are n eigenvalue, eigenvector pairs, λ_i, v_i, $i = 1, \ldots, n$. If $\Lambda = D_n(\lambda_i)$ and $V = (v_1, \ldots, v_n)$, then

$$AV = V\Lambda.$$

I.4.2. Symmetric Matrices

If A is symmetric, the eigenvalues of A are real, and the eigenvectors may be assumed to be real. We shall assume that the eigenvectors are scaled to have length one; that is, $\|v_i\| = 1$.

(i) If $\lambda_1 \neq \lambda_2$, then $v_1'v_2 = 0$; that is, v_1 and v_2 are said to be orthogonal. If all eigenvalues of A are distinct, then $V'V = I$, and V is nonsingular.

(ii) If $\lambda_1 = \lambda_2 = , \ldots , = \lambda_r$, then there are r independent associated eigenvectors, v_i, $i = 1, \ldots, r$, and any linear combination of these is also an eigenvector. From these we can construct vectors U_i, $i = 1, \ldots, r$, which are orthogonal, and they are orthogonal to all other eigenvectors. Thus, we may assume $V'V = I$.

(iii) Note that

$$V'AV = \Lambda.$$

I.4.3. Useful Facts

(i) $tr(A) = \sum_i \lambda_i$

(ii) $|A| = \prod_i \lambda_i$

(iii) $r(A)$ = number of nonzero eigenvalues

I.5. Quadratic Forms

I.5.1. Definition

The function

$$q(y) = y'Ay = \sum_i \sum_j a_{ij}y_iy_j$$

is called a *quadratic form*. Without loss of generality, we shall assume that A is symmetric.

I.5.2. Relation to Eigenvalues and Eigenvectors

The stationary points of $q(y) = y'Ay$ on the unit circle, $y'y = 1$ are given by

$$Ay_i = \lambda_i y_i$$

where $q(y_i) = \lambda_i$.

I.5.3. Nonnegative Quadratic Forms

A quadratic form is said to be positive definite if $q(y) > 0$ for all nonzero y. If $q(y) \geq 0$ and $q(y) = 0$ for some $y \neq 0$, then $q(y)$ is called *positive semi-definite*. These same terms are applied to the matrix of the quadratic form.

(i) The eigenvalues of a positive definite (semi-definite) matrix are positive (non-negative with at least one zero).

(ii) If A is positive definite, the contours of $q(y)$ are ellipsoids.
(iii) If B is $n \times p$ of rank r, then $B'B$ is positive definite (semi-definite) if $r = p\,(r < p)$.

I.5.4. Orthogonal Transformations

If V is the matrix of orthogonal eigenvectors and $y = Vz$, then

$$y'Ay = z'V'AVz = z'\Lambda z = \sum_i \lambda_i z_i^2.$$

I.6. Special Matrices

I.6.1. Orthogonal Matrices

An $n \times n$ matrix A is said to be orthogonal if and only if $A^{-1} = A'$. If A is orthogonal, then

(i) $-1 \leq a_{ii} \leq 1$
(ii) The rows (columns) of A have length one and are orthogonal; that is,

$$A'A = AA' = I.$$

(iii) $|A| = \pm 1$

The matrix of eigenvectors V of a symmetric matrix is orthogonal.

I.6.2. Idempotent Matrices

An $n \times n$ matrix, A, is said to be idempotent if $AA = A$. If A is idempotent, then

(i) $r(A) = n$ implies $A = I$.
(ii) If A is symmetric, then A is positive semi-definite.
(iii) The nonzero eigenvalues of A are $+1$. (If A is symmetric, A is idempotent if and only if the nonzero eigenvalues are $+1$.)
(iv) $tr(A) = r(A)$
(v) Let A_i, $i = 1, \ldots, m$ be symmetric, $A = \Sigma_i A_i$, $r(A_i) = r_i$, and $r(A) = r$. If A is idempotent and $r = \Sigma_i r_i$, then A_i is idempotent, $i = 1, \ldots, m$, and $A_i A_j = 0$, $i \neq j$.

I.7. Diagonalization of Matrices

I.7.1. General Matrices

If A is $m \times n$ of rank r, there exist nonsingular matrices P and Q such that

$$PAQ = \begin{bmatrix} I_r & 0 \\ 0 & 0 \end{bmatrix}.$$

I.7.2. Symmetric Matrices

If A is $n \times n$, symmetric, of rank r, there exists an orthogonal matrix V such that

$$V'AV = D_n(\lambda_i)$$

and a nonsingular matrix P such that

$$P'AP = \begin{bmatrix} I_t & 0 & 0 \\ 0 & -I_{r-t} & 0 \\ 0 & 0 & 0 \end{bmatrix}.$$

For example, λ_i, $i = 1, \ldots, n$, and the columns of V could be the eigenvalues and eigenvectors of A, in which case $P = VD_n(|\lambda_i|^{-1/2})$.

I.7.3. Pairs of Symmetric Matrices

If A and B are symmetric and A is positive definite, then there exists a nonsingular matrix P such that $P'AP = I$ and $P'BP$ is diagonal.

If A and B are symmetric, then there exists an orthogonal matrix V such that $V'AV$ and $V'BV$ are diagonal if and only if AB is symmetric.

I.8. Kronecker Products of Matrices

I.8.1. Definition

The Kronecker (direct) product (see, for example, Neudecker, 1969) of two matrices A, $p \times q$, and B, $m \times n$, is the matrix of dimension $pm \times qn$ given by

$$A \otimes B = (a_{ij} B).$$

I.8.2. Properties of Kronecker Product Matrices

(i) $A \otimes (B \otimes C) = (A \otimes B) \otimes C$
(ii) $(A \otimes B)(C \otimes D) = AC \otimes BD$
(iii) $(A + B) \otimes (C + D) = A \otimes C + A \otimes D + B \otimes C + B \otimes D$
(iv) $(A \otimes B)' = A' \otimes B'$
(v) $(A \otimes B)^{-1} = A^{-1} \otimes B^{-1}$
(vi) $tr(A \otimes B) = tr(A)tr(B)$

I.9. Factorization of Matrices

We shall be concerned only with the factorization of symmetric, positive definite matrices in the form

$$A = S'S.$$

Two factorizations are given.

I.9.1. Eigenvalue Factorization

Let V and Λ be the eigenvector and eigenvalue matrices and define

$$S = D(\lambda_i^{1/2})V'.$$

I.9.2. Cholesky Decomposition

If A is $p \times p$, then S is an upper triangular matrix defined by

$$
\begin{aligned}
s_{11} &= \sqrt{a_{11}} && \\
s_{1j} &= a_{1j}/s_{11} && j = 2, \ldots, p \\
s_{ii} &= \left(\sqrt{a_{ii}} - \sum_{k=1}^{i-1} s_{ki}^2\right) && i > 1 \\
s_{ij} &= \left(a_{ij} - \sum_{k=1}^{i-1} s_{ki}s_{kj}\right) \Big/ s_{ii} && j > i \\
s_{ij} &= 0 && i > j.
\end{aligned}
$$

Note that for the Cholesky decomposition

$$|A| = \prod_{i=1}^{p} s_{ii}^2.$$

The matrix S may be obtained by applying row operations on A. These operations are equivalent to premultiplication by $(S')^{-1}$.

I.10. Matrix Inversion

I.10.1. Matrices of Special Structure

(i) Assuming all multiplications are defined,

$$
\begin{aligned}
(I + AB)^{-1} &= I - A(I + BA)^{-1}B \\
|I + AB| &= |I + BA|.
\end{aligned}
$$

(ii) If A is written in partitioned form as

$$A = \begin{bmatrix} A_{11} & A_{12} \\ A_{21} & A_{22} \end{bmatrix}$$

then

$$A^{-1} = \begin{bmatrix} A^{11} & A^{12} \\ A^{21} & A^{22} \end{bmatrix}$$

where

$$
\begin{aligned}
A^{11} &= (A_{11} - A_{12}A_{22}^{-1}A_{21})^{-1} \\
A^{12} &= -A_{11}^{-1}A_{12}A^{22} \\
A^{22} &= (A_{22} - A_{21}A_{11}^{-1}A_{12})^{-1} \\
A^{21} &= -A_{22}^{-1}A_{21}A^{11}.
\end{aligned}
$$

Also,

$$|A| = |A_{11}| \cdot |A_{22} - A_{21}A_{11}^{-1}A_{12}|$$
$$= |A_{22}| \cdot |A_{11} - A_{12}A_{22}^{-1}A_{21}|.$$

I.10.2. The Sweep Operator

An efficient procedure for inverting matrices uses the Sweep Operator. (See Goodnight, 1979, for a detailed discussion.)

Definition. The matrix A is transformed into the matrix B by Sweep(k)—(A is said to be swept on the kth row) if

$$b_{kk} = 1/a_{kk}$$

$$b_{ik} = -a_{ik}/a_{kk} \qquad i \neq k$$

$$b_{kj} = a_{kj}/a_{kk} \qquad j \neq k$$

$$b_{ij} = a_{ij} - a_{ik}a_{kj}/a_{kk} \qquad i \neq k, \; j \neq k.$$

We write $B = \text{Sweep}(k)A$.

Properties.

(i) *Reversible.*

$$\text{Sweep}(k)\text{Sweep}(k)A = A$$

(ii) *Commutative.*

$$\text{Sweep}(k)\text{Sweep}(r)A = \text{Sweep}(r)\text{Sweep}(k)A$$

Partitioned form. If A is written in partitioned form as

$$A = \begin{bmatrix} A_{11} & A_{12} \\ A_{21} & A_{22} \end{bmatrix}$$

where A_{11} is $r \times r$ and A is swept on the first r rows, then

$$B = \prod_{i=1}^{r} \text{Sweep}(i)A$$
$$= \begin{bmatrix} B_{11} & B_{12} \\ B_{21} & B_{22} \end{bmatrix}$$

where

$$B_{11} = A_{11}^{-1}$$
$$B_{12} = A_{11}^{-1}A_{12}$$
$$B_{21} = -A_{21}A_{11}^{-1}$$
$$B_{22} = A_{22} - A_{21}A_{11}^{-1}A_{12}.$$

Sweeping on all rows yields A^{-1}. The application to solutions of linear equations is suggested by the partitioned form and is examined in the next section.

I.11. Solution of Linear Equations

We shall be primarily concerned with solving the normal equations that arise in the least squares or maximum likelihood analyses of linear models. In general, these equations are of the form

$$X'X\theta = X'Y.$$

I.11.1. Solution of the Normal Equations

Assuming $X'X$ is nonsingular, write

$$A = \begin{bmatrix} X'X & X'Y \\ Y'X & Y'Y \end{bmatrix}.$$

If $X'X$ is $p \times p$, then

$$B = \prod_{i=1}^{p} \text{Sweep}(i)A$$
$$= \begin{bmatrix} (X'X)^{-1} & \hat{\theta} \\ -\hat{\theta}' & RSS \end{bmatrix}$$

where

$$\hat{\theta} = (X'X)^{-1}X'Y$$

and the residual sum of squares

$$RSS = Y'[I - X(X'X)^{-1}X']Y.$$

I.11.2. Adding or Deleting Variables

If we write

$$X = (X_1 \; X_2)$$

where X_1 has r columns, then

$$A = \begin{bmatrix} X_1'X_1 & X_1'X_2 & X_1'Y \\ X_2'X_1 & X_2'X_2 & X_2'Y \\ Y'X_1 & Y'X_2 & Y'Y \end{bmatrix}$$

and

$$\begin{aligned} B &= \prod_{i=1}^{r} \text{Sweep}(i)A \\ &= \prod_{i=r+1}^{p} \text{Sweep}(i) \prod_{j=1}^{p} \text{Sweep}(j)A \\ &= \begin{bmatrix} (X_1'X_1)^{-1} & (X_1'X_1)^{-1}X_1'X_2 & \hat{\theta}_1 \\ -X_2'X_1(X_1'X_1)^{-1} & X_2'MX_2 & X_2'MY \\ -\hat{\theta}_1' & Y'MX_2 & Y'MY \end{bmatrix} \end{aligned}$$

where

$$\hat{\theta}_1 = (X_1'X_1)^{-1}X_1'Y$$

and

$$M = I - X_1(X_1'X_1)^{-1}X_1'.$$

Thus, for fitting the model with only the columns in X_1, $\hat{\theta}_1$ gives the vector of estimates, and the residual sum of squares is given by

$$RSS_1 = Y'MY.$$

To add variables from X_2 to the model or to delete variables from X_1 from the model, we need only sweep on the appropriate rows of B. Note that the effect on the residual sum of squares for the new model after sweeping on row k is determined from B as

$$RSS(\text{new}) = RSS(\text{old}) - (b_{k,p+1})^2/b_{kk}$$

where b_{kk} is the diagonal element and $b_{k,p+1}$ is the element in the last column of the kth row. Note that, if $X'X$ is singular—say, of rank r—the sweep procedure will terminate after r sweeps with

$$X_2'MX_2 = 0$$
$$X_2'MY = 0.$$

However, sweeping again on one of the first r rows (deleting that variable) may allow us to sweep on one of the rows $r + 1$ through p.

II. OPTIMIZATION

We shall be interested in the optimization of functions of several variables that may be subjected to linear equality constraints. The variables may appear in the functions as elements of vectors, matrices, and determinants of matrices. Rules for differentiation are given, and classical results on constrained and unconstrained optimization are stated.

II.1. Differentiation of Matrices and Determinants

II.1.1. Definition

Let A be a matrix whose elements a_{ij} depend on a parameter θ. Then, define

$$dA/d\theta = d[a_{ij}(\theta)]/d\theta = (da_{ij}/d\theta).$$

That is, the derivative of a matrix is defined as the matrix of derivatives of the elements.

II.1.2. Differentiation of Determinants

Let A_{ij} denote the cofactor of a_{ij}. Then

$$|A| = \sum_{j=1}^{p} a_{ij}A_{ij}$$

for any i. Assuming the elements of A are functionally independent, we have the following results:

(i) $d|A|/da_{ij} = A_{ij}$

(ii) $d|A|/d\theta = \sum_i \sum_j (d|A|/da_{ij})da_{ij}/d\theta$

$$= \sum_i \sum_j A_{ij}da_{ij}/d\theta$$

$$= tr[(A_{ij})'(da_{ij}/d\theta)]$$

(iii) If A is symmetric,

$$d|A|/da_{ij} = \begin{cases} A_{ii} & i = j \\ 2A_{ij} & i \neq j \end{cases}$$

(iv) $d\ln|A|/d\theta = (1/|A|)d|A|/d\theta = tr(A^{-1}dA/d\theta)$

(v) $dtr(AB)/d\theta = tr[(dA/d\theta)B] + tr(AdB/d\theta)$

(vi) $dA^{-1}/d\theta = -A^{-1}(dA/d\theta)A^{-1}$

II.2. Differentiation of a Function with Respect to a Vector

II.2.1. Definition

Let $t' = (t_1, t_2, \ldots, t_n)$. Then, if $f(t)$ is a function of t, we define the gradient vector

$$\nabla f = df(t)/dt = \begin{bmatrix} \partial f/\partial t_1 \\ \vdots \\ \partial f/\partial t_n \end{bmatrix}$$

and

$$df(t)/dt' = [df(t)/dt]'.$$

Also, we define the Hessian matrix—that is, the matrix of second derivatives—as

$$\begin{aligned} H_f &= d^2f(t)/dt'dt = d[df(t)/dt]/dt' \\ &= (\partial^2 f/\partial t_j \partial t_i). \end{aligned}$$

Note that H_f is symmetric.

II.2.2. Examples

(i) If $f(t) = t'At + b't + c$, then

$$\begin{aligned} \nabla_f &= 2At + b \\ H_f &= 2A. \end{aligned}$$

(ii) The linear approximation of $f(t)$ at $t = t^*$ is

$$L(t) = f(t^*) + [\nabla f(t^*)]'(t - t^*)$$

and the quadratic approximation is

$$q(t) = f(t^*) + [\nabla f(t^*)]'(t - t^*) + 1/2(t - t^*)'H_f(t^*)(t - t^*).$$

Here $\nabla f(t^*)$ and $H_f(t^*)$ denote the gradient vector and the Hessian matrix evaluated at $t = t^*$.

II.3. Differentiation of a Function with Respect to a Matrix

II.3.1. Definition

Let X be an $n \times m$ matrix with elements x_{ij}, $i = 1, \ldots, n, j = 1, \ldots, m$. Then if $f(X)$ is a function of the elements of X, we define

$$df/dX = (\partial f/\partial x_{ij}, \; i = 1, \ldots, n, \; j = 1, \ldots, m).$$

II.3.2. Examples

(i) If $f(X) = a'Xb$, then

$$df/dX = ab'.$$

(ii) If $f(X) = tr(AXB)$ where A is $p \times n$ and B is $m \times p$, then

$$df/dX = A'B'.$$

(iii) If X is symmetric and $f(X) = a'Xb$, then

$$df/dX = ab' + ba' - D(ab')$$

where $D(ab')$ denotes the diagonal matrix whose diagonal elements are those of ab'.

(iv) If X is symmetric and $f(X) = tr(AXB)$, then

$$df/dX = A'B' + BA - D(BA).$$

(v) If X and A are symmetric and $f(X) = tr(AXAX)$, then

$$df/dX = 2AXA.$$

II.4. Optimization of a Function

II.4.1. The Unconstrained Case

We shall assume that the function $f(t)$ is twice differentiable.

Definition. The stationary points of $f(t)$ are the solutions to the system of equations

$$\nabla f = 0.$$

Results.

(i) If t^* represents a point at which $f(t)$ takes on a relative minimum or maximum, then t^* is a stationary point of $f(t)$.

(ii) If $H_f(t^*)$ is positive (negative) definite, then t^* defines a local minimum (maximum) of $f(t)$.

II.4.2. The Constrained Case

We shall consider only the case of linear constraints that we write in the form

$$Gt = g$$

where G is $q \times n$ of rank q.

The reduction method. Let $G = (G_1 | G_2)$ where G_1 is nonsingular, and let $t' = (t_1' \; t_2')$. Then, solving the constraints for t_1 yields

$$t_1 = G_1^{-1}(g - G_2 t_2).$$

Let

$$f^*(t_2) = f[G_1^{-1}(g - G_2 t_2), t_2]$$

then the stationary points of $f(t)$ subject to $Gt = g$ are determined by the stationary points of $f^*(t_2)$ with associated values of t_1 determined by the constraints. Constrained maximization (minimization) of $f(t)$ subject to $Gt = g$ is equivalent to the unconstrained maximization (minimization) of $f^*(t_2)$.

The LaGrange multiplier method. The LaGrange function is defined by

$$F(t, \lambda) = f(t) + \lambda'(Gt - g)$$

where λ is a q vector of LaGrange Multipliers. The stationary points of $F(t, \lambda)$ are given by

$$\nabla_t F = \nabla f + G'\lambda = 0$$

and

$$\nabla_\lambda F = Gt - g = 0.$$

If (t^*, λ^*) is a stationary point of $F(t, \lambda)$, then t^* defines a local minimum (maximum) of $f(t)$ subject to $Gt = g$ if $H_f(t^*)$ is positive (negative) definite.

APPENDIX B
STATISTICAL FACTS

This appendix contains a brief summary of some of the basic results in the classical theory of estimation and hypothesis testing. It is assumed that the reader is familiar with the concepts. The results are stated here for ease of reference as well as to establish the notation used in the book. (For more detail, see, for example, Mood, Graybill, and Boes, 1974).

I. ESTIMATION

I.1. The Likelihood Function

Let Y be a random vector whose density depends on a parameter vector, θ. The density of Y—say, $f(Y)$—when viewed as a function of θ—say, $L(\theta)$—is called the *likelihood function*.

I.2. Maximum Likelihood Estimation

Maximization of $L(\theta)$ with respect to θ will be denoted by

$$\max_{\theta} L(\theta).$$

Solutions to this problem will be functions of Y—say, $\theta(Y)$—and are called *maximum likelihood estimators*. Evaluation of $\theta(Y)$ for a specific vector yields a maximum likelihood estimate that we shall denote by $\hat{\theta}$ or $\tilde{\theta}$. In this book, the maxima will occur at stationary points of $L(\theta)$ or, equivalently, of the logarithmic transform, $\text{Ln}L(\theta)$. In general, the maximizer may not have a closed form expression but must be determined numerically. In such cases, it is difficult to establish exact distributional results. In Chapters 2 through 7 this is not a problem, but in the last three chapters we shall generally have to appeal to large sample distributions for our estimators.

It is known that the maximum likelihood estimator is asymptotically distributed as

$$\hat{\theta} \sim N(\theta, I^{-1})$$

where I is the Information Matrix whose (i, j)th element is

$$I_{ij} = -E[\partial^2 \text{Ln}L(\theta)/\partial\theta_i\partial\theta_j].$$

A lower bound on the variance of unbiased estimates of θ_i is given by the ith diagonal element of I^{-1}. The Cramer-Rao lower bound is given by the reciprocal of I_{ii}.

I.3. Constrained Estimation

In some cases, not all values of θ are acceptable as maximizers; hence, we must solve the constrained maximization problem

$$\max L(\theta)$$
$$\text{subject to} \quad \theta \in \Theta$$

where Θ is a subset of Euclidean space. Sometimes these restrictions are automatically satisfied. For example, the estimate of σ^2 based on a random sample from $N(\mu, \sigma^2)$ is assured to be nonnegative. In other cases, the maximization must be constrained. The simplest situation is that in which the feasible values are defined by equality constraints that may be used to reduce the number of parameters in the model. In this book, the situation is especially simple since the constraints are linear.

In other cases, the constraint space may be defined by inequalities—for example, $\theta_i \geq 0$, $\theta_i \geq \theta_j$, and so on. Such problems come under the heading of mathematical programming, which is beyond the scope of this book. (See Hadley, 1964, for a good discussion.) We encounter some special cases of this problem in Chapters 8 through 10.

I.4. Complete, Sufficient Statistics

Let $T_i(Y)$, $i = 1, \ldots, p$, be a set of statistics such that the conditional distribution of Y given the $T_i(Y)$ does not depend on the parameters $\theta_1, \ldots, \theta_n$. Then $T_i(Y)$, $i = 1, \ldots, p$, are jointly sufficient for $\theta_1, \ldots, \theta_n$. We shall be interested in the minimal set of sufficient statistics that are complete in the sense that there is no function of T with expected value zero. We exclude functions that are zero almost everywhere.

I.5. The Exponential Family

Suppose the density of Y can be written in the form

$$f(Y, \theta) = a(\theta)b(Y)\exp\left[\sum_{j=1}^{p} c_j(\theta)T_j(Y)\right]$$

where θ is a p vector of parameters, and the range on the variables does not depend on θ. Then the density belongs to the exponential family, and, subject to some mild conditions, the statistics $T_j(Y), j = 1, \ldots, p$ are complete, sufficient statistics for θ. An extension of this result, due to Gautschi (1959), shows that the family of distributions given by

$$f(Y, \theta, \mu) = a(\theta, \mu)b(Y)\exp\left[\sum_{j=1}^{p} C_j(\theta)T_j(Y) + T_{p+1}(Y)d(\theta, \mu) + T_{p+1}^2(Y)h(\theta)\right]$$

is also complete.

The following two theorems stress the importance of statistics with this property.

I.6. The Rao-Blackwell Theorem

If $g(Y)$ is an unbiased estimator of a scalar parameter function—say, $h(\theta)$—and T_i, $i = 1, \ldots, p$, are jointly sufficient for θ, there exists an estimator $P(T)$ depending on the data only through the sufficient statistics such that $E[P(T)] = h(\theta)$, and $\text{Var}[P(T)] \leq \text{Var}[g(Y)]$. Further, this estimator can be constructed by determining the expected value of $g(Y)$ conditional on the T_i; that is, $P(T) = E[g(Y)|T_1, \ldots, T_p]$.

I.7. The Lehmann-Scheffé Theorem

If T_i, $i = 1, \ldots, p$, are complete, sufficient statistics for θ, then $P(T)$ from the Rao-Blackwell Theorem is unique and hence gives the minimum variance, unbiased estimate of $h(\theta)$.

II. TESTS OF HYPOTHESES AND CONFIDENCE REGIONS

The tests in this book will generally be derived using the likelihood ratio principle. Let $L(\theta)$ denote the likelihood function where θ is the vector of parameters restricted by $\theta \in \Theta$. Suppose we wish to test the hypothesis H_0: $\theta \in \Theta_1$ versus the alternative H_A: $\theta \notin \Theta_1$ where Θ_1 is a subset of Θ. Then the test statistic is given by the ratio

$$\gamma(Y) = \frac{\max_{\theta \in \Theta_1} L(\theta)}{\max_{\theta \in \Theta} L(\theta)}.$$

The numerator of γ is the maximum value of the likelihood function, subject to the constraints of the hypothesis and any natural constraints on the parameters included in the model statement. The denominator maximization is constrained only by the latter. It follows that $0 \leq \gamma \leq 1$, with small values of γ suggesting rejection of H_0. In practice, we may use any monotone function of γ as our test statistic.

If we reject H_0 in favor of H_A when $\gamma(Y) \leq \gamma^*$, then the size of the test, α, is defined by

$$\text{Prob}[\gamma(Y) \leq \gamma^* \mid \theta \in \Theta_1] \leq \alpha.$$

The probability of rejecting H_0 as a function of θ is called the *power of the test* and is denoted by

$$\prod(\theta) = \text{Prob}[\gamma(Y) \leq \gamma^* \mid \theta].$$

In general, the exact, small sample distribution of $\gamma(Y)$ is difficult to obtain. An approximate result, based on large sample theory, is given by

$$-2 \text{ Ln } \gamma(Y) \mathrel{\dot{\sim}} \chi^2(v)$$

under H_0. Here, v denotes the number of independent constraints in the hypothesis.

We shall develop confidence regions on θ (or functions of θ) by inverting test statistics; that is, a confidence region on θ—say, $R(Y)$—will be a region defined by the data, Y, which has the property that, if $\theta^* \in R(Y)$, the hypothesis H_0: $\theta = \theta^*$ would be accepted based on the data Y.

APPENDIX C
TABLES

TABLE C-1 Upper α probability points of central F distribution with n_1 d.f. in numerator and n_2 d.f. in denominator. Entries are $F_{\alpha:n_1,n_2}$.

α	n_2 \ n_1	1	2	3	4	5	6	7	8	9	10	12	15	20	30	60	120	∞
.10		39.9	49.5	53.6	55.8	57.2	58.2	58.9	59.4	59.9	60.2	60.7	61.2	61.7	62.3	62.8	63.1	63.3
.05		161	200	216	225	230	234	237	239	241	242	244	246	248	250	252	253	254
.025	1	648	800	864	900	922	937	948	957	963	969	977	985	993	1000	1010	1010	1020
.01		4,050	5,000	5,400	5,620	5,760	5,860	5,930	5,980	6,020	6,060	6,110	6,160	6,210	6,260	6,310	6,340	6,370
.005		16,200	20,000	21,600	22,500	23,100	23,400	23,700	23,900	24,100	24,200	24,400	24,600	24,800	25,000	25,200	25,400	25,500
.10		8.53	9.00	9.16	9.24	9.29	9.33	9.35	9.37	9.38	9.39	9.41	9.42	9.44	9.46	9.47	9.48	9.49
.05		18.5	19.0	19.2	19.2	19.3	19.3	19.4	19.4	19.4	19.4	19.4	19.4	19.5	19.5	19.5	19.5	19.5
.025	2	38.5	39.0	39.2	39.2	39.3	39.3	39.4	39.4	39.4	39.4	39.4	39.4	39.4	39.5	39.5	39.5	39.5
.01		98.5	99.0	99.2	99.2	99.3	99.3	99.4	99.4	99.4	99.4	99.4	99.4	99.4	99.5	99.5	99.5	99.5
.005		199	199	199	199	199	199	199	199	199	199	199	199	199	199	199	199	199
.10		5.54	5.46	5.39	5.34	5.31	5.28	5.27	5.25	5.24	5.23	5.22	5.20	5.18	5.17	5.15	5.14	5.13
.05		10.1	9.55	9.28	9.12	9.01	8.94	8.89	8.85	8.81	8.79	8.74	8.70	8.66	8.62	8.57	8.55	8.53
.025	3	17.4	16.0	15.4	15.1	14.9	14.7	14.6	14.5	14.5	14.4	14.3	14.3	14.2	14.1	14.0	13.9	13.9
.01		34.1	30.8	29.5	28.7	28.2	27.9	27.7	27.5	27.3	27.2	27.1	26.9	26.7	26.5	26.3	26.2	26.1
.005		55.6	49.8	47.5	46.2	45.4	44.8	44.4	44.1	43.9	43.7	43.4	43.1	42.8	42.5	42.1	42.0	41.8
.10		4.54	4.32	4.19	4.11	4.05	4.01	3.98	3.95	3.93	3.92	3.90	3.87	3.84	3.82	3.79	3.78	3.76
.05		7.71	6.94	6.59	6.39	6.26	6.16	6.09	6.04	6.00	5.96	5.91	5.86	5.80	5.75	5.69	5.66	5.63
.025	4	12.2	10.6	9.98	9.60	9.36	9.20	9.07	8.98	8.90	8.84	8.75	8.66	8.56	8.46	8.36	8.31	8.26
.01		21.2	18.0	16.7	16.0	15.5	15.2	15.0	14.8	14.7	14.5	14.4	14.2	14.0	13.8	13.7	13.6	13.5
.005		31.3	26.3	24.3	23.2	22.5	22.0	21.6	21.4	21.1	21.0	20.7	20.4	20.2	19.9	19.6	19.5	19.3

continued

TABLE C-1 (continued)

α	n_2 \ n_1	1	2	3	4	5	6	7	8	9	10	12	15	20	30	60	120	∞
.10		4.06	3.78	3.62	3.52	3.45	3.40	3.37	3.34	3.32	3.30	3.27	3.24	3.21	3.17	3.14	3.12	3.11
.05		6.61	5.79	5.41	5.19	5.05	4.95	4.88	4.82	4.77	4.74	4.68	4.62	4.56	4.50	4.43	4.40	4.37
.025	5	10.0	8.43	7.76	7.39	7.15	6.98	6.85	6.76	6.68	6.62	6.52	6.43	6.33	6.23	6.12	6.07	6.02
.01		16.3	13.3	12.1	11.4	11.0	10.7	10.5	10.3	10.2	10.1	9.89	9.72	9.55	9.38	9.20	9.11	9.02
.005		22.8	18.3	16.5	15.6	14.9	14.5	14.2	14.0	13.8	13.6	13.4	13.1	12.9	12.7	12.4	12.3	12.1
.10		3.78	3.46	3.29	3.18	3.11	3.05	3.01	2.98	2.96	2.94	2.90	2.87	2.84	2.80	2.76	2.74	2.72
.05		5.99	5.14	4.76	4.53	4.39	4.28	4.21	4.15	4.10	4.06	4.00	3.94	3.87	3.81	3.74	3.70	3.67
.025	6	8.81	7.26	6.60	6.23	5.99	5.82	5.70	5.60	5.52	5.46	5.37	5.27	5.17	5.07	4.96	4.90	4.85
.01		13.7	10.9	9.78	9.15	8.75	8.47	8.26	8.10	7.98	7.87	7.72	7.56	7.40	7.23	7.06	6.97	6.88
.005		18.6	14.5	12.9	12.0	11.5	11.1	10.8	10.6	10.4	10.2	10.0	9.81	9.59	9.36	9.12	9.00	8.88
.10		3.59	3.26	3.07	2.96	2.88	2.83	2.78	2.75	2.72	2.70	2.67	2.63	2.59	2.56	2.51	2.49	2.47
.05		5.59	4.74	4.35	4.12	3.97	3.87	3.79	3.73	3.68	3.64	3.57	3.51	3.44	3.38	3.30	3.27	3.23
.025	7	8.07	6.54	5.89	5.52	5.29	5.12	4.99	4.90	4.82	4.76	4.67	4.57	4.47	4.36	4.25	4.20	4.14
.01		12.2	9.55	8.45	7.85	7.46	7.19	6.99	6.84	6.72	6.62	6.47	6.31	6.16	5.99	5.82	5.74	5.65
.005		16.2	12.4	10.9	10.1	9.52	9.16	8.89	8.68	8.51	8.38	8.18	7.97	7.75	7.53	7.31	7.19	7.08
.10		3.46	3.11	2.92	2.81	2.73	2.67	2.62	2.59	2.56	2.54	2.50	2.46	2.42	2.38	2.34	2.31	2.29
.05		5.32	4.46	4.07	3.84	3.69	3.58	3.50	3.44	3.39	3.35	3.28	3.22	3.15	3.08	3.01	2.97	2.93
.025	8	7.57	6.06	5.42	5.05	4.82	4.65	4.53	4.43	4.36	4.30	4.20	4.10	4.00	3.89	3.78	3.73	3.67
.01		11.3	8.65	7.59	7.01	6.63	6.37	6.18	6.03	5.91	5.81	5.67	5.52	5.36	5.20	5.03	4.95	4.86
.005		14.7	11.0	9.60	8.81	8.30	7.95	7.69	7.50	7.34	7.21	7.01	6.81	6.61	6.40	6.18	6.06	5.95
.10		3.36	3.01	2.81	2.69	2.61	2.55	2.51	2.47	2.44	2.42	2.38	2.34	2.30	2.25	2.21	2.18	2.16
.05		5.12	4.26	3.86	3.63	3.48	3.37	3.29	3.23	3.18	3.14	3.07	3.01	2.94	2.86	2.79	2.75	2.71
.025	9	7.21	5.71	5.08	4.72	4.48	4.32	4.20	4.10	4.03	3.96	3.87	3.77	3.67	3.56	3.45	3.39	3.33
.01		10.6	8.02	6.99	6.42	6.06	5.80	5.61	5.47	5.35	5.26	5.11	4.96	4.81	4.65	4.48	4.40	4.31
.005		13.6	10.1	8.72	7.96	7.47	7.13	6.88	6.69	6.54	6.42	6.23	6.03	5.83	5.62	5.41	5.30	5.19
.10		3.29	2.92	2.73	2.61	2.52	2.46	2.41	2.38	2.35	2.32	2.28	2.24	2.20	2.15	2.11	2.08	2.06
.05		4.96	4.10	3.71	3.48	3.33	3.22	3.14	3.07	3.02	2.98	2.91	2.84	2.77	2.70	2.62	2.58	2.54
.025	10	6.94	5.46	4.83	4.47	4.24	4.07	3.95	3.85	3.78	3.72	3.62	3.52	3.42	3.31	3.20	3.14	3.08
.01		10.0	7.56	6.55	5.99	5.64	5.39	5.20	5.06	4.94	4.85	4.71	4.56	4.41	4.25	4.08	4.00	3.91
.005		12.8	9.43	8.08	7.34	6.87	6.54	6.30	6.12	5.97	5.85	5.66	5.47	5.27	5.07	4.86	4.75	4.64
.10		3.18	2.81	2.61	2.48	2.39	2.33	2.28	2.24	2.21	2.19	2.15	2.10	2.06	2.01	1.96	1.93	1.90
.05		4.75	3.89	3.49	3.26	3.11	3.00	2.91	2.85	2.80	2.75	2.69	2.62	2.54	2.47	2.38	2.34	2.30
.025	12	6.55	5.10	4.47	4.12	3.89	3.73	3.61	3.51	3.44	3.37	3.28	3.18	3.07	2.96	2.85	2.79	2.72
.01		9.33	6.93	5.95	5.41	5.06	4.82	4.64	4.50	4.39	4.30	4.16	4.01	3.86	3.70	3.54	3.45	3.36
.005		11.8	8.51	7.23	6.52	6.07	5.76	5.52	5.35	5.20	5.09	4.91	4.72	4.53	4.33	4.12	4.01	3.90

continued

TABLE C-1 (continued)

α	n_2 \ n_1	1	2	3	4	5	6	7	8	9	10	12	15	20	30	60	120	∞
.10		3.07	2.70	2.49	2.36	2.27	2.21	2.16	2.12	2.09	2.06	2.02	1.97	1.92	1.87	1.82	1.79	1.76
.05		4.54	3.68	3.29	3.06	2.90	2.79	2.71	2.64	2.59	2.54	2.48	2.40	2.33	2.25	2.16	2.11	2.07
.025	15	6.20	4.77	4.15	3.80	3.58	3.41	3.29	3.20	3.12	3.06	2.96	2.86	2.76	2.64	2.52	2.46	2.40
.01		8.68	6.36	5.42	4.89	4.56	4.32	4.14	4.00	3.89	3.80	3.67	3.52	3.37	3.21	3.05	2.96	2.87
.005		10.8	7.70	6.48	5.80	5.37	5.07	4.85	4.67	4.54	4.42	4.25	4.07	3.88	3.69	3.48	3.37	3.26
.10		2.97	2.59	2.38	2.25	2.16	2.09	2.04	2.00	1.96	1.94	1.89	1.84	1.79	1.74	1.68	1.64	1.61
.05		4.35	3.49	3.10	2.87	2.71	2.60	2.51	2.45	2.39	2.35	2.28	2.20	2.12	2.04	1.95	1.90	1.84
.025	20	5.87	4.46	3.86	3.51	3.29	3.13	3.01	2.91	2.84	2.77	2.68	2.57	2.46	2.35	2.22	2.16	2.09
.01		8.10	5.85	4.94	4.43	4.10	3.87	3.70	3.56	3.46	3.37	3.23	3.09	2.94	2.78	2.61	2.52	2.42
.005		9.94	6.99	5.82	5.17	4.76	4.47	4.26	4.09	3.96	3.85	3.68	3.50	3.32	3.12	2.92	2.81	2.69
.10		2.88	2.49	2.28	2.14	2.05	1.98	1.93	1.88	1.85	1.82	1.77	1.72	1.67	1.61	1.54	1.50	1.46
.05		4.17	3.32	2.92	2.69	2.53	2.42	2.33	2.27	2.21	2.16	2.09	2.01	1.93	1.84	1.74	1.68	1.62
.025	30	5.57	4.18	3.59	3.25	3.03	2.87	2.75	2.65	2.57	2.51	2.41	2.31	2.20	2.07	1.94	1.87	1.79
.01		7.56	5.39	4.51	4.02	3.70	3.47	3.30	3.17	3.07	2.98	2.84	2.70	2.55	2.39	2.21	2.11	2.01
.005		9.18	6.35	5.24	4.62	4.23	3.95	3.74	3.58	3.45	3.34	3.18	3.01	2.82	2.63	2.42	2.30	2.18
.10		2.79	2.39	2.18	2.04	1.95	1.87	1.82	1.77	1.74	1.71	1.66	1.60	1.54	1.48	1.40	1.35	1.29
.05		4.00	3.15	2.76	2.53	2.37	2.25	2.17	2.10	2.04	1.99	1.92	1.84	1.75	1.65	1.53	1.47	1.39
.025	60	5.29	3.93	3.34	3.01	2.79	2.63	2.51	2.41	2.33	2.27	2.17	2.06	1.94	1.82	1.67	1.58	1.48
.01		7.08	4.98	4.13	3.65	3.34	3.12	2.95	2.82	2.72	2.63	2.50	2.35	2.20	2.03	1.84	1.73	1.60
.005		8.49	5.80	4.73	4.14	3.76	3.49	3.29	3.13	3.01	2.90	2.74	2.57	2.39	2.19	1.96	1.83	1.69
.10		2.75	2.35	2.13	1.99	1.90	1.82	1.77	1.72	1.68	1.65	1.60	1.54	1.48	1.41	1.32	1.26	1.19
.05		3.92	3.07	2.68	2.45	2.29	2.18	2.09	2.02	1.96	1.91	1.83	1.75	1.66	1.55	1.43	1.35	1.25
.025	120	5.15	3.80	3.23	2.89	2.67	2.52	2.39	2.30	2.22	2.16	2.05	1.94	1.82	1.69	1.53	1.43	1.31
.01		6.85	4.79	3.95	3.48	3.17	2.96	2.79	2.66	2.56	2.47	2.34	2.19	2.03	1.86	1.66	1.53	1.38
.005		8.18	5.54	4.50	3.92	3.55	3.28	3.09	2.93	2.81	2.71	2.54	2.37	2.19	1.98	1.75	1.61	1.43
.10		2.71	2.30	2.08	1.94	1.85	1.77	1.72	1.67	1.63	1.60	1.55	1.49	1.42	1.34	1.24	1.17	1.00
.05		3.84	3.00	2.60	2.37	2.21	2.10	2.01	1.94	1.88	1.83	1.75	1.67	1.57	1.46	1.32	1.22	1.00
.025	∞	5.02	3.69	3.12	2.79	2.57	2.41	2.29	2.19	2.11	2.05	1.94	1.83	1.71	1.57	1.39	1.27	1.00
.01		6.63	4.61	3.78	3.32	3.02	2.80	2.64	2.51	2.41	2.32	2.18	2.04	1.88	1.70	1.47	1.32	1.00
.005		7.88	5.30	4.28	3.72	3.35	3.09	2.90	2.74	2.62	2.52	2.36	2.19	2.00	1.79	1.53	1.36	1.00

This table is abridged from "Tables of percentage points of the inverted beta distribution," *Biometrika,* Vol. 33 (1943). It is here published with the kind permission of the editor of *Biometrika*.

TABLE C-2 Upper 5 percent points of the studentized range. The entries are $q_{.05:m,n}$, where $P[Q < q_{.05:m,n}] = .95$.

n \ m	2	3	4	5	6	7	8	9	10	11	12	13	14	15	16	17	18	19	20
1	17.97	26.98	32.82	37.08	40.41	43.12	45.40	47.36	49.07	50.59	51.96	53.20	54.33	55.36	56.32	57.22	58.04	58.83	59.56
2	6.08	8.33	9.80	10.88	11.74	12.44	13.03	13.54	13.99	14.39	14.75	15.08	15.38	15.65	15.91	16.14	16.37	16.57	16.77
3	4.50	5.91	6.82	7.50	8.04	8.48	8.85	9.18	9.46	9.72	9.95	10.15	10.35	10.52	10.69	10.84	10.98	11.11	11.24
4	3.93	5.04	5.76	6.29	6.71	7.05	7.35	7.60	7.83	8.03	8.21	8.37	8.52	8.66	8.79	8.91	9.03	9.13	9.23
5	3.64	4.60	5.22	5.67	6.03	6.33	6.58	6.80	6.99	7.17	7.32	7.47	7.60	7.72	7.83	7.93	8.03	8.12	8.21
6	3.46	4.34	4.90	5.30	5.63	5.90	6.12	6.32	6.49	6.65	6.79	6.92	7.03	7.14	7.24	7.34	7.43	7.51	7.59
7	3.34	4.16	4.68	5.06	5.36	5.61	5.82	6.00	6.16	6.30	6.43	6.55	6.66	6.76	6.85	6.94	7.02	7.10	7.17
8	3.26	4.04	4.53	4.89	5.17	5.40	5.60	5.77	5.92	6.05	6.18	6.29	6.39	6.48	6.57	6.65	6.73	6.80	6.87
9	3.20	3.95	4.41	4.76	5.02	5.24	5.43	5.59	5.74	5.87	5.98	6.09	6.19	6.28	6.36	6.44	6.51	6.58	6.64
10	3.15	3.88	4.33	4.65	4.91	5.12	5.30	5.46	5.60	5.72	5.83	5.93	6.03	6.11	6.19	6.27	6.34	6.40	6.47
11	3.11	3.82	4.26	4.57	4.82	5.03	5.20	5.35	5.49	5.61	5.71	5.81	5.90	5.98	6.06	6.13	6.20	6.27	6.33
12	3.08	3.77	4.20	4.51	4.75	4.95	5.12	5.27	5.39	5.51	5.61	5.71	5.80	5.88	5.95	6.02	6.09	6.15	6.21
13	3.06	3.73	4.15	4.45	4.69	4.88	5.05	5.19	5.32	5.43	5.53	5.63	5.71	5.79	5.86	5.93	5.99	6.05	6.11
14	3.03	3.70	4.11	4.41	4.64	4.83	4.99	5.13	5.25	5.36	5.46	5.55	5.64	5.71	5.79	5.85	5.91	5.97	6.03
15	3.01	3.67	4.08	4.37	4.59	4.78	4.94	5.08	5.20	5.31	5.40	5.49	5.57	5.65	5.72	5.78	5.85	5.90	5.96
16	3.00	3.65	4.05	4.33	4.56	4.74	4.90	5.03	5.15	5.26	5.35	5.44	5.52	5.59	5.66	5.73	5.79	5.84	5.90
17	2.98	3.63	4.02	4.30	4.52	4.70	4.86	4.99	5.11	5.21	5.31	5.39	5.47	5.54	5.61	5.67	5.73	5.79	5.84
18	2.97	3.61	4.00	4.28	4.49	4.67	4.82	4.96	5.07	5.17	5.27	5.35	5.43	5.50	5.57	5.63	5.69	5.74	5.79
19	2.96	3.59	3.98	4.25	4.47	4.65	4.79	4.92	5.04	5.14	5.23	5.31	5.39	5.46	5.53	5.59	5.65	5.70	5.75
20	2.95	3.58	3.96	4.23	4.45	4.62	4.77	4.90	5.01	5.11	5.20	5.28	5.36	5.43	5.49	5.55	5.61	5.66	5.71
24	2.92	3.53	3.90	4.17	4.37	4.54	4.68	4.81	4.92	5.01	5.10	5.18	5.25	5.32	5.38	5.44	5.49	5.55	5.59
30	2.89	3.49	3.85	4.10	4.30	4.46	4.60	4.72	4.82	4.92	5.00	5.08	5.15	5.21	5.27	5.33	5.38	5.43	5.47
40	2.86	3.44	3.79	4.04	4.23	4.39	4.52	4.63	4.73	4.82	4.90	4.98	5.04	5.11	5.16	5.22	5.27	5.31	5.36
60	2.83	3.40	3.74	3.98	4.16	4.31	4.44	4.55	4.65	4.73	4.81	4.88	4.94	5.00	5.06	5.11	5.15	5.20	5.24
120	2.80	3.36	3.68	3.92	4.10	4.24	4.36	4.47	4.56	4.64	4.71	4.78	4.84	4.90	4.95	5.00	5.04	5.09	5.13
∞	2.77	3.31	3.63	3.86	4.03	4.17	4.29	4.39	4.47	4.55	4.62	4.68	4.74	4.80	4.85	4.89	4.93	4.97	5.01

continued

From E. S. Pearson and H. O. Hartley, "Biometrika Tables for Statisticians," Vol. 1, pp. 176–177, published by the Biometrika Trustees, Cambridge University Press, London, 1954. Reproduced with the permission of the authors and the publisher. Corrections of ± 1 in the last figure, supplied by Dr. James Pachares, have been incorporated in some entries.

TABLE C-2 (continued) Upper 10 percent points of the studentized range. The entries are $q_{.10:m,n}$, where $P[Q < q_{.10:m,n}] = .90$.

n \ m	2	3	4	5	6	7	8	9	10	11	12	13	14	15	16	17	18	19	20
1	8.93	13.44	16.36	18.49	20.15	21.51	22.64	23.62	24.48	25.24	25.92	26.54	27.10	27.62	28.10	28.54	28.96	29.35	29.71
2	4.13	5.73	6.77	7.54	8.14	8.63	9.05	9.41	9.72	10.01	10.26	10.49	10.70	10.89	11.07	11.24	11.39	11.54	11.68
3	3.33	4.47	5.20	5.74	6.16	6.51	6.81	7.06	7.29	7.49	7.67	7.83	7.98	8.12	8.25	8.37	8.48	8.58	8.68
4	3.01	3.98	4.59	5.03	5.39	5.68	5.93	6.14	6.33	6.49	6.65	6.78	6.91	7.02	7.13	7.23	7.33	7.41	7.50
5	2.85	3.72	4.26	4.66	4.98	5.24	5.46	5.65	5.82	5.97	6.10	6.22	6.34	6.44	6.54	6.63	6.71	6.79	6.86
6	2.75	3.56	4.07	4.44	4.73	4.97	5.17	5.34	5.50	5.64	5.76	5.87	5.98	6.07	6.16	6.25	6.32	6.40	6.47
7	2.68	3.45	3.93	4.28	4.55	4.78	4.97	5.14	5.28	5.41	5.53	5.64	5.74	5.83	5.91	5.99	6.06	6.13	6.19
8	2.63	3.37	3.83	4.17	4.43	4.65	4.83	4.99	5.13	5.25	5.36	5.46	5.56	5.64	5.72	5.80	5.87	5.93	6.00
9	2.59	3.32	3.76	4.08	4.34	4.54	4.72	4.87	5.01	5.13	5.23	5.33	5.42	5.51	5.58	5.66	5.72	5.79	5.85
10	2.56	3.27	3.70	4.02	4.26	4.47	4.64	4.78	4.91	5.03	5.13	5.23	5.32	5.40	5.47	5.54	5.61	5.67	5.73
11	2.54	3.23	3.66	3.96	4.20	4.40	4.57	4.71	4.84	4.95	5.05	5.15	5.23	5.31	5.38	5.45	5.51	5.57	5.63
12	2.52	3.20	3.62	3.92	4.16	4.35	4.51	4.65	4.78	4.89	4.99	5.08	5.16	5.24	5.31	5.37	5.44	5.49	5.55
13	2.50	3.18	3.59	3.88	4.12	4.30	4.46	4.60	4.72	4.83	4.93	5.02	5.10	5.18	5.25	5.31	5.37	5.43	5.48
14	2.49	3.16	3.56	3.85	4.08	4.27	4.42	4.56	4.68	4.79	4.88	4.97	5.05	5.12	5.19	5.26	5.32	5.37	5.43
15	2.48	3.14	3.54	3.83	4.05	4.23	4.39	4.52	4.64	4.75	4.84	4.93	5.01	5.08	5.15	5.21	5.27	5.32	5.38
16	2.47	3.12	3.52	3.80	4.03	4.21	4.36	4.49	4.61	4.71	4.81	4.89	4.97	5.04	5.11	5.17	5.23	5.28	5.33
17	2.46	3.11	3.50	3.78	4.00	4.18	4.33	4.46	4.58	4.68	4.77	4.86	4.93	5.01	5.07	5.13	5.19	5.24	5.30
18	2.45	3.10	3.49	3.77	3.98	4.16	4.31	4.44	4.55	4.65	4.75	4.83	4.90	4.98	5.04	5.10	5.16	5.21	5.26
19	2.45	3.09	3.47	3.75	3.97	4.14	4.29	4.42	4.53	4.63	4.72	4.80	4.88	4.95	5.01	5.07	5.13	5.18	5.23
20	2.44	3.08	3.46	3.74	3.95	4.12	4.27	4.40	4.51	4.61	4.70	4.78	4.85	4.92	4.99	5.05	5.10	5.16	5.20
24	2.42	3.05	3.42	3.69	3.90	4.07	4.21	4.34	4.44	4.54	4.63	4.71	4.78	4.85	4.91	4.97	5.02	5.07	5.12
30	2.40	3.02	3.39	3.65	3.85	4.02	4.16	4.28	4.38	4.47	4.56	4.64	4.71	4.77	4.83	4.89	4.94	4.99	5.03
40	2.38	2.99	3.35	3.60	3.80	3.96	4.10	4.21	4.32	4.41	4.49	4.56	4.63	4.69	4.75	4.81	4.86	4.90	4.95
60	2.36	2.96	3.31	3.56	3.75	3.91	4.04	4.16	4.25	4.34	4.42	4.49	4.56	4.62	4.67	4.73	4.78	4.82	4.86
120	2.34	2.93	3.28	3.52	3.71	3.86	3.99	4.10	4.19	4.28	4.35	4.42	4.48	4.54	4.60	4.65	4.69	4.74	4.78
∞	2.33	2.90	3.24	3.48	3.66	3.81	3.93	4.04	4.12	4.21	4.28	4.35	4.41	4.47	4.52	4.57	4.61	4.65	4.69

From James Pachares, "Table of the upper 10% points of the Studentized range," *Biometrika,* Vol. 46, Pts. 3 and 4 (1959), pp. 461–466. Reproduced with the permission of the author and the Biometrika Trustees.

REFERENCES

Addelman, S. (1974), "Computing the Analysis of Variance Table for Experiments Involving Qualitative Factors and Zero Amounts of Quantitative Factors," *The American Statistician, 28,* 21–22.

Aitkin, M. (1978), "The Analysis of Unbalanced Cross-Classifications," *Journal of the Royal Statistical Society,* A, *141,* 195–223.

Alalouf, I. S. (1980), "A Multivariate Approach to a Mixed Linear Model," *Journal of the American Statistical Association, 75,* 194–200.

Anderson, T. W. (1973), "Asymptotically Efficient Estimation of Covariance Matrices with Linear Structure," *Annals of Statistics, 1,* 135–141.

Andrews, D. F. (1974), "A Robust Method for Multiple Linear Regression," *Technometrics, 16,* 523–531.

Appelbaum, M. I., and Cramer, E. M. (1974), "Some Problems in the Non-Orthogonal Analysis of Variance," *Psychological Bulletin, 81,* 335–343.

Atkinson, A. C. (1982), "Regression Diagnostics, Transformations and Constructed Variables," *Journal of the Royal Statistical Society, 44,* 1–22.

Bargmann, R. E. (1977), "Multiple Factor Factorial Experiments," *Proceedings of the Statistical Computations Section,* ASA Annual Meeting, Chicago, Ill., 27–32. Available through the ASA office in Washington, D.C.

Barr, A. J., Goodnight, J. H., Sall, J. P., and Helwig, J. T. (1976), *A User's Guide to SAS 76,* Raleigh, N.C.: SAS Institute.

Belsley, D. A., Kuh, E., and Welsch, R. E. (1980), *Regression Diagnostics.* New York: Wiley.

Bozivich, H., Bancroft, T. A., and Hartley, H. O. (1956), "Power of Analysis of Variance Procedures for Certain Incompletely Specified Models," *Annals of Mathematical Statistics, 27,* 1017–1043.

Bryce, G. R. (1980), *Data Analysis in Rummage—A User's Guide.* Department of Statistics, Brigham Young University, Provo, Utah.

Bryce, G. R., Scott, D. T., and Carter, M. W. (1980), "Estimation and Hypothesis Testing in Linear Models—A Reparameterization Approach to the Cell Means Model," *Communications in Statistics, A9,* 131–150.

Burdick, D. S., and Herr, D. G. (1980), "Counterexamples in Unbalanced Two-Way Analysis of Variance," *Communications in Statistics, A9,* 231–241.

Burdick, D. S., Herr, D. G., O'Fallon, W., and O'Neill, B. V. (1974), "Exact Methods in the Unbalanced, Two-Way Analysis of Variance—A Geometric Approach," *Communications in Statistics, 3,* 581–594.

Carlson, J. E., and Timm, K. H. (1974), "Analysis on Nonorthogonal Fixed-Effects Designs," *Psychological Bulletin, 31,* 335–343.

Chew, V. (1977), *Comparisons among Treatment Means in an Analysis of Variance,* Agriculture Research Service, U.S. Department of Agriculture, Washington, D.C.: U.S. Government Printing Office.

Cook, R. D. (1977), "Detection of Influential Observations in Linear Regression," *Technometrics, 19,* 15–18.

Cook, R. D., and Weisberg, S. (1982), *Residuals and Influence in Regression*. New York: Chapman and Hall.

Corbeil, R. R., and Searle, S. R. (1976), "Restricted Maximum Likelihood (REML) Estimation of Variance Components in the Mixed Model," *Technometrics, 18,* 31–38.

Cox, D. R., and McCullagh, P. (1982), "Some Aspects of Analysis of Covariance," *Biometrics, 38,* 1–17.

Cramer, E. M., and Appelbaum, M. I. (1980), "Nonorthogonal Analysis of Variance—Once Again," *Psychological Bulletin, 87,* 51–57.

Daniel, C., and Wood, F. S. (1980), *Fitting Equations to Data* (2nd ed.). New York: Wiley.

Dixon, W. J. (Ed.), (1977), *BMD Biomedical Computer Programs* (3rd ed.). Berkeley: University of California Press.

Dixon, W. J., and Brown, M. B. (Eds.), (1977), *BMDP Biomedical Computer Programs, P-Series.* Berkeley, Calif.: University of California Press.

Draper, N., and Smith, H. (1981), *Applied Regression Analysis* (2nd ed.). New York: Wiley.

Elston, R. C., and Bush, J. (1964), "The Hypotheses That Can Be Tested When There Are Interactions in an Analysis of Variance Model," *Biometrics, 20,* 681–699.

Finney, D. J. (1948), "Main Effects and Interactions," *Journal of the American Statistical Association, 43,* 566–571.

Fisher, R. A. (1925), "Theory of Statistical Estimation," *Proceedings of the Cambridge Philosophical Society, 22,* 700–725.

Francis, I. (1973), "Comparison of Several Analysis of Variance Programs," *Journal of the American Statistical Association, 68,* 860–865.

Frane, J. W. (1979), "Some Computing Methods for Unbalanced Analysis of Variance and Covariance," *Communications in Statistics, A9,* 151–166.

Freund, R. J. (1980), "The Case of the Missing Cell," *The American Statistician, 34,* 94–98.

Gautschi, W. (1959), "Some Remarks on Herbach's Paper 'Optimum Nature of the F-test for Model II in the Balanced Case.'" *Annals of Mathematical Statistics, 30,* 960–963.

Ghosh, B. K. (1973), "Some Monotonicity Theorems for Chi-Square, F, and t Distributions with Applications," *Journal of the Royal Statistical Society, 35,* 480–492.

Golhar, M., and Skillings, J. (1976), "A Comparison of Several Analysis of Variance Programs with Unequal Cell Size," *Communications in Statistics, B5,* 43–54.

Goodnight, J. H. (1976), "The General Linear Models Procedure," *Proceedings of the First International SAS User's Group,* Raleigh, N.C.: SAS Institute.

Goodnight, J. H. (1979), "A Tutorial on the SWEEP Operator," *The American Statistician, 33,* 149–158.

Goodnight, J. H. (1980), "Tests of Hypotheses in Fixed Effects Linear Models," *Communications in Statistics, A9,* 167–180.

Goodnight, J. H., and Speed, F. M. (1980), "Computing Expected Mean Squares," *Biometrics, 36,* 123–125.

Gosslee, D. G., and Lucas, H. L. (1965), "Analysis of Variance of Disproportionate Data When Interactions Are Present," *Biometrics, 21,* 115–133.

Graybill, F. A. (1961), *An Introduction to Linear Statistical Models.* New York: McGraw-Hill.

Graybill, F. A. (1969), *Introduction to Matrices with Applications in Statistics.* Belmont, Calif.: Wadsworth.

Graybill, F. A. (1976), *Theory and Applications of the Linear Model.* North Scituate, Mass.: Duxbury.

Gundberg, W. R., Jr. (1984) "A History of Results on Independence of Quadratic Forms of Normal Variables," Unpublished Masters Thesis, Department of Mathematics, Arizona State University, Tempe, Arizona 85287.

Gunst, R. F., and Mason, R. L. (1980), *Regression Analysis and Its Application.* New York: Marcel Dekker.

Hadley, G. (1964), *Nonlinear and Dynamic Programming.* Reading, Mass.: Addison-Wesley.

Hartley, H. O. (1967), "Expectations, Variances and Covariances of ANOVA Mean Squares by 'Synthesis'," *Biometrics, 23,* 105–114.

Hartley, H. O., and Rao, J. N. K. (1967), "Maximum Likelihood Estimation for the Mixed Analysis of Variance Model," *Biometrics, 54,* 93–108.

Hartley, H. O., Rao, J. N. K., and Lamotte, L. R. (1978), "A Simple 'Synthesis'-Based Method of Variance Component Estimation," *Biometrics, 34,* 233–242.

Harville, D. A. (1976), "Confidence Intervals and Sets for Linear Combinations of Fixed and Random Effects," *Biometrics, 32,* 403–407.

Harville, D. A. (1977), "Maximum Likelihood Approaches to Variance Component Estimation and to Related Problems," *Journal of the American Statistical Association, 72,* 320–338.

Hemmerle, W. J. (1974), "Non-Orthogonal Analysis of Variance Using Iterative Improvement and Balanced Residuals," *Journal of the American Statistical Association, 69,* 772–778.

Hemmerle, W. J. (1979), "Recognizing Balance with Unbalanced Data," *Communications in Statistics, A9,* 201–212.

Hemmerle, W. J., and Hartley, H. O. (1973), "Computing Maximum Likelihood Estimates for the Mixed Analysis of Variance Model Using the *W* Transformation," *Technometrics, 15,* 819–831.

Henderson, C. R. (1953), "Estimation of Variance and Covariance Components," *Biometrics, 9,* 226–252.

Herr, D. (1975), "A Geometric Characterization of Connectedness in a Two-Way Design," *Biometrika, 63,* 93–100.

Hoaglin, D. C., and Welsch, R. E. (1978), "The Hat Matrix in Regression and ANOVA," *The American Statistician, 32,* 17–22.

Hochberg, Y. (1975), "An Extension to the *T*-Method to General Unbalanced Models of Fixed Effects," *Journal of the Royal Statistical Society,* B, *37,* 426–433.

Hocking, R. R. (1973), "A Discussion of the Two-Way Mixed Model," *The American Statistician, 27,* 148–151.

Hocking, R. R. (1976), "The Analysis and Selection of Variables in Linear Regression," *Biometrics, 32,* 1–49.

Hocking, R. R. (1981), "An Analysis of Methods Used in ANOVA with Missing Cells," *Proceedings of the SAS Users Group,* Orlando, Fla., 94–99.

Hocking, R. R. (1983a), "A Diagnostic Tool for Mixed Models with Application to Negative Estimates of Variance Components," *Proceedings of the SAS Users Group,* New Orleans, La., 711–716.

Hocking, R. R. (1983b), "Developments in Linear Regression Methodology: 1959–1982" (with discussion), *Technometrics, 25,* 219–245.

Hocking, R. R., Hackney, O. P., and Speed, F. M. (1978), "The Analysis of Linear Models with Unbalanced Data," in *Contributions to Survey Sampling and Applied Statistics—Papers in Honor of H. O. Hartley,* H. A. David (Ed.). New York: Academic Press.

Hocking, R. R., and Kutner, M. H. (1975), "Some Analytical and Numerical Comparisons of Estimators for the Mixed AOV Model," *Biometrics, 31,* 19–28.

Hocking, R. R., and Pendleton, O. J. (1983), "The Regression Dilemma," *Communications in Statistics, 12,* 497–527.

Hocking, R. R., and Speed, F. M. (1975), "A Full Rank Analysis of Some Linear Model Problems," *Journal of the American Statistical Association, 70,* 706–712.

Hocking, R. R., and Speed, F. M. (1980), "The Cell Means Model for the Analysis of Variance," presented at the Annual Technical Conference of ASQC and ASA, Cincinnati, Ohio.

Hocking, R. R., Speed, F. M., and Coleman, A. T. (1980), "Hypotheses To Be Tested with Unbalanced Data," *Communications in Statistics, A9,* 117–130.

Hocking, R. R., Speed, F. M., and Lynn, M. J. (1976), "A Class of Biased Estimators in Linear Regression," *Technometrics, 18,* 425–438.

Hoerl, A. E., and Kennard, R. W. (1970a), "Ridge Regression: Applications to Non-Orthogonal Problems," *Technometrics, 12,* 69–82.

Hoerl, A. E., and Kennard, R. W. (1970b), "Ridge Regression: Biased Estimation for Non-Orthogonal Problems," *Technometrics, 12,* 55–67.

Huber, P. J. (1981), *Robust Statistics.* New York: Wiley.

Hultquist, R. A., and Graybill, F. A. (1965), "Minimal Sufficient Statistics for the Two-Way Classification Mixed Model Design," *Journal of the American Statistical Association, 60,* 182–192.

John, P. W. M. (1971), *Statistical Design and Analysis of Experiments.* New York: Macmillan.

Kempthorne, O. (1952), *The Design and Analysis of Experiments.* New York: Wiley.

Kempthorne, O. (1975), "Fixed and Mixed Models in the Analysis of Variance," *Biometrics, 31,* 473–486.

Kennedy, W. J., Jr., and Gentle, J. E. (1980), *Statistical Computing,* New York: Marcel Dekker.

Khuri, A. I. (1981), "Simultaneous Confidence Intervals for Functions of Variance Components on Random Models," *Journal of the American Statistical Association, 76,* 878–885.

Khuri, A. I. (1982), "Interval Estimation of Fixed Effects and of Functions of Variance Components in Balanced Mixed Models," *Technical Report No. 173,* Department of Statistics, University of Florida, Gainsville, Fla.

Khuri, A. I. (1983), "Direct Products: A Powerful Tool for the Analysis of Balanced Data," *Communications in Statistics, 11,* 2903–2910.

Kramer, C. Y. (1956), "Extension of Multiple Range Tests to Group Means with Unequal Numbers of Replication," *Biometrics, 13,* 13–18.

Kshirsagar, A. M. (1972), *Multivariate Analysis.* New York: Marcel Dekker.

Kshirsagar, A. M. (1983), *A Course in Linear Models.* New York: Marcel Dekker.

Kutner, M. H. (1971), "Maximum Likelihood Analysis of Balanced Incomplete Block Models," unpublished Ph.D. dissertation, Texas A & M University, College Station, Tex.

Kutner, M. H. (1974), "Hypothesis Testing in Linear Models, (Eisenhart Model 1)," *The American Statistician, 29,* 98–100; 133–134.

LaMotte, L. R. (1970), "A Class of Estimators of Variance Components," *Technical Report No. 10,* Department of Statistics, University of Kentucky, Lexington, Ky.

Littell, R. C., and Lynch, R. O. (1983), "Power Comparisons of the Weighted Squares of Means Method and the Method of Fitting Constants in Unbalanced Designs," *Proceedings of the SAS Users Group,* New Orleans, La., 757–764.

Low, L. Y. (1969), "Estimators of Variance Components in the Balanced Incomplete Block," *Journal of the American Statistical Association, 64,* 1014–1030.

Mandel, J. (1971), "A New Analysis of Variance Model for Non-Additive Data," *Technometrics, 13,* 1–18.

McGuire, J. U., Jr. (1975), *Least-Squares Analysis of Data,* Agriculture Research Service, U.S. Department of Agriculture, Washington, D.C.: U.S. Government Printing Office.

Miller, R. G. (1981), *Simultaneous Statistical Inference.* New York: Springer-Verlag.

Monlezun, C. J. (1979), "Two-Dimensional Plots for Interpreting Interactions in the Three-Factor Analysis of Variance Model," *American Statistician, 33,* 63–69.

Mood, A., Graybill, F. A., and Boes, D. C. (1974), *Introduction to the Theory of Statistics.* New York: McGraw-Hill.

Murray, L. W., and Smith, D. W. (to be published 1984), "Estimability, Testability, and Connectedness in the Analysis of Linear Models," *Communications in Statistics.*

Nelder, J. A. (1954), "The Interpretation of Negative Components of Variance," *Biometrika, 41,* 544–548.

Neter, J., Wasserman, W., and Kutner, M. H. (1982), *Applied Linear Statistical Models.* Homewood, Ill.: Richard D. Irwin.

Neudecker, H. (1969), "Some Theorems on Matrix Differentiation with Special Reference to Kronecker Matrix Products," *Journal of the American Statistical Association, 64,* 953–963.

Nie, N. H., Hull, C. H., and Jenkins, J. G. (1975), *SPSS: Statistical Package for the Social Sciences.* New York: McGraw-Hill.

O'Neill, R., and Wetherill, G. B. (1971), "The Present State of Multiple Comparison Methods," *Journal of the Royal Statistical Society, 2,* 218–241.

Ostle, B. (1963), *Statistics in Research.* Ames: Iowa State University Press.

Owen, D. B. (1968), "A Survey of Properties and Applications of the Noncentral *t*-Distribution," *Technometrics, 10,* 445–478.

Patnaik, P. B. (1949), "The Noncentral Chi-Squared and *F* Distributions and Their Applications," *Biometrika, 36,* 202–232.

Patterson, H. D., and Thompson, R. (1971), "Recovery of Interblock Information When Block Sizes Are Unequal," *Biometrika, 58,* 545–554.

Ramsey, P. H. (1978), "Power Differences between Pairwise Multiple Comparisons," *Journal of the American Statistical Association, 73,* 479–485.

Rao, C. R. (1967), "Least Squares Theory Using an Estimated Dispersion Matrix and Its Application to Measurement of Signals," *Proceedings of the Fifth Berkeley Symposium in Mathematical Statistics and Probability.* Berkeley: University of California Press, 355–372.

Rao, C. R. (1970), "Estimation of Heteroscedastic Variances in Linear Models," *Journal of the American Statistical Association, 65,* 161–172.

Rao, C. R. (1973), *Linear Statistical Inference and Its Applications*. New York: Wiley.

Rao, J. N. K. (1968), "On Expectations, Variances, and Covariances of ANOVA Mean Squares by 'Synthesis'," *Biometrics, 24,* 963–978.

Rubin, D. B. (1972), "A Non-Iterative Algorithm for Least Squares Estimation of Missing Values in Any Analysis of Variance Design," *Applied Statistics, 21,* 136–141.

Satterthwaite, F. E. (1946), "An Approximate Distribution of Estimates of Variance Components," *Biometrika Bulletin, 2,* 110–114.

Scheffé, H. (1953), "A Method for Judging All Contrasts in the Analysis of Variance," *Biometrika, 40,* 87–104.

Scheffé, H. (1959), *The Analysis of Variance*. New York: Wiley.

Searle, S. R. (1968), "Another Look at Henderson's Methods of Estimating Variance Components," *Biometrics, 24,* 749–788.

Searle, S. R. (1971a), *Linear Models*. New York: Wiley.

Searle, S. R. (1971b), "Topics in Variance Component Estimation," *Biometrics, 27,* 176.

Searle, S. R. (1972), "Using the *R*( )-Notation for Reductions in Sums of Squares When Fitting Linear Models," presented at the Spring Regional Meetings of ENAR, Ames, Iowa.

Searle, S. R. (1976), "Comments on ANOVA Calculations for Messy Data," *Proceedings of the First International SAS Users Group,* Raleigh, N.C.: SAS Institute, 298–308.

Seber, G. A. F. (1977), *Linear Regression Analysis*. New York: Wiley.

Smith, D. W. (1971), "A New Approach to the Estimation of Variance Components," unpublished Ph.D. dissertation, Texas A & M University, College Station, Tex.

Smith, D. W., and Hocking, R. R. (1978), "Maximum Likelihood Analysis of the Mixed Linear Model: The Balanced Case," *Communications in Statistics, A7,* 1253–1266.

Snedecor, G. W., and Cochran, W. G. (1967), *Statistical Methods*. Ames: Iowa State University Press.

Snee, R. D. (1982), "Nonadditivity in a Two-Way Classification: Is It Interaction or Nonhomogeneous Variance?" *Journal of the American Statistical Association, 77,* 515–519.

Speed, F. M. (1979), "Choice of Sums of Squares for the Estimation of Components of Variance," presented at the Annual Meeting of the American Statistical Association, Washington, D.C.

Speed, F. M., and Hocking, R. R. (1974), "Computation of Expectations, Variances and Covariances of ANOVA Mean Squares," *Biometrics, 30,* 157–169.

Speed, F. M., and Hocking, R. R. (1976), "The Use of the *R*( )-Notation with Unbalanced Data," *The American Statistician, 30,* 30–33.

Speed, F. M., Hocking, R. R., and Hackney, O. P. (1978), "Methods of Analysis of Linear Models with Unbalanced Data," *Journal of the American Statistical Association, 73,* 105–112.

Speed, F. M., and Monlezun, C. J. (1979), "Exact F Tests for the Method of Unweighted Means in a 2^k Experiment," *The American Statistician, 33,* 15–18.

Spjøtvoll, E., and Stoline, M. R. (1973), "An Extension of the T-Method of Multiple Comparison to Include the Cases with Unequal Sample Sizes," *Journal of the American Statistical Association, 68,* 975–978.

Steel, R. G. D., and Torrie, J. H. (1960), *Principles and Procedures of Statistics*. New York: McGraw-Hill.

Stroup, W. W., Evans, J. W., and Anderson, R. L. (1980), "Maximum Likelihood Estimation of Variance Components in a Completely Random BIB Design," *Communications in Statistics, A9,* 725–756.

Thompson, W. A., and Moore, J. R. (1963), "Non-negative Estimates of Variance Components," *Technometrics, 5,* 441–450.

Tukey, J. W. (1949), "One Degree of Freedom for Nonadditivity," *Biometrics, 5,* 232–242.

Urquhart, N. S., and Weeks, D. L. (1978), "Linear Models in Messy Data: Some Problems and Alternatives," *Biometrics, 34,* 696–705.

Urquhart, N. S., Weeks, D. L., and Henderson, C. R. (1973), "Estimation Associated with Linear Models: A Revisitation," *Communciations in Statistics, 1,* 303–330.

Vandaele, W. H., and Chowdhury, S. R., (1971), "A Revised Method of Scoring," *Statistica Neerlandica, 25,* 101–112.

Wansbeck, T., and Kapteyn, A. (1982), "A Simple Way to Obtain the Spectral Decomposition of Variance Components Models for Balanced Data," *Communications in Statistics, 11,* 2105–2112.

Weeks, D. L., and Graybill, F. A. (1961), "Minimal Sufficient Statistics for the Balanced Incomplete Block Design under an Eisenhart Model II," *Sankya*, A, *23*, 261–298.

Weisberg, S. (1980), *Applied Linear Regression*. New York: Wiley.

Welch, B. L. (1956), "On Linear Combinations of Several Variances," *Journal of the American Statistical Association, 51*, 132–148.

Williams, J. S. (1962), "A Confidence Interval for Variance Components," *Biometrika, 49*, 278–281.

Yates, F. (1934), "The Analysis of Multiple Classifications with Unequal Numbers in the Different Classes," *Journal of the American Statistical Association, 29*, 51–66.

Yates, F. (1940), "The Recovery of Inter-Block Information in Balanced Incomplete Block Designs," *Annals of Eugenics, 10*, 317–325.

INDEX